AF531435

Computer Aided Environmental Management

Prof. S.A. ABBASI,
PhD, DSc, FIE, PE.
Director
Centre for Pollution Control and Energy Technology,
Pondicherry University, Pondicherry—605 014.

Dr. FAISAL I. KHAN,
BSc (Engg.), ME, PhD, MIE
Assistant Professor, Chemical Engineering Group,
BITS, Pilani, Rajasthan—333 031.

DISCOVERY PUBLISHING HOUSE
NEW DELHI- 1100 02.

First Published-2000
Reprinted: 2013
ISBN 81-7141-562-8

Published by

DISCOVERY PUBLISHING HOUSE
4831/24, Ansari Road, Prahlad Street,
Darya Ganj, New Delhi-110002 (India)
Phone: 3279245 • Fax: 91-11-3253475
E-mail:dph@indiatimes.com

Printed at: Dynamic printers, Delhi

FOREWORD

Environmental management is a vast subject because environment is a vast subject, too. If we make its definition flexible, 'environment' may encompass not only the entire world but the solar system as well. Environmental management, likewise, tends to encompass a mind-buggling variety of concepts, methodologies, techniques, and tools.

Add the dimension of computers to all this – and we may have a truly bewildering branch of study before us!

When writing a book on computer-aided environmental management (CAEM), one would understandably be at a loss to decide what topics to include and what not to. The authors have tackled this problem by focussing on several key areas of CAEM. An attempt has been made to describe these areas with the clarity and the depth necessary to do jutice to this complex branch of knowledge. The authors have also tried to demonstrate the application of different facets of CAEM in solving some of the most poignant real-life problems of our times. These include water quality modelling, dispersion and control of hazardous gases, benefit-cost analysis of biowaste treatment systems, design of greenbelts, forecasting of accidents (fires, explosions, and toxic releases) in chemical industries etc.

The authors' approach has been to balance scholastic value with practical utility. I hope this would enable the book to stimulate the growth of CAEM in India, as indeed in other parts of the world.

I congratulate Discovery Publishing House, New Delhi, and its leader Mr. Tilak Wassan, for their initiative in bringing out this significant publication on one of the latest and most important branches of knowledge.

Prof. V. T. Patil

Vice Chancellor, Pondicherry (Central) University

PREFACE

In the domain of science and technology 'new' fields emerge from time to time. Recent examples are environmental science, biotechnology, and materials science. The domain of humanities and social sciences also gets its share of 'shining' new fields. Business administration is an example.

These 'new' fields become hubs of great activity. We find in them a new charisma, a fresh surge of energy and creativity. New fields also generate new avenues of employment.

But, come to think of it, most 'new' fields are hardly new. Environmental science is nothing but ecology with its frontiers expanded. Biotechnology is as old as the days when man learnt to set curd, make pickle, or distill wine. Materials science is even older – it began to roll with the first ever wheel! Likewise a lot of business administration is formed from the segments of the discipline of commerce.

What happens is that some new discovery, a technological breakthrough, or realization of new potential applications in a classical branch of knowledge stimulate a new thrust of enquiery. Certain new vistas begin to come in view. While pursuing these leads some other branches of knowledge are tread upon. A synthesis takes place. And a new 'field' emerges! This is how biotechnology came to the fore : classical biology acquired a new meaning with discoveries in molecular biology; It led to genetic engineering which, on synthesis with other classical branches such as microbiology and enzyme engineering metamorphosed into biotechnology! Environmental science came up as a response to *growing* threat of pollution. We have emphasized the word 'growing' because pollution was always there and was known as such. But when it began to *grow alarmingly* a new branch of science came up to study and control it.

With the passage of time each 'new' field develops its own techniques and tools, concepts and protocols, lexicon and metaphor. It carves its own niche, becomes established, matures, and becomes ready to 'breed' newer disciplines!

Computer-aided environmental management (CAEM) is an exciting 'new' field. True to the tradition discribed above, CAEM has its roots in by-now well-established branches of knowledge – environmental science and computer science. But the exceeding vastness of the two parant branches makes CAEM a bewilderingly large zone of activity. And it keeps growing by the minute! In this book we have tried to present the defining traits of CAEM as it is applied to several of the 'hot' problems of the present day.

We are thankful to the All India Council of Technical Education (AICTE), New Delhi, for instituting a CAEM unit in Centre for Pollution Control and Energy Technology. A good deal of work described in this book has been done under the aegis of this unit.

S.A. Abbasi

F. I. Khan

CONTENTS

Chapter 1

COBAS: COST BENEFIT ANALYSIS OF ANAEROBIC DIGESTER SYSTEMS

P.C. Nipaney* and E. V. Ramasamy[1]

In this chapter, we present an overview of the importance and the present state of knowledge of anaerobic digestion-based biowaste treatment processes. We then present a computer-automated tool developed by us, named COBAS (Cost Benefit Analysis of anaerobic digestion Systems). The tool enables the user to swiftly and rationally chose which type of digester to use for best results under given feed type and environmental conditions.

ANAEROBIC DIGESTION - THE PAST AND THE PRESENT

For nearly a century since it was first formally used in the 1880 (McCarty, 1982), anaerobic digestion had enjoyed only limited application as a process for stabilising high strength biodegradable wastes while the bulk of biowaste treatment was done by aerobic digestion (Kirby, 1980; Van den berg, 1984). Unlike the processes based on aerobic digestion, anaerobic digestion consumed less energy, in fact produced energy in the form of methane-rich biogas, and also generated easier-to-dispose sludges. But two factors severely limited the application

* Senior Manager, Safety & Environment, Jindal Stripes Ltd. Raigarh

1 Scientist-in-charge, Environmental Biotechnology lab, Centre for Pollution Control & Energy Technology, Pondycherry University

of anaerobic digestion - its extreme slowness and its tendency to easily get upset by toxicants or minor changes in pH/nutrient balance (Clark and Speece, 1970; Ghosh *et al*, 1981; Speece and Parkin, 1984).

In India as well as in many developing countries anaerobic digestion has also been providing a means of small scale energy generation via biogas digesters. The popularity of this application has also been limited essentially due to the slow rate and process instability of anaerobic digestion; the slow rate means large digester volumes-consequently greater costs and space requirements-and process instability means lack of assurance of steady energy supply (Abbasi and Abbasi 1996).

But this situation is poised to change dramatically as a result of a string of breakthroughs which have occurred in recent years and are likely to occur in the future. Introduction of anaerobic filter, upflow anaerobic sludge blanket reactor, expanded/fluidised bed anaerobic reactor, and phase separated reactors (Figures 1.1–1.4) have brought down the hydrualic retention time (HRT) of anaerobic digesters from 35-40 days of typical unstirred reactors (like conventional biogas digesters) or 15–20 days of typical continuously stirred (and heated) tank reactors (Figure 1.5) to a few hours (Tables 1.1–1.5); even a few minutes such as fluidised bed anaerobic reactors. Such drastic reduction in HRT has a dramatically favourable impact in terms of smaller digester sizes and consequently lesser digester costs. Further it has opened the possibility of treating high-volume low-strength wastes such as sewage by anaerobic process; such wastes earlier could be speedily treated only by the then faster-than-anaerobic,aerobic processes. The other major breakthrough is occurring in our understanding of the nutrient requirements and uptake mechanisms of anaerobic and facultative microbes. (Abbasi and Krihnakumari 1996;Abbasi and Nipaney 1984; Ramasamy and Abbasi 1997)

It is being realised that proper regulation of nutrient composition and supply may have as dramatic improvements in biodegradation and energy (methane)production as changes in the digester configurations. Yet another pursuit which may have far-reaching impact on the applicability of anaerobic digestion is the control of 'dry digestion' or digestion of feeds with solids content higher than common at present. (Abbasi and Ramasamy 1998).

Once the two conventional disadvantages of anaerobic digestion-slow rate and process instability-are overcome the two major advantages-ability to generate energy and stable sludges-may overwhelmingly change the global biowaste management scenario from the one presently dominated by aerobic digestion to the one dominated by anaerobic digestion.

BIOGAS SYSTEMS AND INDIA

For a country like India where energy continues to be precious, with oil prices continuing to risc unlike in the West, anaerobic digestion has far greater relevance than it has to many other regions of the world. By their ability to produce energy, the modern anaerobic digesters are capable of treating sewage and other biological wastes at much lesser costs than the energy-intensive (or cumbersome if one considers oxidation ponds) aerobic digestion processes. This aspect holds great hope for India in particular and the third world countries in general where monitory constraints preclude wider application of known but expensive technologies such as aerobic activated sludge processes which are the mainstay of biodegradable waste management in developed countries.Anaerobic digestion also continues to be of great relevance to us as a source of clean energy, especially for the semi-urban and the rural milieu. The recent and future breakthroughs have the potential of overcoming the technological and (consequently) economic barriers that have so far prevented the popular acceptance of biogas technology and full utilisation of its potential as waste-stabilisers-cum-energy producers.

A survey reveals an increasing trend, not only abroad but India as well of using high-rate anaerobic digesters for treating high strength biowaste. (Abbasi and Ramasamy 1996,1998) Large scale digester systems are becoming especially popular in the chemical, biochemical and agricultural sectors. High-rate digesters also appear to be the best bet for aquatic weed utilisation and family size dairy waste utilisation systems (Abbasi and Abbasi 1996, Abbasi and Ramasamy, 1998). The situation is becoming more and more common when managers/ administrators/scientists are faced with the question - which digester - type of what capacity to be installed under which operating conditions in the given regional setting, to get the highest benefit-to-cost (B C) ratio. As a favourable BC is a direct result of a positive energy balance, a digester of given process type, running on a given feedstock, under given operating conditions must produce more energy than it consumes, to be economically viable. In several cases, especially the ones involving high-rate digestion, the digester capacity may have a pronounced effect on the B C ratios. The large number of process options and the large number of other variables influencing the benefit-cost ratios in biogas systems, make it very difficult to find out the options that would be cost-effective and the ones, which would not. One possible way is to construct the various digester-types of various capacities and to actually operate them for several months with the given feedstock under various operating conditions before finding out the most cost-effective options. But this would involve enormous resource inputs. Further, the data thus generated would be appropriate only to the given feedstock and massive long-term trials will be needed for each potential feedstock.

The present work makes an attempt to circumvent this problem by developing a rational decision criteria on the basis of computer-aided modelling and simulation for use by scientists/managers/ administrators in quickly assessing the economic viability of any proposed digester as also to determine the optimal set of process-capacity-operating conditions for any feedstock which would give the highest benefit to cost ratio. A computer program has been developed on the basis of detailed engineering design and costing of four most promising high-rate digester types namely continuously stirred tank reactor (CSTR), anaerobic filter (AF), upflow anaerobic sludge blanket reactor (UASB) and diphasic digester (DD). The installation and operating costs of various capacities at typical hydraulic retention times (HRTs), VS

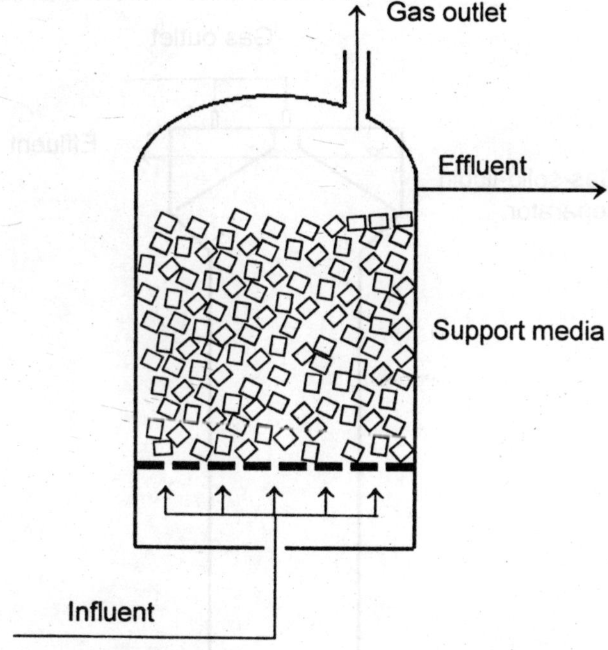

Figure 1.1 Upflow anaerobic (fixed-bed) filter reactor (UAF)

loading rates, ambient temperature conditions, and feedstock biodegradabilities have been worked out and balanced with several energy output options.

The benefit-cost analysis as above shall be of utility to scientists as well as managers/decision makers to obtain quick appraisal of the economic viability of any combination of feedstock, digester type, and capacity.

ELEMENTS OF THE CAEM PROGRAMME

1. Working out the construction and operation (C & O) costs of CSTRs of 1, 2, and 3 TPD (tonnes per day) capacities operating at HRTs (hydraulic retention times) of 10, 15 and 20 days, at ambient temperatures of 10, 20, and 30°C.
2. Working out the C and O costs of AF and UASB digesters of 1, 2, and 3 TPD operating at HRTs of 1, 2, and 4 days

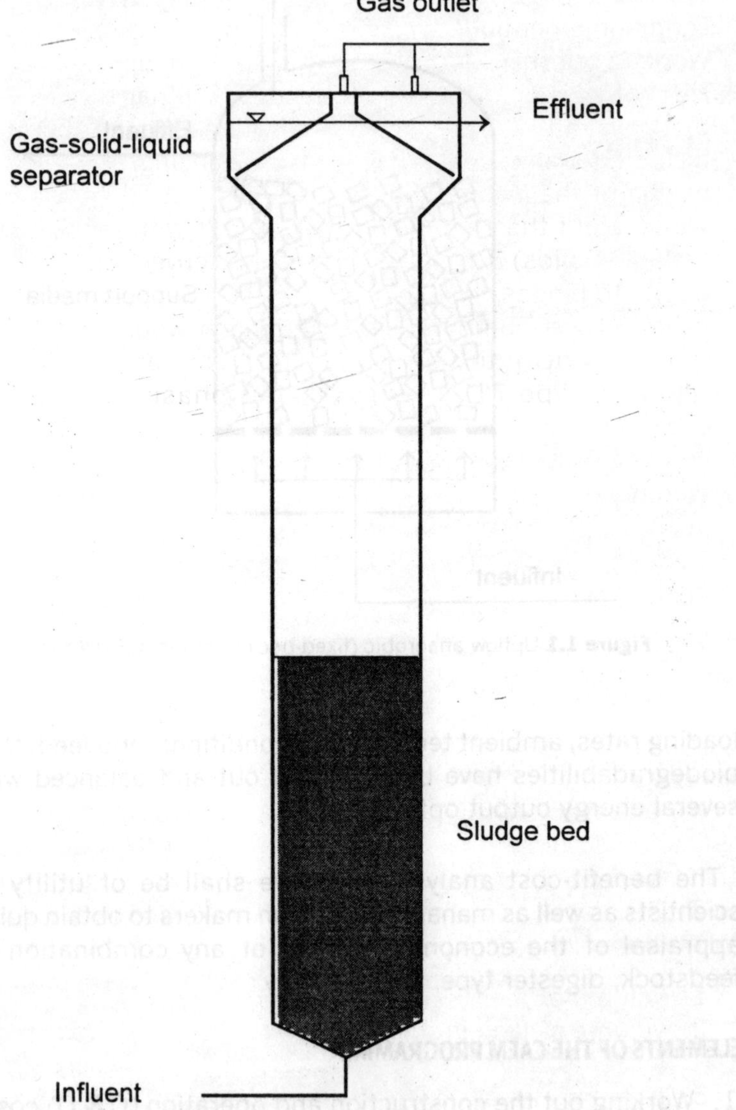

Figure 1.2. Upflow anaerobic sludge bed (UASB) reactor

(for AF) and 12, 18 and 24 hours (for UASB) and other conditions mentioned in (1) above.

3. Working out the C and O costs of Diphasic digesters with HRTs of acid phase 1/2/4 days and of methane phase 4/6/8 days, and other conditions mentioned in (1) above.
4. Including processing costs such as mincing/shredding/mixing of the feedstocks.
5. Working out the energy outputs of feeds with percent VS (Volatile Solids) 3/5/7 and percent VS conversions of 55/65/75 to biogas.
6. Developing a computer program which would give C/B (Cost-Benefit) ratios when fed with information such as Digester type (CSTR/AF/UASB/Diphasic), Digester capacity, HRT, VS(%), Biodegradability (%VS producing biogas), Ambient temperature, Price levels, etc.
7. Working out regression equations and incorporating them in the programme, correlating CB ratios with digester capacities so that gas guesstimates can be made for installation of any desired size.

In pursuance of the above protocol, a systems model has been developed which is summarised in the flow chart depicted in Figure 1.6 (a & b). The flow charts for the digester construction, digester heating, pumping, and benefit-cost calculations are presented in figures 1.7, 1.8 (a, b, c), 1.9, and 1.10 respectively. The notations used in the flow charts are listed vide table 1.6 . The program named COBAS (Cost Benefit Analysis of anaerobic digestion Systems) works out benefit-cost ratios when fed with the following basic information:

1. volume of feed
2. volume of feed in second digester (in case of diphasic systems only)
3. retention time
4. retention time in second digester (in case of diphasic systems only)
5. rate of gas production (anticipated or known)
6. height of the digester
7. volatile solids present
8. volatile solids destruction (anticipated or known)

9. ambient temperature
10. temperature maintained inside the digester
11. rate of interest
12. duration of benefit expected.

Upto ten different sets of such digestion options can be simultaneously processed by the programme. Of these upto three different sets can be simultaneously processed using microcomputers. The output gives five best benefit-cost ratios and the set of options that lead to the five best values. Details of the programme, including source codes, are available elsewhere (Abbasi and Nipaney, 1993). Typical simulation results are presented on the following pages.

THE PROGRAMME AND TYPICAL SIMULATION RESULTS

The programme

The programme COBAS can process up to 10 sets of variables; it can thus evaluate up to ten options (up to ten different sets of process variables) simultaneously. In order to run the programme at its maximum capacity and get output within a reasonable time a midi or maxi computer will be needed (Table 1.7). For speedily processing up to three options, with up to nine variables differing in each option, a personal computer (Pentium) is adequate having a RAM of > 4 MB.

The full codes of COBAS have been detailed elsewhere (Abbasi and Nipaney 1993).

Typical Simulation Results

Case I: Benefit-cost analysis of a digester proposed to be run in locations varying widely in climate. In this illustrative example, cases of CSTR type digesters of identical capacity and operating conditions are considered to be run at three distinct locations having average ambient temperatures 5°C, 15° and 30° C respectively.

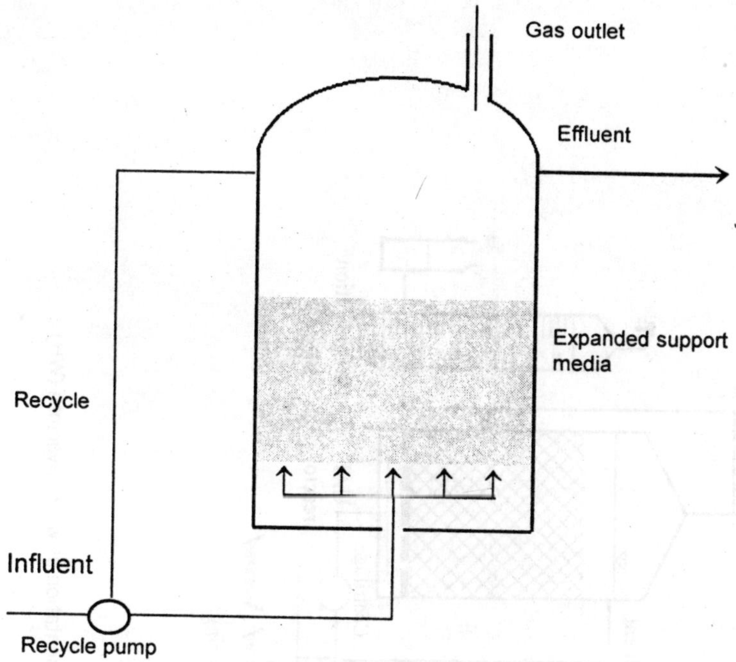

Figure 1.3. Expanded/Fluidised · bed reactor

The input set is given in Table 1.8 and the results of benefit-cost analysis are presented in Tables 1.9–1.10. It is seen that the digesters running with the given feed quantity and characteristics in locations with ambient temperature 15°C and 30°C will have B:C ratio marginally greater than unity while the digester running in the third location will be uneconomical.

Case II: Benefit-cost analysis when the same quantity and quality of feed is processed by different types of digesters at three different ambient temperatures. In this exercise a feed with VS 6%, VS conversion rate 65%, and quantity 1 TPD (tonnes per day) is processed using four digester types CSTR, AF, UASB, and Diphasic – each constructed either with RCC or with bricks.

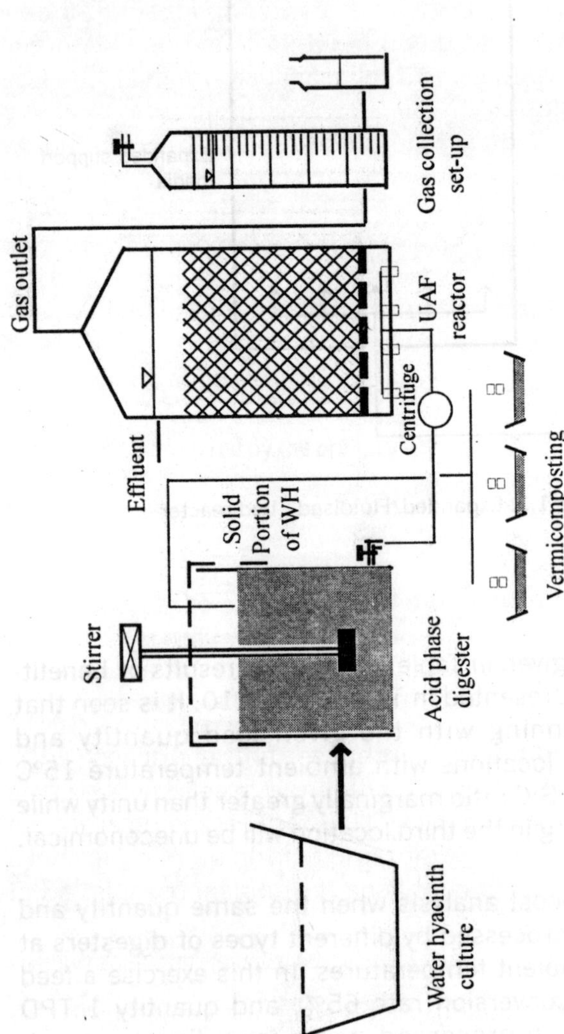

Figure 1.4. Di-phasic anaerobic fermentation of water hyacinth (WH)

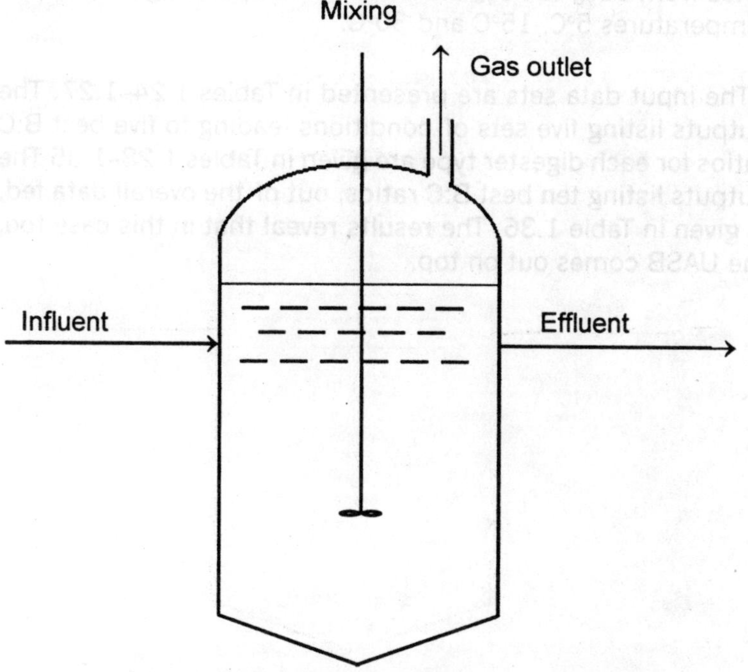

Figure 1.5. Continuosly stirred tank reactor (CSTR)

The input data given in Tables 1.11—1.14. The outputs listing three sets of conditions leading to the three best B:C rations for each digester type are given in Tables 1.15–1.22. The output listing ten sets of conditions giving the ten best B:C rations, out of the overall data feed, is given in Table 1.23. It is seen that the most economic system comes out to be a UASB digester of RCC construction running on an RT of 18 hours at the next two best B:C ratios are for the identical digesters running in the lower temperature conditions. Anaerobic filter of RCC construction operated at RTs of 3 days occupies the next three positions vis a vis the benefit-cost balance, followed by AF BRICK and CSTR.

Case III: Benefit-cost analysis of digesters running on different feeds varying in VS content and VS conversion rates. In this simulation digesters of 3 TPD capacity running on feeds varying in VS content from 4% to 6% and varying in their VS conversion rates from 55% to 75% are considered operating at ambient temperatures 5°C, 15°C and 30°C.

The input data sets are presented in Tables 1.24–1.27. The outputs listing five sets of conditions leading to five best B:C ratios for each digester type are given in Tables 1.28–1.35 The outputs listing ten best B:C ratios, out of the overall data fed, is given in Table 1.36. The results reveal that in this case too, the UASB comes out on top.

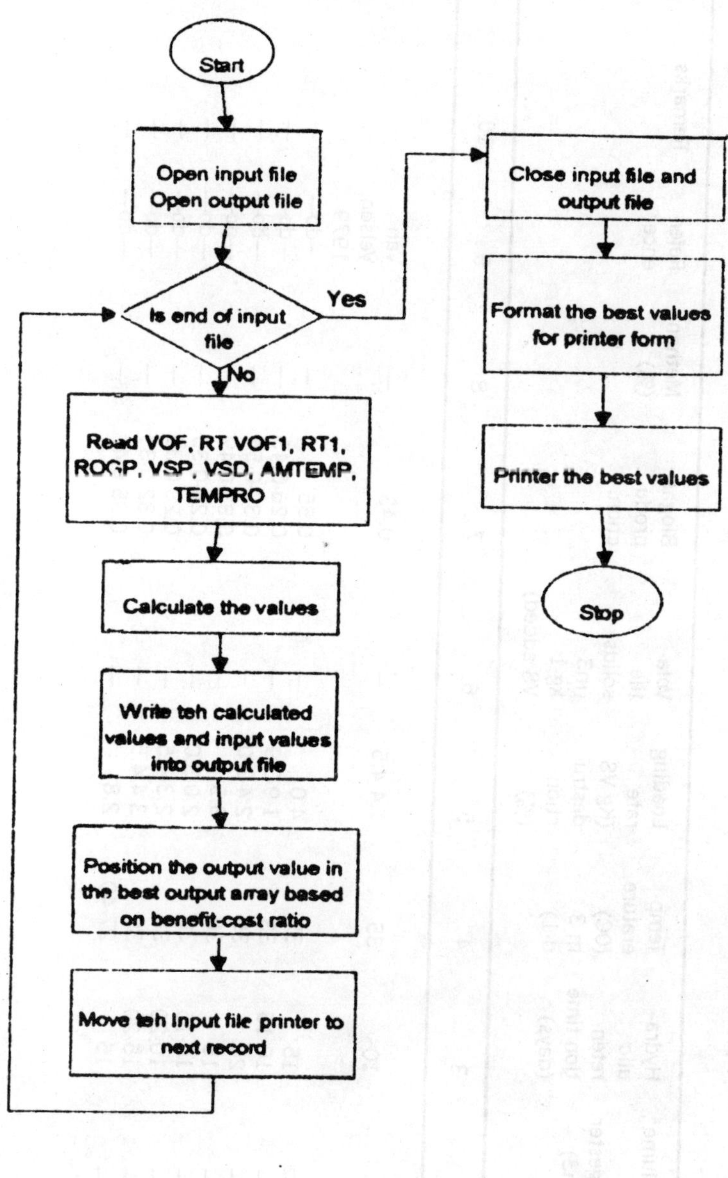

Fig. 1.6a System flow-chart of the model depicting the sub-system which address the individual reactors

Table 1.1. Continuously stirred tank reactor (CSTR)

Type of feed	Volume of digester (m3)	Hydraulic retention time (days)	Temperature (OC) m·3 d·1)	Loading rate (Kg VS destruction (%)	Volatile solids (m3 kg·1 VS added)	Biogas production	Methane (%)	References	Remarks
1	2	3	4	5	6	7	8	9	10
Piggery wastes	—	10	35	1.4·4.5	—	0.43	—	Van Velson, 1979	—
—do—	—	15	35	4.0	—	0.55	—	—do—	—
—do—	—	10·15	33	1.9·3.9	—	0.26·0.45	—	—do—	—
—do—	—	20	35	2.4·3.0	—	0.37·0.54	—	—do—	—
—do—	—	15	30	3.5	—	0.33·0.49	—	—do—	—
—do—	—	10·15	30	2.0·3.0	—	0.24·0.33	—	—do—	—
—do—	—	10·20	30	2.3·4.5	—	0.28·0.39	—	—do—	—
—do—	—	15·20	30	3.4·4.5	—	0.32·0.33	—	—do—	—
—do—	—	15	27·40	2.8	—	0.35·0.40	—	—do—	—

1	2	3	4	5	6	7	8	9	10
Fattening pigs manure + straw & silage	60	22	33·38	5.0	—	0.49	60	—do—	(i) Heating: internal serpentine coils (ii) Stirring: horizontal mechanical axle motor
Dairy cattle manure + silage waste	32	15.3	34·40	5.6	33	0.14	56	—do—	(i) Heating: external heat exchanger (ii) Stirring: gas recirculation
Distillary wastes	—	10	35	2.7	55	0.55	65	—do—	(i) Heating: not necessary since tempe-rature of feed was about 400 C (ii) Stirring: gaslift

1	2	3	4	5	6	7	8	9	10
Liquid manure from cow, heifers, calves straw vegetable material	100 (partly mixed)	10·25	30·50	2.0·5.0	35·40	0.38·0.42	59·62	—do—	(i) Heating:two external heat exchanger (ii) Stirring: Internal propeller, installed in the central tube, acting downwards centrifugal pump
Piggery wastes	18.6	10·15	—	6.0	—	0.47	70·71 1980	Kirby,	—
Dairy manure + Barley straw	0.13	25	35	5.2	29	0.84	65	Hills, 1980	Heating: Circulation of hot water in a jacket surrounding the unit.

1	2	3	4	5	6	7	8	9	10
	0.13	20	35	6.5	28	0.93	64	—do—	—do—
	0.13	15	35	8.7	26	0.96	61	—do—	—do—
	0.13	10	35	12.5	24	0.83	58	—do—	—do—
Primary sludge	—	17	35	1.4	—	0.37	68	Chynoweth & Srjvastava, 1980	—
Macrocystic pyrifera (raw kelp)	2++	18	35	1.6	50.8	0.478	58.2	,,	Heating: Keeping the digesters in a constant temparature chamber Stirring: intermittantly (15 min hourly) by magnetic stirrer at 130 rpm
—do—	2++	10	35	1.6	38.6	0.37	58.8	,,	—do—
Baseline treated kelp	2++	10	35	1.6	35.6	0.35	59.3	,,	—do—
—do—	2++	18	35	1.6	36.2	0.35	59.8	,,	—do—

1	2	3	4	5	6	7	8	9	10
Baseline treated kelp	2++	18	35	1.6	35.6	0.40	58.7	,,	—do—
Juice (4:1)									
—do—	2++	10	35	1.6	24.0	0.32	64.1	,,	—do—
—do—	2++	18	35	1.6	42.0	0.35	60.3	,,	—do—
(3:2)									
—do—	2++.	10	35	1.6	23.3	0.18	66.1	,,	—do—
Raw kelp (without nutrient)	2++	12	35	1.6	45.1	0.41	57.8	,,	—do—
Raw kelp (with nutrient)	2++	12	35	1.6	42.6	0.39	59.8	,,	—do—
Cow manure	35	15	—	—	—	0.28@	55	Jewell et al, 1981	—
	35	10	—	—	—	0.21@	60	—do—	—
Cattle manure	10	24	35	2.5-2.8	— 0.25	0.20·	60	Hansen, 1983	—do—
Screened dairy	0.003	16	22	2.1	–12.2(3.3)***	0.124	63.5	Lo et al	—

1	2	3	4	5	6	7	8	9	10
manure								1985	
	0.003	15	22	1.9	–13.8(2.9)***	0.149	64.6	—do—	—
	0.003	12	22	2.9	–11.8(3.4)***	0.112	63.5	—do—	—
	0.003	10	22	2.8	–14.3(2.8)***	0.102	64.6	—do—	—
	0.003	10	22	2.9	–13.8(2.9)***	0.115	62.7	—d0—	—
Water hyacinth + sewage sludge	—	31·8	—	1.6-6.4	—	—	0.25	Hayes et al 1985	
Brown Algae Laminar-ia hyper-borea	10	24	35	1,7	—	0.53	52.8	Hanssen,et al,1987	Heating fitted with an automatic temperature control Stirring : agitation system wtih continuously adjustable speed, stirring for 10 minutes every 2 hour

								Process: semicontinu-ous, feeding once a day	
Laminaria saccharina	10	24	35	1.7	.	0.45	51.1	—do—	—do—
Ascophy-llum nodosum	10	24	35	1.8	.	0.22	50.0	—do—	—do—
Palm oil Mill Effluent	0.015	14	32	3.2*	50.8*	0.19	53	Ng, et al 1987	—
		21	32	2.1*	50.6*	0.25	56	—do—	—
Petro chemical industry wastewater (DMT)	2.5 L	10	35	3.3**	.	.	.	Sharma et al 1994	—

++ Volume of digester 2dm3, semicontinuos.
@ At standard condition, 20 0C and 1 atm.
* As COD.
** COD g l·1.
*** Value given in the parenthesis are VS (%) in the feed.

Table 1.2. Anaerobic filters (AF)

Type of feed	Volume of digester (litres)	Filter media	Hydraulic retention time (days) d-1)	Temperature (0C)	Loading rate (Kg COD m-3	COD reduction (%)	Biogas production (m3 kg-1 COD added)	Methane (%)	References	Remarks
1	2	3	4	5	6	7	8	9	10	11
Pig waste	200	—	2-8.5	26-30	3.74-15.65	—	—	—	Chavedej, 1980	Active liqu d volume =127 l
Silage Effluent	8.4	Lime Stone chips	3	28 + 3oC	4.7	76	0.49*	84	Barry & Colleran, 1982	—
Water hyaci-nth + Bermuda grass + MSW	6.4	Plastic raschig rings	5.2	35	6.1	94	0.60**	68	Ghosh & Henry, 1982	Active liquid volume = 2.5 l
Cane molases	8.1	—	9	—	7.0	50-70	—	—	Szendrey, 1983a	—

1	2	3	4	5	6	7	8	9	10	11
stillage —	—	—	1·10	—	1·10	80·95	—	—	Van den Berg, & Kennedy, 1983	—
Sows and weaness manure	3500	Plastic matrix	5.5	25	3.7·4.7	66	0.26	86	Demuynck et al,1984 exchanger Stirring : none	Heating: external heat
Pig slurry	18	clay, coral, mussel shells, plastic	6	33+2oC	5	69·73.3	0.39·0.46**	85·87	Wilkie & Colleran, 1984	Active liguid volume =12.4 L
Sulphite pulp mills	10	—	6.2	—	0.44	69	—	—	Geller & Eattsching 1985	—
	10	—	4.5	—	0.63	79	—	—	—do—	—
	10	—	2.1	—	1.3	90	—	—	—do—	—
Pig slurry super	3500	Poly·propy·lene	6	25	2.2	66	0.26	86	Wilkie and Colleran, 1986	Empty bed volume = 2800 litres

1	2	3	4	5	6	7	8	9	10	11
natant		cascade minirings								
—do—	3500	—	6	25	4.3	66	0.26	86	—do—	—do—
—do—	3500	—	3	25	8.4	52	0.20	87	—do—	—do—
—do—	3500		3	35	9.9	60	0.25	87	—do—	—do—
Distil-lary water + water (1:1)	9	Stone rubble	30	24.5-31.0	2.05	77	—	72	Gadre & Godbule 1986	Active volume of the digester = 3.6 l
	9		20	24.5-31.0	3.02	74	—	65	—do—	—do—
	9		15	24.5-31.0	3.46	62	—	60	—do—	—do—
Raw Dis tillary water	9	Stone rubble	30	24.5-31.0	3.56	72	—	65	—do—	—do—
	9		20	24.5-31.0	5.74	60	—	63	—do—	—do—
	9		15	24.5-31.0	6.81	55		59	—do—	—do—
Mining wastewater	1.0	dolomitic pebbles	20 h	—	3.3	—	—	—	Marce et al 1987	Loading rate as mgL-1
Cattle slurry	20	Filter-pak CR 50 rings	5-14	35	—	—	0.26	—	Peck & Hawkes, 1987	—
Molasses wastewater	1,0	Pumice stones	—	35	10.0	15	—	—	Hilton and Archer, 1988	—

1	2	3	4	5	6	7	8	9	10	11
Olive mill wastewater	11.5	Polyurethane foam	2	—	6	65·70	—	—	Rossi et al 1989	Total volume in dm3 LR as KgTOC
Swine slurry	15	wood chips PVC Clay	4.5 4.5 4.5	30 30 30	1.9 3.1 1.4	— — —	212 144 32	72 77 68	Sorlini et al 1990	Loading rate Ld·1 Gas yield Lkg·1VS
Water hyacinth	73.63	—	10	28·30	—	—	130	64	Delgado et al, 1992	Gas yield as methane L K5·1 VS
Olive mill effluent	—	—	2,1	—	40	70·80	—	—	Andreoni et al, 1993	—
Date pro-cessing Industry waste	—	Plastic pipes	—	—	1.6·25	90	0.4·0.53	—	Boller, 1993	—
Palm oil mill effluent (POME)	23	—	15·6	—	1.2·11.4	90	20·165#	60	Borja and Banks, 1994 expressed as dm3	Vol.of reactor
Tuna process-ing waste	–1	PVC rasching rings	4·1	37	11·13	75	—	—	Veiga etal, 1994	—

1	2	3	4	5	6	7	8	9	10	11
Sea food processing wastewater	—	—	6.6	—	1.3	65	1.3	—	Prasertsan et al, 1994	—
Dairy waste	4.3	Polypropy lere ras ching rings	24-40 h	35-37	9	70	—	—	Monroy et al, 1994	Vol. of the reac tor expre ssed as m3
Synthetic wastewater	0.9	Plastic tubes	3-9	20	0.27-0.82	81-90	—	54-78	Hanaki et al, 1994	—
Synthetic wastewater clay pellets	—	Fire expanded	1.1	—	—	92.8	6.4@	—	Chua and Fung, 1995	Reactor content was re cycled
-do-	—	-do-	2	—	—	74.1	2.13@	—	-do-	-do-

* Gas yield from per Kg COD de
** Gas yield calculated for STP
@ Gas yield interms of Liters per liter per day
\# Gas yield interms of dm3 d-1
h HRT interms of hours
$ in terms of TOC%

Table 1.3. Upflow anaerobic sludge bed (UASB)

Type of feed	Volume of digester (m3)	Hyadraulic retention time (days)	Temperature (0C)	Loading rate (Kg COD m·3 d·1)	COD destruc·tion (%)	Biogas yield (m3kg·1 COD added)	Methane (%)	Reference	Remarks
1	2	3	4	5	6	7	8	9	10
Sugar beet sap,un·soured	0.064	1·2	30	4·5	95	—	—	Lettinga & Van Velson, 1974	—
Sugar beet sap, soured (closed circuit)	0.018	0.5·1	30	8·10	84·95	—	—	—do—	—
Sugar beet sap soured (two step)	0.018	1	30	8·9	97	—	—	—do—	—

1	2	3	4	5	6	7	8	9	10
Beet sugar waste	—	0.7·2.1	25	5·8	95	0.47	83	Ghosh, 1982	—
Thermal sludge conditi-oning liquor	—	2	35	5.9	71	0.18*	—	Hall & Jovanovic, 1983	—
Raw sewage	6	1·0.3	20	0.04	70·75	0.12·0.16	—	Grin,et al 1983	—
	—	0.5·6	—	5·30	85·95	—	—	Van den Berg Kennedy, 1983	—
Paper mill effluent	70	2.5	—	9·9.5	70 0.41	70·80		Habets and Knelissen 1985	Gas yield in m3 kg·1 COD removal
Maize starch waste water	800	0.8	40	15	90·95	180·200	—	Hulstaff pol & Lettinga, 1986	—
Acid water	20	0.25·0.5	—	—	—	—	—	Rinzema & Schultz 1987	

1	2	3	4	5	6	7	8	9	10
from edible oil refinery Sewage	3.7	10·18 h	24·26	660mg·1	73	140	—	Nobre et al 1987	Gas yield in Nl kg·1 COD added
Raw domastic sewage	35	5.2 h	23·24	0.43·0.52@	66	—	—	Schellinkout et al, 1988	
·do·	120 L	4 h	19·28	0.627@	74	—	—	Barbosa and Sant Anna Jr. 1989	
Domestic sewage	6	0.3	9.5·1Q	—	—	0.17·0.24	30·55	Lettinga et al,1980	—
Glucose	3L	—	—	166.44 gd·1	99.7	8.5 Ld·1	—	Palns et al 1990	
Cheese production	4	46 h	35	31@	90	—	—	Rico Gutie rrez et al, 1991	
Thermo mechanical pulping wastewater	—	0.9 h	55	81	61	—	—	Rintala and Lepisto, 1992	—

1	2	3	4	5	6	7	8	9	10
Methanolic waste	—	120	—	32@	70	—	—	Bhatti et al, 1993	—
Synthetic waste	5.5 L	—	55	87	91	—	—	Uemura and Harada, 1993	—
Slaughter house waste	—	4·1.5 h	30	15	80·95	—	—	Sayed et al, 1993	—
Domestic wastewater	1.2	—	>12	—	60	300**	78	Bogte e al, 1993	—
Synthetic coffee waste	—	24 h	>55	4	>75	—	—	Shi and Forster, 1993	—
Synthetic waste	1.7 L	—	30	0.18·2	98	—	—	Visser el al, 1993	Loading rate reported as g CODg·1 VSSd·1
Synthetic waste	1.2 L	—	—	13.8·39.6@	57.8·96.7	—	—	Shen et al, 1993	—
Sewage	67.5	—	—	—	80	—	—	Vieria et al, 1994	—
Synthetic starch/ sucrose waste	—	—	—	0.5@	75	—	—	Harada et al, 1994	—

1	2	3	4	5	6	7	8	9	10
Synthetic resin pro duction wastewater	—	—	—	4.5	71	2.94#	—	Peng et al, 1994	Seeding material without acclimati sation
·do·	—	—	—	4.5	78.	3.5#	—	—do—	Seeding was accli matisd for 30 days
Low strength wastewater with Ethanol/ whey	—	—	30	0.3·6.8@	95	—	—	Kato et al, 1994	—
Wastewater with formate	—	—	37	10·20 29·65@	97·98 86·90	—	—	Chui et al, 1994	—
Malt whisky wastewater	—	2.1	—	15	90	—	—	Goodwin & Stuart 1994	—
Tannery effluent	10 L	12 h	—	—	70	0.15·0.25	80	Rajamani et al, 1995	—
Domestic wastewater	2 L	4.6 h	20	3	53·76	—	—	Soto et al, 1995	—

1	2	3	4	5	6	7	8	9	10
Tannery wastewater	4 L Ld-1	5·7	29·34	1.5	39·50	0.1-0.3**	—	Van Gro eviestijn, 1995	—
Wheetstraw pulp black liquor	—	—	—	10	40·46	510$	—	He et al, 1995	—
Domastic wastewater	5 mld	—	—	—	75.3	0.15	75·80	Khan, 1995	

* = As methane yield
L = Volume of the reactor expressed as liters
h = HRT in terms of hours
@ = Loading rate interms of g COD per liter per day
# = Biogas yield interms of dm3 per dm3 of wastewater per day
** = Biogas yield interms of liters per day
$ = Biogas yield interms of liters per Kg COD added

Table 1.4. Performance data for fluidized and expanded bed reactors

Type of waste	Waste strength g/L*	Suspended solids g/L	Reactor size L	Reactor temperature C	Loading rate g.L-1.day-1	Conversion %	Reference
1	2	3	4	5	6	7	8
Liquor from heat-treated sewage digestor sludge	4·9	—	7	35	4.2·26.5	48·75	Bell et al 1980
Whey	10	—	0.5	35	10·20	87	Switzenbaum 1980
Raw sewage	0.15·0.3	<0.1	1.0	20	0.65·35	0·85	Jewell et al 1981
Synthetic waste with cellulose	0·3·1.8	—	0.5	30	2	74·83	Morris and Jewell 1981
					5	75·83	·do·
					8	40·83	·do·
Thermal sludge conditioning liquor	11	<0.5	29	35	0·30	52	Hall 1982
Whey	55*	—	50	35	17·37	65·84	Hickey 1982

1	2	3	4	5	6	7	8
Whey permeate	10·30	—	60*	30·35	8·24	80·90	Li et al 1982
Black-liquor Condensate	1.4	—	1	22	10	80	Norman 1982
Synthetic	9.7	—	—	36	19.7#	72	Sutton and Huss, 1984
refining wastewater							
Synthetic wastewater	1.1	—	—	35	4.5 (VS)	25·93	Toldra et al, 1986
Synthetic waste with glucose	—	18.6 mg	—	35	2 4·3.6#	—	Yoda et al, 1987
Wine industry waste	7.8@	—	—	35	6.20#	91.1	Converti et al, 1990
Brewary waste water	—	—	—	25	27·30#	85	Liang et al, 1993

Brewary waste water	—	—	1	35	—	95	Borjra et al, 1993
Synthetic wastewater	—	—	—	—	30	83·91	Rinzema, 1993
Pickled-plum effluent	—	—	—	—	11.1	84.6	Tanemura et al, 1994
Ice-cream wastewater	—	—	—	35	15.6	94.4	Borja et al, 1995

Note : Results are based on chemical oxygen demand (COD), biological oxygen demand (BOD), or total volatile solids (TVS). Loading rates given are mostly the highest reported or achievable. Loading rates and conversions are given in COD. BOD or TVS depending on units used for waste strength.

* Measured as COD

\# Loading rate expressed in Kg COD m^{-3} d^{-1}

@ Kg COD m^{-3}

Table 1.5. Diphasic digestion (DD)

Type of feed	Volume of di-gester (litre)	Hydraulic retention time(days)	Tempera-ture(0C)	Loading rate (kg vs m-3d-1)	Type of methane reactor	Volatile solids destruc-tion(%)	Biogas production (m3 kg-1 vs added)	Methane (%)	Reference	Remarks
1	2	3	4	5	6	7	8	9	10	11
Glucose	AP=10L	AP=0.67	37	—	CSTR	—	AP=0.012	AP=5	Massey &	without biomass
	MP=10L	MP=2.8	37	—	CSTR	—	MP=0.065	MP=71 1978	Pohland,	recycle
—do—	—do—	AP=0.8	37	—	CSTR	—	AP=0.032	AP=20	—do—	—do—
		MP=3.4	37	—	CSTR	—	MP=0.072	MP=70		
—do—	—do—	AP=1.0	37	—	CSTR	—	—	—	—do—	—do—
		MP=4.0	37	—	CSTR	—	—	64		
—do—	—do—	AP=1.0	37	—	CSTR	—	—	—	—do—	with biomass
		MP=4.9	37	—	CSTR	—	—	70		recycle
—do—	—do—	AP=0.6	37	—	CSTR	—	—	—	—do—	—do—
		MP=4.0	37	—	CSTR	—	—	66		
—do—	—do—	AP=0.7	37	—	CSTR	—	—	—	—do—	—do—
		MP=2.0	37	—	CSTR	—	—	66		
—do—	—do—	AP=0.5	37	—	CSTR	—	—	—	—do—	—do—
		MP=2.0	37	—	CSTR	—	—	65		

1	2	3	4	5	6	7	8	9	10	11
Indust- rial waste	—do—	AP=1 MP=4	37 37	— —	CSTR CSTR	— —	— —	— 68	—do—	—do—
		AP=1.4 MP=5.6	37 37	— —	CSTR CSTR	— —	— —	— 73	—do—	—do—
Water hyacin- th + Ber- muda gra- ss + MSW	AP=25 MP=26.7	AP=0.8 MP=2.33	35-40	AP=20 MP=1.39 =1.54 =2.07	Anaero- bic fil- ters	—	0.43 0.58 0.35	73 72 71	Ghosh & Henry,1982	— —do—
Beet sugar waste	MP=7m3 (UASB)	AP=0.2- 0.4 MP=0.7- 2.1	35	AP=17- 29 MP=3-8	Upflow anaero- bic slu- dge bed reactor	95*	0.47*	83	Ghosh,	— 1982
Sewage sludge	—	AP=2.73 MP=4.58	35 35	AP=8.64	— —	80 —	— 0.538	— 68.1	Ghosh, 1982	—
Chopped straw	30	18***	—	—	CSTR	45-52	0.16- 0.19*@	—	Barry,1983	

1	2	3	4	5	6	7	8	9	10	11
NaOH-treated straw	30	18***	—	—	CSTR	75-82	0.25-0.27*@	—	—do—	—
Grass clippings	—	12-16***	—	—	CSTR	62-71	0.25-0.33*@	—	—do—	—
Cabbage leaves	—	16***	—	—	CSTR	—	0.29-0.35*@	—	—do—	
Coffee pulp juice	AP=12 MP=1.5	AP=1	35	—	Anaerobic filters	—	—	—	Calzada et al,1984	—
		MP=5	35	MP =				—		—
		=4	35	5.2-8.8				89-92	—do—	—
		=3	35	6.5-11.0	—	—	—	91-97	—do—	—
		=2	35	8.7-14.7				85-92	—do—	—
				13.0-22.0				89-91		
—do—	MP=12	MP=3	35	8.5	—	—		81	—do—	—
		=2	35	15.2	—	—		82	—do—	—
Synthetic feed	AP=2.5 MP=10	AP=0.7-1.9 MP=2.9-7.7	35 —	AP=4.9-54.1 MP=0.98-12.2*	CSTR	AP=9.8-22.6* MP=97.9-99.8*	0.42-0.50*	60-81	Anderson et al 1986	—

1	2	3	4	5	6	7	8	9	10	11
Water Hyacin-th	—	AP=2.1 MP=5.7	35·37 35·37	— —	CSTR —	— —	— b	— —	Saraswat & Khanna, 1986	Water hyacinth was treated with alkali for 24 hours
—do—	—	AP=2 MP=3.2	35·37 35·37	— —	CSTR —	— c	— —	— —do—		Without pretreat-ment of water hycianth
Palm oil mill effluent	AP=0.87 MP=12 & 15	AP=1 MP=10 20 30	32	3.4* 1.8* 1.2*	CSTR CSTR CSTR	63.9* 67.3* 74.1*	0.44** 0.75** 0.98**	60 61 57	Ng et al, 1987	—
Glucose	AP=4 MP=16	AP=0.25 MP=1·4	35	—	—	—	—	—	Schwitzguebel & Peringer, 1987	—
Ethanol fermen-tation waste water	AP=45dm3 MP=11dm3	AP:2 MP:17h	37	AR:15.1* MP:34.5*	Fixed bed loop reactor	—	—	—	Alvasidis et al 1989	—

1	2	3	4	5	6	7	8	9	10	11
Ice cream wastewater bed		AP=IOL MP	AR=1 MR=1	4.5* dised	Flui·	—	0.157**	75	David et al 1990	—
Synthetic meat extract waste water		MP=13.5 MP=2	37	MP=8.84*	Flui· dised bed	—	—	—	Dinopoulou et al 1990	—
Vegetable waste	24	30	RT	—	Floating doom bio· gas plant	—	465@	69.5	Ranade et al, 1991	—
—do—	200	30	—	6.6	—do—	593@	65.3		—do—	
Thermo chemical pulp water (cellulosic wastewater)	—	21 h	—	11	—	90(COD)	—	—	Vinas et al 1993	—
Coffee waste	0.45(MR) 2(AR)	—	37	2$	—	—	1.7 L#	—	Ikbal et al	— 1994

1	2	3	4	5	6	7	8	9	10	11
Icecream waste water	AP=3 MP-2L	AR=0.3 MR=1	35	AR 17.3*g MR 5.0*g	Anaerobic filter	— —	0.2 0.35	AR 34 MR 64	Borja and Banks 1995b	—
Municipal solid waste (MSW)	AP=1 MP=2.5	AR=6 MR=6	29-35	—	Anaerobic packed bed	80** 90	10	40 · 50	Bhoyar & Bhide 1996	

AP = Acid Phase; MP = Methane Phase

- * As COD
- ** As COD destroyed
- *** Overall (Total) hydraulic retention time
- *a As methane yield
- b Gas yield 50 l kg-1 WH(dry) d-1
- c Gas yield 32.5 l kg-1 WH(dry) d-1

AR Acid phase reactor
MR Methane phase reactor

- h HRT expressed as hours
- @ Biogas yield interms of liters per Kg TVS
- # Biogas yield interms of liter per gram TOC
- $ Loading rate as gram TOC per litre per day
- \+ Biogas yield interm of liter per Kg MSW

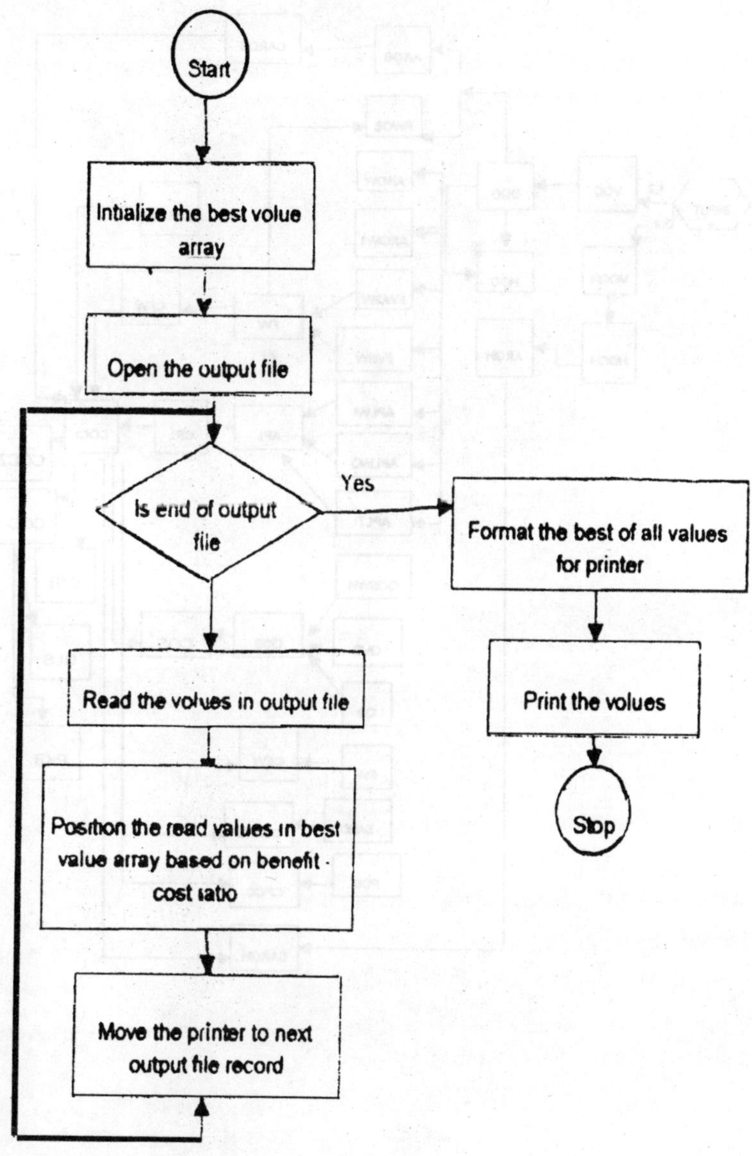

Fig. 1.6b System flow-chart of the model depicting the sub-system which address the individual reactors

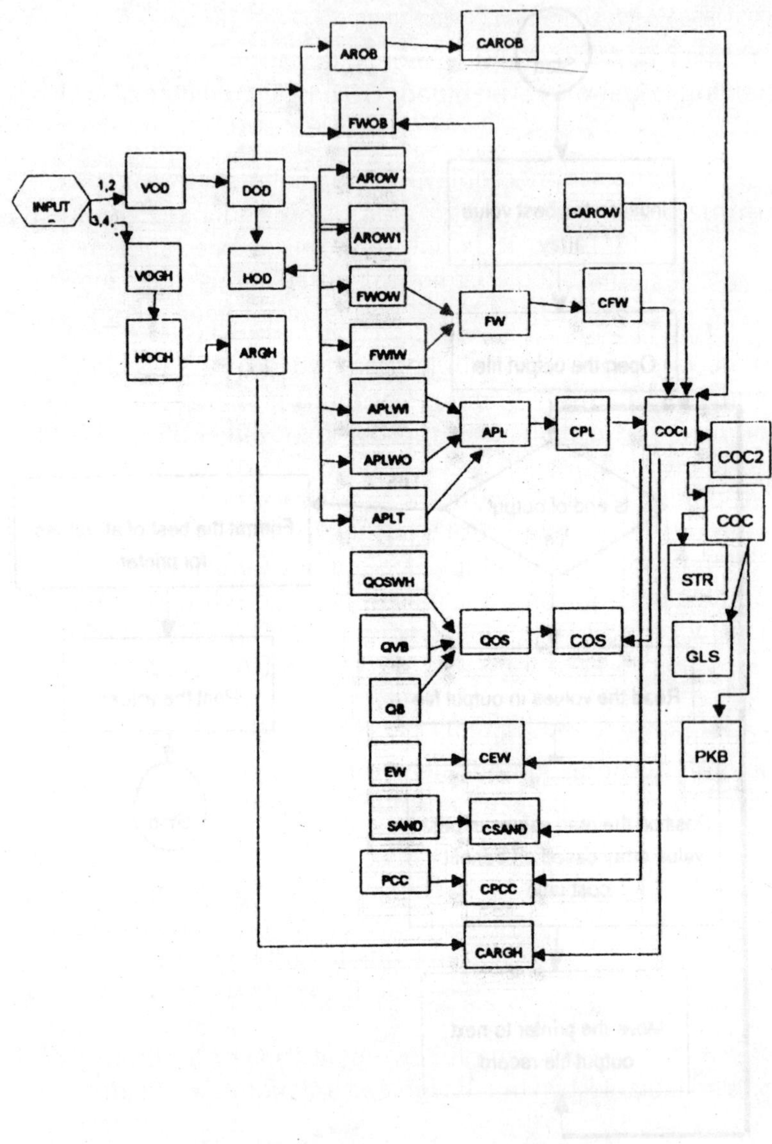

Input are: 1: VOF; 2: RT; 3:ROGP; 4:VSP; 5:VSD; 6:AMTEMP; 7:TEMPQ; 8:ROI; 9:Y; 10:VOF1; 11:RT1;; 12:HOD, if applicable

Fig. 1.7 System flow-chart for civil engineering design of the digester and its costing.

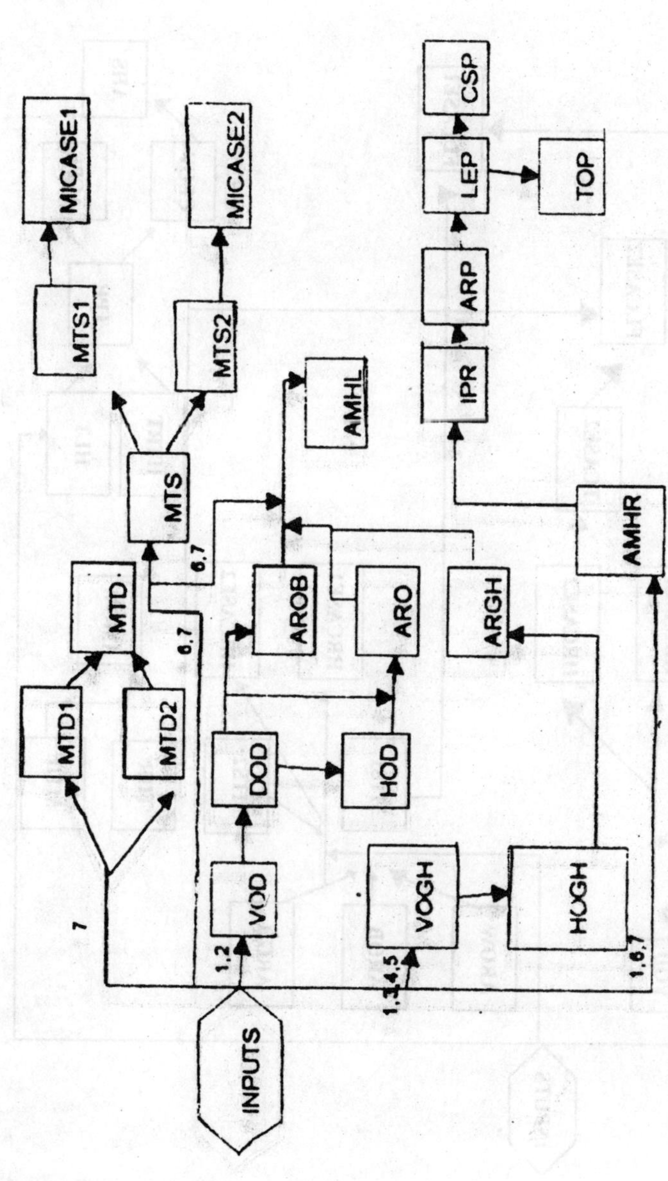

Inputs are: 1:VOF; 2:RT; 3:ROGP; 4:VSP; 5:VSD; 6:AMTEMP; 7: TEMPQ; 8:ROI; 9:Y; 10:VOF1; 11:RT1; 12:HOD, if applicable

Fig. 1.8a System flow-chart for heating system design

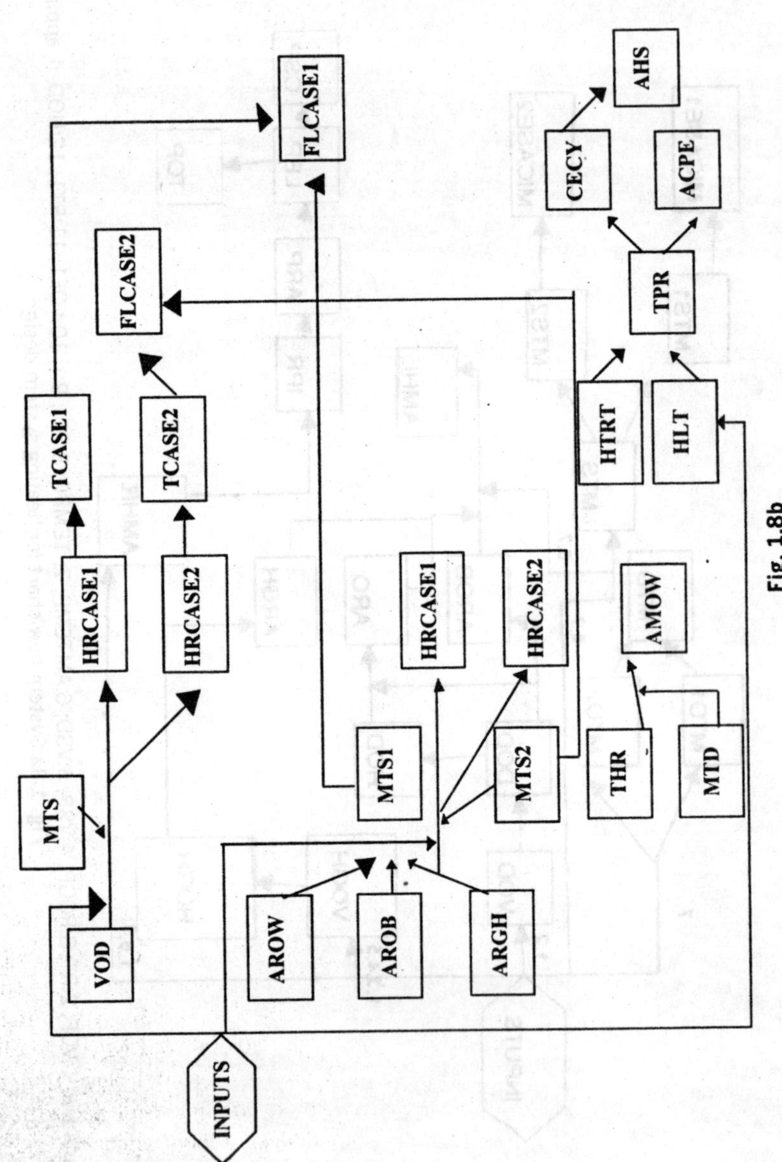
INPUTS
VOD
MTS
HRCASE1
HRCASE2
TCASE1
TCASE2
FLCASE2
FLCASE1
AROW
AROB
ARGH
MTS1
MTS2
HRCASE1
HRCASE2
THR
MTD
AMOW
HTRT
HLT
TPR
CECY
ACPE
AHS

Fig. 1.8b

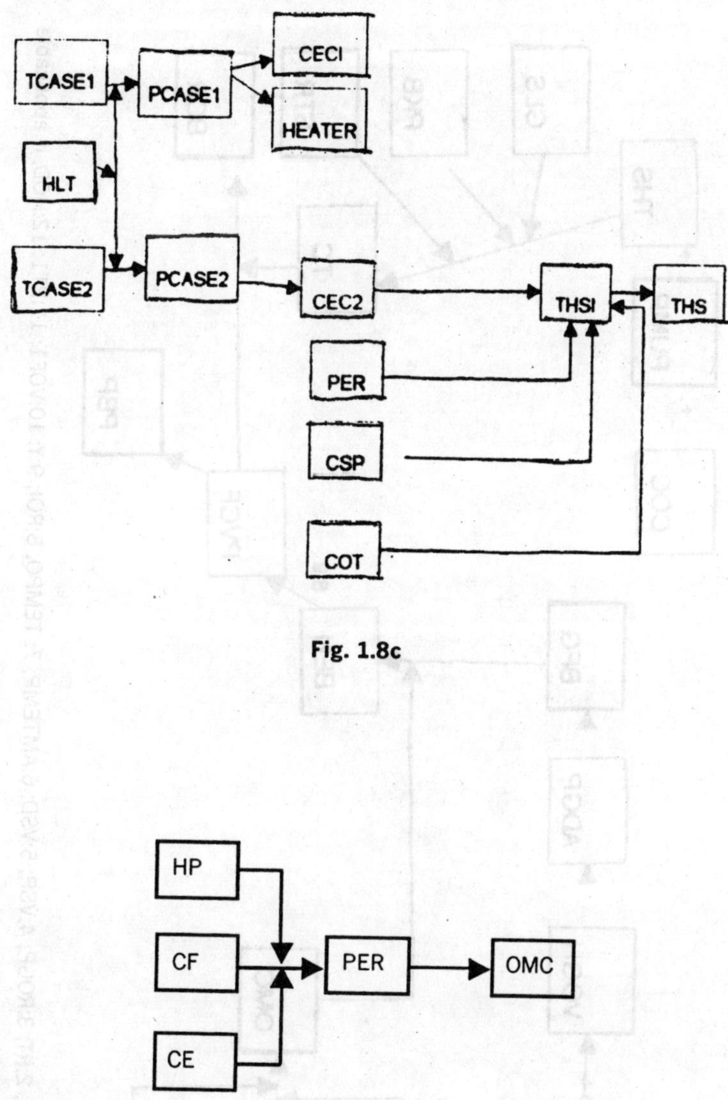

Fig. 1.8c

Inputs are: 1:VOF; 2:RT; 3:ROGP; 4:VSP; 5:VSD; 6:AMTEMP; 7: TEMPQ; 8:ROI; 9:Y; 10:VOF1; 11:RT1; 12:HOD, if applicable

Fig. 1.9 System flow-chart for pumping system design of the digester and its costing.

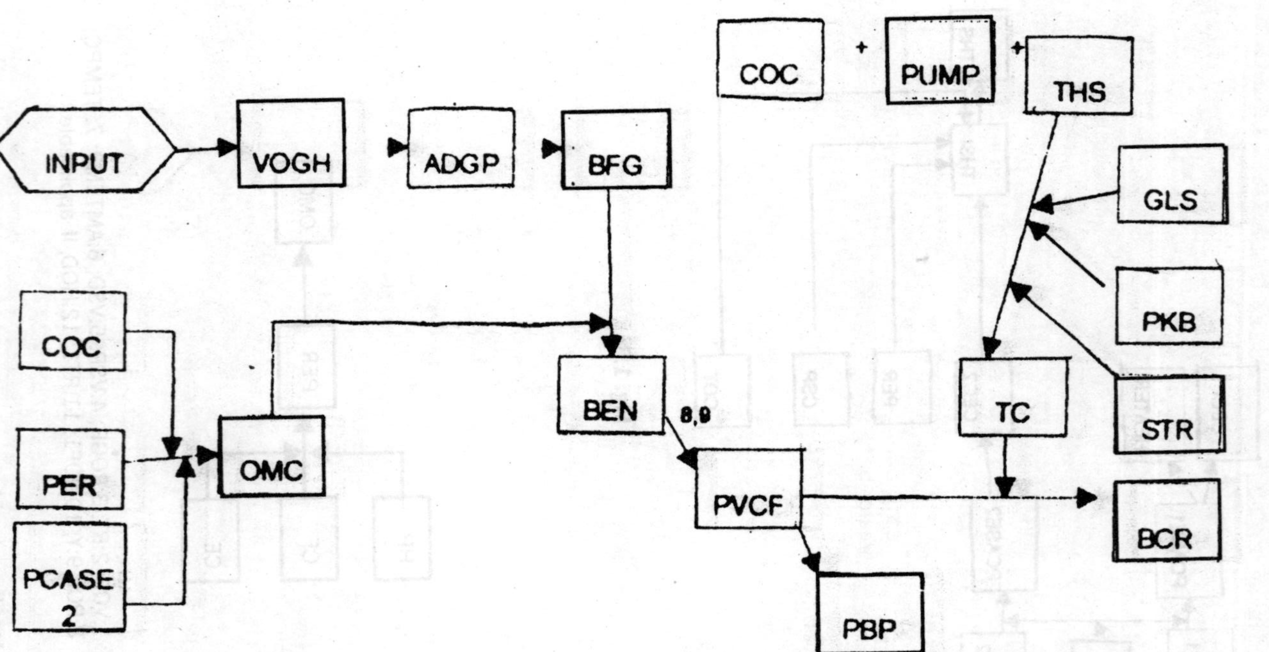

Inputs are: 1:VOF; 2:RT; 3:ROGP; 4:VSP; 5:VSD; 6:AMTEMP; 7: TEMPQ; 8:ROI; 9:Y; 10:VOF1; 11:RT1; 12:HOD, if applicable

Fig. 1.10: System flow-chart for cost-benefit ratio and pay-back period

Table: 1.6 Notations

AMHL	:	Amount of heat loss from the digester of CSTR, AF, UASB and DD(AP), watts
AMHL1	:	Amount of heat loss from the methane phase of diphasic digester, watts
AMHR	:	Amount of heat required to maintain desired temperature in CSTR, AF, UASB and DD(AP), watts
AMHR1	:	Amount of heat required to maintain desired temperature in DD(MP), watts
AMOW	:	Amount of water required for circulation in CSTR, AF, UASB and DD(AP), litres
AMOW1	:	Amount of water required for circulation if DD(MP), litres
AMTEMP	:	Ambient temperature, OC
AOB	:	Area of brick wall in CSTR, AF, UASB and DD(AP), m2
AOB1	:	Area of brick work in DD(MP), m2
AOGP	:	Amount of gas produced in CSTR, AF and UASB, m3
AOGP1	:	Amount of gas produced in methane phase of diphasic digester, m3
APL	:	Total area of plastering, m2
APLGHI	:	Area of plastering of inner wall of gas holder, m2, in CSTR, AF and DD(AP)
APLGHO	:	Area of plastering of outer wall of gas holder, m2, in CSTR, AF and DD(AP)
APLGHT	:	Area of plastering of top of gas holder, m2, in CSTR, AF and DD(AP)
APLT	:	Area of plastering of top of digester wall in CSTR only, m2
APLT1	:	Area of plastering for the top of the digester wall, m2
APLWI	:	Area of plastering of inner wall of digester, m2, in CSTR, AF and DD(AP)
APLWI1	:	Area of plastering for the inner wall of the digester DD(MP), m2
APLWO	:	Area of plastering of outer wall of digester, m2, in CSTR, AF and DD(AP)
APLWO1	:	Area of plastering for the outer wall of the digester in DD(MP), m2
ARGH	:	Area of the gas holder, m2, in CSTR, AF, UASB and DD(AP)
ARGH1	:	Area of the gas holder, m2, in DD(MP)
AROB	:	Area of base, m2, in CSTR, AF UASB and DD(AP)
AROB1	:	Area of the base of the digester, m2, in DD(MP)
AROW	:	Area of wall for calculating heat losses, m2, in CSTR, AF and DD(AP)
AROW1	:	Area of the wall of the digester through which heat loss occurs, m2, in DD(MP)
AROW2	:	Area of wall for RCC work, m2, in CSTR, AF and DD(AP)
AROW3	:	Area of the wall of the digester for RCC structure, m2, in DD(MP)
ARP	:	Surface area of the heating pipe in CSTR, AF, UASB and DD(AP), m
ARP1	:	Surface area of the heating pipe in DD(MP), m2
AVTH	:	Average thickness of the brick wall in CSTR, AF, UASB DD(AP), m

AVTH1	:	Average thickness of the brick wall in DD(MP), m
BCR	:	Benefit cost ratio
BEN	:	Benefit, Rs
CARGH	:	Cost of gas holder of the digester, Rs
CAROB	:	Cost of base of the digester, Rs
CAROW	:	Cost of wall of the digester, Rs
CEC1	:	Cost of electricity consumption in first half of heating of the slurry in CSTR, AF, UASB and DD(AP), Rs
CEC2	:	Cost of electricity in the first half of the heating in DD(MP), Rs
CEC3	:	Cost of electricity consumption in second half of heating of the slurry in CSTR, AF, UASB and DD(AP), Rs
CEC4	:	Cost of electricity in the second half of the heating in DD(MP), Rs
CECY	:	Cost of electricity consumption in an year, Rs
CECY1	:	Annual consumption of electricity
CF	:	Heat transfer coefficient of floor, concrete, W/m2 O
CEW	:	Cost of earth work, Rs
CFW	:	Cost of frame work, Rs
CGH	:	Heat transfer coefficient of steel, Mild steel, W/m2 OC(gas holder)
COB	:	Cost of brick work in CSTR, AF, UASB and DD(AP), Rs
COB1	:	Cost of brick work in DD(MP), Rs
COC	:	Cost of construction with 15% unforseen expenditure, Rs
COC1	:	Cost of construction, without any unforseen expenditure, Rs
COC2	:	Cost of construction with 10% petty expenditure, Rs
COS	:	Cost of total steel bars, Rs
COT	:	Cost of water heating tank, Rs
CP	:	Heat transfer coefficiency of the heating pipe
CPCC	:	Cost of PCC, Rs
CPL	:	Cost of plastering of the digester, Rs
CSAND	:	Cost of sand filling, RS
CSP	:	Cost of steel pipe used for hot water circulation, Rs
CT	:	Heat transfer coefficiency of water heating tank
CW	:	Heat transfer coefficient of wall, concrete, W/m2 OC
DOD	:	Diameter of the digester, m, in CSTR, AF, UASB and DD(AP)
DOD1	:	Dia of the digester, m, in DD(MP)
DOGH	:	Diameter of the gas holder, m, in CSTR, AF, UASB and DD(AP)
DOGH1	:	Dia of the gas holder, m, in DD(MP)
DOS	:	Density of the slurry, Kg m-3
EW	:	Volume of earth work, m3, in CSTR, AF, UASB and DD(AP)
EW1	:	Volume of earth work, m3, in DD(MP)
FLCASE1	:	Flow of water during first half of the heating in CSTR, AF, UASB and DD(AP), kg sec-1
FLCASE2	:	Flow rate in the first half of the heating in DD(MP), kg sec-1
FLCASE3	:	Flow of water during second half of the heating in CSTR, AF, UASB and DD(AP), kg sec-1
FLCASE4	:	Flow rate in the second half of the heating in DD(MP), kg sec-1
FW	:	Total frame work, m1
FWB	:	Frame work for the base, m2, in CSTR, AF, UASB and DD(AP)
FWB1	:	Frame work for the base of the digester, m2, in DD(MP)

FWGHI : Frame work for the inner wall of gas holder, m2, in CSTR, AF and DD(AP)
FWGHO : Frame work for the outer wall of gas holder, m2, in CSTR, AFUASB and DD
FWGHT : Frame work for the top of gas holder, m2, in CSTR, AF UASB and DD
FWIW : Frame work for the inner wall of digester, m2, in CSTR, AF, UASB and DD(AP)
FWIW1 : Frame work for the inner wall of the digester, m2, in DD(MP)
FWOW : Frame work for the outer wall of digester, m2, in CSTR, AF, UASB and DD(AP)
FWOW1 : Frame work for the outer wall of the digester, m2, in DD(MP)
GLS : Cost of gas liquid solid separator, Rs
HEATER : Cost of heater, Rs
HLT : Heat losses from water heating tank in CSTR, AF, US and DD(AP), watts
HLT1 : Heat loss from the water heating tank in DD(MP), watts
HLCASE1 : Heat losses in the first half of the heating of the slurry in CSTR, AF, UASB and DD(AP), watts
HLCASE2 : Heat loss in the first half of the heating of the slurry in DD(MP), watts
HLCASE3 : Heat losses in the second half of the heating of the slurry in CSTR, AF, UASB and DD(AP), watts
HLCASE4 : Heat loss in the second half of the heating of the slurry in DD(MP), watts
HOD : Height of the digester, m, in CSTR, AF, UASB and DD(AP)
HOD1 : Height of the digester, m, in DD(MP)
HOGH : Height of the gas holder, m, in CSTR, AF, UASB and DD(AP)
HOGH1 : Height of the gas holder, m, in DD(MP)
HRCASE1 : Heat required in the first half of the heating of the slurry in CSTR, AF, UASB and DD(AP), watts
HRCASE2 : Heat required in the first half of the heating of the slurry in DD(MP), watts
HRCASE3 : Heat required in the second half of the heating of the slurry in CSTR, AF, UASB and DD(AP)
HRCASE4 : Heat required in the second half of the heating of the slurry DD(MP)
HTRT : Heat required to maintain the temperature of heating tank
LEP : Length of heating pipe, m, in CSTR, AF, UASB and DD(AP)
LEP1 : Length of the heating pipe, m, in DD(MP)
MTCASE1 : Mean temperature difference during the first half of heating of the slurry in CSTR, AF, UASB and DD(AP)
MTCASE2 : Mean temperature difference during the second half of heating of the slurry in DD(MP)
MTD : Mean temperature difference
MTD1 : Mean temperature difference between ambient to mean temperature difference (MTD)
MTD2 : Mean temperature difference between MTD to desired
MTS : Mean temperature difference of the slurry

MTS1 : Mean temperature difference of the slurry between ambient to MTS
MTS2 : Mean temperature difference of the slurry between MTS to temperature to be maintained
NB : Number of bricks in CSTR, AF, UASB and DD(AP)
NB1 : Number of bricks in DD(MP)
OMC : Operation and maintenance cost, Rs
PBP : Pay back period
PCASE1 : Total Power (heater) required for first half of heating in CSTR, AF, UASB and DD(AP)
PCASE2 : Power required in the first half of the heating in DD(MP)
PCASE3 : Total power (heater) required for second half of heating in CSTR, AF, UASB and DD(AP)
PCASE4 : Power required in the second half of the heating in DD(MP)
PCC : Volume of PCC, m3, in CSTR, AF, UASB and DD(AP)
PCC1 : Volume of PCC, m3, in DD(MP)
PER : Cost of pump energy, Rs
PKB : Cost of packing bed in Anaerobic filters, Rs
PRICE : Price of the biogas in terms of kerosene, Rs
PUMP : Installation cost of pumping system, Rs
PVCF : Present value of cash flow, Rs
QB : Quantity of base bars with 20% extra quantity, kg, in CSTR, AF, UASB and DD(AP)
QB1 : Quantity of base bars, kg, in CSTR, AF, UASB and DD(AP)
QB2 : Quantity of base bars, kg, in DD(MP)
QB3 : Quantity of base bars with 20% extra quantity, kg, in DD(MP)
QD : Quantity of steel bars for gas holder with 20% extra quantity, kg, in AF, UASB and DD(AP)
QD1 : Quantity of steel bars in gas holder, kg, in AF, UASB and DD(AP)
QOS : Total quantity of steel bars, kg
QOSWH : Quantity of steel bars in horizontal position, kg, in CSTR, AF, UASB and DD(AP)
QOSWH1 : Quantity of the steel bars in horizontal position of the digester, kg, in DD(MP)
QOSWH2 : Quantity of steel bars in horizontal position with 20 quantity, kg, in CSTR, AF, UASB and DD(AP)
QOSWH3 : Quantity of the steel bars of horizontal position with 20% extra quantity, kg, in DD(MP)
QVB : Quantity of vertical bars, kg, in CSTR, AF, UASB and DD(AP)
QVB1 : Quantity of vertical bars, kg, in DD(MP)
QVB2 : Quantity of vertical bars with 20% extra quantity, kg, in CSTR, AF, UASB and DD(AP)
QVB3 : Quantity of vertical bars with 20% extra quantity, kg, in DD(MP)
ROGP : Rate of gas production expected, l Kg-1 feed
ROI : Rate of interest, %
RT : Retention time, days, in CSTR, AF, UASB and DD(AP)
RT1 : Retention time, days, in DD(MP)
SAND : Volume of Sand filling , m3, in CSTR, AF, UASB and DD(AP)
SAND1 : Volume of sand filling, m3, in DD(MP)
SPH : Specific heat of the slurry

STR	:	Cost of stirring, RS
TC	:	Total installation cost, Rs
TCASE1	:	Total heat required in the first half of the heating the slurry in CSTR, AF, UASB and DD(AP)
TCASE2	:	Total heat required in the first half of heating in DD(MP)
TCASE3	:	Total heat required in the second half of the heating of slurry in CSTR, AF, UASB and DD(AP)
TCASE4	:	Total heat required in the second half of heating in DD(MP)
TEMPRQ	:	Temperature required to be maintained inside the digester, OC
THR	:	Total heat required for heating the digester, watts
THS	:	Cost of heating system with insulation, Rs
THS1	:	Cost of heating system with out insulation, Rs
TOP	:	Number of turns of the heating pipe in CSTR, AF, UASB and DD(AP)
TOP1	:	Number of turns in DD(MP)
TPR	:	Total power required, watts, in CSTR, AF, UASB and DD(AP)
TPR1	:	Total power required, watts, in DD(MP)
VBR	:	Number of vertical bars in CSTR, AF, UASB and DD(AP)
VBR1	:	Number of vertical bars in DD(MP)
VOD	:	Volume of the digester, m3, in CSTR, AF, UASB and DD(AP)
VOD1	:	Volume of the digester, m3, in DD(MP)
VOF	:	Volume of feed, m3 or TPD, in CSTR, AF, UASB and DD(AP)
VOF1	:	Volume of the feed, m3 or TPD, in DD(MP)
VOGH	:	Volume of the gas holder, m3, in CSTR, AF, UASB and DD(AP)
VOGH1	:	Volume of the gas holder, m3, in DD(MP)
VSD	:	Volatile solids destruction expected, %
VSP	:	Volatile solids present in the feed, %
WW	:	Width of the brick wall, m, in CSTR, AF, UASB and DD(AP)
WW1	:	Width of the brick wall, m, in DD(MP)
Y	:	Duration of benefits expected, years.

Table: 1.7 Processing time needed by different capacities of computers for running COBAS

Type of Computer	CPU Speed	Number of options (Dimensions of array) dealt simultaneously	
		Three	Ten
PC-XT	100 Kips	90 hrs*	**
Mini (AT-286)	500 Kips	18 hrs	66 hrs
Super Mini	1 Mips	9 hrs	33 hrs
Midi	10 Mips	54 min	3 hrs
Maxi or Super	100 Mips	5 min	20min

* This is the time required if all the twelve process variables are different in each of the three sets of options; for nine and five different process variables in each set of option the running time would be 197 minutes and 20 minutes respectively.

** Not possible to process on a PC-XT

Table : 1.8 Inputs used for CSTR type digesters

1. Volume of feed, TPD	3		
2. Retention time, Days	20		
3. Rate of gas production, M^3/kg VS	0.3		
4. Volatile solids present, %	6		
5. Volatile solids destruction, %	65		
6. Ambient temperature, (°C)	5	15	30
7. Temperature maintained inside the Digester (°C)	35		
8. Rate of interest	0.12		
9. Duration of benefits expected (years)	20		

Table : 1.9 Bestout 1.Txt
The best three values of cst-rcc construction

SL. NO.	TYPE OF REAC: TOR	TYPE OF CONSTR-UCTION	VOF (TPD)	VOF1 (TPD)	ROGP (M3)	VSD (%)	AMTEMP (oC)	ROI (%)	VOD (M3)	COST OF CONSTR-UCTION	BENEFIT COST RATIO	PAY-BACK PERIOD (YEARS
			RT *(D)*	RT1 *(D)*	VSP *(%)*	HOD *(M)*	TEMPRQ *(oC)*	Y *(YEAR)*				
1	CST	RCC	3.00	0.00	0.30	65.00	30.00	0.12	60.00	81575.22	1.14	6.56
			20.00	0.00	6.00	0.00	35.00	20.00				
2	CST	RCC	3.00	0.00	0.30	65.00	15.00	0.12	60.00	91144.60	1.02	7.33
			20.00	0.00	6.00	0.00	35.00	20.00				
3	CST	RCC	3.00	0.00	0.30	65.00	5.00	0.12	60.00	97524.19	0.95	7.85
			20.00	0.00	6.00	0.00	35.00	20.00				

Table : 1.10
Bestout2.Txt
The best three values of cst-brick construction

SL. NO.	TYPE OF REAC-TOR	TYPE OF CONSTR-UCTION	VOF (TPD) / RT *(D)*	VOF1 (TPD) / RT1 *(D)*	ROGP (M3) / VSP *(%)*	VSD (%) / HOD *(M)*	AMTEMP (oC) / TEMPRQ *(oC)*	ROI (%) / Y *(YEAR)*	VOD (M3)	COST OF CONSTR-UCTION	BENEFIT COST RATIO	PAY-BACK PERIOD (YEARS
1	CST	BRICK	3.00	0.00	0.30	65.00	30.00	0.12	60.00	178773.78	0.44	16.88
			20.00	0.00	6.00	0.00	35.00	20.00				
2	CST	BRICK	3.00	0.00	0.30	65.00	15.00	0.12	60.00	178773.78	0.42	17.91
			20.00	0.00	6.00	0.00	35.00	20.00				
3	CST	BRICK	3.00	0.00	0.30	65.00	5.00	0.12	60.00	178773.78	0.40	18.60
			20.00	0.00	6.00	0.00	35.00	20.00				

Table: 1.11 Input used for CSTR type Digesters

1.	Volume of feed,TPD	1		
2.	Retention time, Days	20		
3.	Rate of gas production M³/kg VS	0.3		
4.	Volatile solids present (%)	6		
5.	Volatile solids destruction (%)	6.5		
6.	Ambient temperature (°C)	5	15	30
7.	Temperature maintained inside the digester (°C)	3.5		
8.	Rate of interest	0.12		
9.	Duration of benefits expected (years)	2.0		

Table: 1.12 Inputs used for anaerobic filters

1.	Volume of feed, TDP	1		
2.	Retention time, days	3		
3.	Height of digester, m	4		
4.	Rate of gas production, M³/kg VS	0.3		
5.	Volatile solids present (%)	6		
6.	Volatile solids destruction,%	65		
7.	Ambient temperature (°C)	5	15	30
8.	Temperature inside the digester (°C)	35		
9.	Rate of Interest	0.12		
10.	Duration of benefits expected, years	20		

Table: 1.13 Inputs used for UASB digesters

1.	Volume of feed, TPD	1		
2.	Retention time, hours	18		
3.	Height of digester, m	6		
4.	Rate of gas production, M³/kg VS	0.3		
5.	Volatile solids present (%)	6		
6.	Volatile solids destructioin (%)	65		
7.	Ambient temperature (°C)	5	15	30
8.	Temperature inside the digester (°C)	35		
9.	Rate of interest	0.12		
10.	Duration of benefits expected, years	20		

Table: 1.14 Inputs used for diphasic digester

1. Volume of feed in acid phase, TPD	1		
2. Volume of feed in methane phase, TPD	1		
3. Retention time in acid phase, days	2		
4. Retention time in methane phase, days	6		
5. Rate of gas production M³/kg VS	0.3		
6. Volatile solids present (%)	6		
7. Volatile soilds destruction (%)	65		
8. Ambient temperature (°C)	5	15	30
9. Temperature inside the digester (°C)	35		
10. Rate of interest	0.12		
11. Duration of benefits expected, years	20		

TABLE : 1.15
BESTOUT1.TXT
The best three values of cst-rcc construction

SL. NO.	TYPE OF REAC-TOR	TYPE OF CONSTR-UCTION	VOF (TPD) / RT *(D)*	VOF1 (TPD) / RT1 *(D)*	ROGP (M3) / VSP *(%)*	VSD (%) / HOD *(M)*	AMTEMP (oC) / TEMPRQ *(cC)*	ROI (%) / Y *(YEAR)*	VOD (M3)	COST OF CONSTR-UCTION	BENEFIT COST RATIO	PAY-BACK PERIOD (YEARS)
1	CST	RCC	1.00	0.00	0.30	65.00	30.00	0.12	20.00	41594.02	0.64	11.67
			20.00	0.00	6.00	0.00	35.00	20.00				
2	CST	RCC	1.00	0.00	0.30	65.00	15.00	0.12	20.00	45203.89	0.59	12.68
			20.00	0.00	6.00	0.00	35.00	20.00				
3	CST	RCC	1.00	0.00	0.30	65.00	5.00	0.12	20.00	47610.47	0.56	13.36
			20.00	0.00	6.00	0.00	35.00	20.00				

TABLE : 1.16
BESTOUT2.TXT
The best three (if exists) values for cst - brick construction

SL. NO.	TYPE OF REAC-TOR	TYPE OF CONSTR-UCTION	VOF (TPD)	VOF1 (TPD)	ROGP (M3)	VSD (%)	AMTEMP (oC)	ROI (%)	VOD (M3)	COST OF CONSTR-UCTION	BENEFIT COST RATIO	PAY-BACK PERIOD (YEARS)
			RT *(D)*	RT1 *(D)*	VSP *(%)*	HOD *(M)*	TEMPRQ *(oC)*	Y *(YEAR)*				
1	CST	BRICK	1.00	0.00	0.30	65.00	30.00	0.12	20.00	71268.18	0.33	22.38
			20.00	0.00	6.00	0.00	35.00	20.00				
2	CST	BRICK	1.00	0.00	0.30	65.00	15.00	0.12	20.00	71268.18	0.32	23.67
			20.00	0.00	6.00	0.00	35.00	20.00				
3	CST	BRICK	1.00	0.00	0.30	65.00	5.00	0.12	20.00	71268.18	0.30	24.54
			20.00	0.00	6.00	0.00	35.00	20.00				

TABLE : 1.17
BESTOUT3.TXT
The best three values of af-rcc construction

SL. NO.	TYPE OF REACTOR	TYPE OF CONSTRUCTION	VOF (TPD)	VOF1 (TPD)	ROGP (M3)	VSD (%)	AMTEMP (oC)	ROI (%)	VOD (M3)	COST OF CONSTRUCTION	BENEFIT COST RATIO	PAY-BACK PERIOD (YEARS)
			RT (D)	*RT1 (D)*	*VSP (%)*	*HOD (M)*	*TEMPRQ (oC)*	*Y (YEAR)*				
1	AF	RCC	1.00	0.00	0.30	65.00	30.00	0.12	3.00	21760.55	1.63	4.59
			3.00	0.00	6.00	4.00	35.00	20.00				
2	AF	RCC	1.00	0.00	0.30	65.00	15.00	0.12	3.00	23840.19	1.49	5.03
			3.00	0.00	6.00	4.00	35.00	20.00				
3	AF	RCC	1.00	0.00	0.30	65.00	5.00	0.12	3.00	25226.61	1.40	5.32
			3.00	0.00	6.00	4.00	35.00	20.00				

TABLE : 1.18
BESTOUT4.TXT
The best three values of af-brick construction

SL. NO.	TYPE OF REAC-TOR	TYPE OF CONSTR-UCTION	VOF (TPD) / RT *(D)*	VOF1 (TPD) / RT1 *(D)*	ROGP (M3) / VSP *(%)*	VSD (%) / HOD *(M)*	AMTEMP (oC) / TEMPRQ *(oC)*	ROI (%) / Y *(YEAR)*	VOD (M3)	COST OF CONSTR-UCTION	BENEFIT COST RATIO	PAY-BACK PERIOD (YEARS)
1	AF	BRICK	1.00	0.00	0.30	65.00	30.00	0.12	3.00	35420.57	0.97	7.73
			3.00	0.00	6.00	4.00	35.00	20.00				
2	AF	BRICK	1.00	0.00	0.30	65.00	15.00	0.12	3.00	37796.88	0.91	8.25
			3.00	0.00	6.00	4.00	35.00	20.00				
3	AF	BRICK	1.00	0.00	0.30	65.00	5.00	0.12	3.00	39381.09	0.87	8.59
			3.00	0.00	6.00	4.00	35.00	20.00				

TABLE : 1.19
BESTOUT5.TXT
The best three (if exists) values of uasb-rcc construction

SL. NO.	TYPE OF REAC-TOR	TYPE OF CONSTR-UCTION	VOF (TPD) / RT *(D)*	VOF1 (TPD) / RT1 *(D)*	ROGP (M3) / VSP *(%)*	VSD (%) / HOD *(M)*	AMTEMP (oC) / TEMPRQ *(oC)*	ROI (%) / Y *(YEAR)*	VOD (M3)	COST OF CONSTR-UCTION	BENEFIT COST RATIO	PAY-BACK PERIOD (YEARS)
1	UASB	RCC	1.00	0.00	0.30	65.00	30.00	0.12	1.33	14352.21	2.51	2.97
			18.00	0.00	6.00	6.00	35.00	20.00				
2	UASB	RCC	1.00	0.00	0.30	65.00	15.00	0.12	1.33	15883.07	2.27	3.29
			18.00	0.00	6.00	6.00	35.00	20.00				
3	UASB	RCC	1.00	0.00	0.30	65.00	5.00	0.12	1.33	16903.65	2.13	3.50
			18.00	0.00	6.00	6.00	35.00	20.00				

TABLE : 1.20
BESTOUT6.TXT
The best three (if exists) values of uasb-brick construction

SL. NO.	TYPE OF REAC-TOR	TYPE OF CONSTR-UCTION	VOF (TPD)	VOF1 (TPD)	ROGP (M3)	VSD (%)	AMTEMP (oC)	ROI (%)	VOD (M3)	COST OF CONSTR-UCTION	BENEFIT COST RATIO	PAY-BACK PERIOD (YEARS)
			RT *(D)*	RT1 *(D)*	VSP *(%)*	HOD *(M)*	TEMPRQ *(oC)*	Y *(YEAR)*				
1	UASB	BRICK	1.00	0.00	0.30	65.00	30.00	0.12	1.33	61309.69	0.48	15.43
			18.00	0.00	6.00	6.00	35.00	20.00				
2	UASB	BRICK	1.00	0.00	0.30	65.00	15.00	0.12	1.33	63722.99	0.47	16.03
			18.00	0.00	6.00	6.00	35.00	20.00				
3	UASB	BRICK	1.00	0.00	0.30	65.00	5.00	0.12	1.33	65331.85	0.45	16.44
			18.00	0.00	6.00	6.00	35.00	20.00				

TABLE : 1.21
BESTOUT7.TXT
The best three values of diphasic digester - rcc construction

SL. NO.	TYPE OF REACTOR	TYPE OF CONSTRUCTION	VOF (TPD)	VOF1 (TPD)	ROGP (M3)	VSD (%)	AMTEMP (oC)	ROI (%)	VOD (M3)	COST OF CONSTRUCTION (Rs)	BENEFIT COST RATIO	PAY-BACK PERIOD (YEARS)
			RT *(D)*	RT1 *(D)*	VSP *(%)*	HOD *(M)*	TEMPRQ *(oC)*	Y *(YEAR)*	VODI *(m3)*			
2	DIPHASE	RCC	1.00	1.00	0.30	65.00	15.00	0.12	2.00	71539.62	0.56	13.40
1	DIPHASE	RCC	1.00	1.00	0.30	65.00	30.00	0.12	2.00	67185.96	0.59	12.59
			2.00	6.00	6.00	0.00	35.00	20.00	6.00			
2	DIPHASE	RCC	1.00	1.00	0.30	65.00	15.00	0.12	2.00	71539.62	0.56	13.40
			2.00	6.00	6.00	0.00	35.00	20.00	6.00			
3	DIPHASE	RCC	1.00	1.00	0.30	65.00	5.00	0.12	2.00	74442.06	0.54	13.95
			2.00	6.00	6.00	0.00	35.00	20.00	6.00			

TABLE : 1.22
BESTOUT8.TXT
The best three values of diphasic digester - brick construction

SL. NO.	TYPE OF REAC-TOR	TYPE OF CONSTR-UCTION	VOF (TPD) RT *(D)*	VOF1 (TPD) RT1 *(D)*	ROGP (M3) VSP *(%)*	VSD (%) HOD *(M)*	AMTEMP (oC) TEMPRQ *(oC)*	ROI (%) Y *(YEAR)*	VOD (M3)	COST OF CONSTR-UCTION (Rs)	BENEFIT COST RATIO	PAY-BACK PERIOD (YEARS)
1	DIPHASE	BRICK	1.00 2.00	1.00 6.00	0.30 6.00	65.00 0.00	30.00 35.00	0.12 20.00	2.00	78622.39	0.51	14.73
2	DIPHASE	BRICK	1.00 2.00	1.00 6.00	0.30 6.00	65.00 0.00	15.00 35.00	0.12 20.00	2.00	83138.81	0.48	15.57
3	DIPHASE	BRICK	1.00 2.00	1.00 6.00	0.30 6.00	65.00 0.00	5.00 35.00	0.12 20.00	2.00	86149.76	0.46	16.14

TABLE : 1.23
BSTOFALL.TXT
The best ten values of all type digesters

SL. NO.	TYPE OF REACTOR	TYPE OF CONSTRUCTION	VOF (TPD) / RT *(D)*	VOF1 (TPD) / RT1 *(D)*	ROGP (M3) / VSP *(%)*	VSD (%) / HOD *(M)*	AMTEMP (oC) / TEMPRQ *(oC)*	ROI (%) / Y *(YEAR)*	VOD (M3)	COST OF CONSTRUCTION (Rs)	BENEFIT COST RATIO	PAY-BACK PERIOD (YEARS)
1	UASB	RCC	1.00	0.00	0.30	65.00	30.00	0.12	1.33	14352.21	2.51	2.97
			18.00	0.00	6.00	6.00	35.00	20.00				
2	UASB	RCC	1.00	0.00	0.30	65.00	15.00	0.12	1.33	15883.07	2.27	3.29
			18.00	0.00	6.00	6.00	35.00	20.00				
3.	UASB	RCC	1.00	0.00	0.30	65.00	5.00	0.12	1.33	16903.65	2.13	3.50
			18.00	0.00	6.00	6.00	35.00	20.00				
4.	AF	RCC	1.00	0.00	0.30	65.00	30.00	0.12	3.00	21760.55	1.63	4.59
			3.00	0.00	6.00	4.00	35.00	20.00				
5.	AF	RCC	1.00	0.00	0.30	65.00	15.00	0.12	3.00	23840.19	1.49	5.03
			3.00	0.00	6.00	4.00	35.00	20.00				
6.	AF	RCC	1.00	0.00	0.30	65.00	5.00	0.12	3.00	25226.61	1.40	5.32
			3.00	0.00	6.00	4.00	35.00	20.00				
7.	AF	BRICK	1.00	0.00	0.30	65.00	30.00	0.12	3.00	35420.57	0.97	7.73
			3.00	0.00	6.00	4.00	35.00	20.00				
8.	AF	BRICK	1.00	0.00	0.30	65.00	15.00	0.12	3.00	37796.88	0.91	8.25
			3.00	0.00	6.00	4.00	35.00	20.00				
9.	AF	BRICK	1.00	0.00	0.30	65.00	5.00	0.12	3.00	39381.09	0.87	8.59
			3.00	0.00	6.00	4.00	35.00	20.00				
10.	CST	RCC	1.00	0.00	0.30	65.00	30.00	0.12	20.00	41594.02	0.64	11.67
			3.00	0.00	6.00	4.00	35.00	20.00				

Table: 1.24 Input used for CSTR type digesters

1. Volume of feed, TPD	3		
2. Retention time, Days	20		
3. Rate of gas production, M³/kg VS	0.3		
4. Volatile solids present, %	4	5	6
5. Volatile solids destruction, %	55	65	75
6. Ambient temperature, (°C)	5	15	30
9. Temperature maintained inside the Digester (°C)	35		
8. Rate of interest	0.12		
9. Duration of benefits expected (years)	20		

Table: 1.25 Inputs used for anaerobic filters

1. Volume of feed, TPD	3		
2. Retention time, Days	3		
3. Height of digester, m	4		
4. Rate of gas production, M³/kg VS	0.3		
5. Volatile solids present, (%)	4	5	6
6. Volatile solids destruction, %	55	65	75
7. Ambient temperature, (°C)	5 15	30	
8. Temperature inside the digester (°C)	35		
9. Rate of interest	0.12		
10.Duration of benefits expected, years	20		

Table: 1.26 Inputs used for UASB digesters

1. Volume of feed, TPD	3		
2. Retention time, hours	18		
3. Height of digester, m	6		
4. Rate of gas production, M³/kg VS	0.3		
5. Volatile solids present, (%)	4	5	6
6. Volatile solids destruction, (%)	55	65	75
7. Ambient temperature, (°C)	5	15	30
8. Temperature inside the digester (°C)	35		
9. Rate of interest	0.12		
10. Duration of benefits expected, years	20		

Table: 1.27 Inputs used for diphasic digester

1. Volume of feed in acid phase, TPD	3		
2. Volume of feed in methane phase, TPD	3		
3. Retention time in acid phase, days	2		
4. Retention time in methane phase, days	6		
5. Rate of gas production, M^3/kg VS	0.3		
6. Volatile solids present, (%)	4	5	6
7. Volatile solids destruction, (%)	55	65	75
8. Ambient temperature, (^oC)	5	15	30
9. Temperature inside the digester (^oC)	35		
10. Rate of interest	0.12		
11. Duration of benefits expected, years	20		

TABLE : 1.28
BESTOUT1.TXT
The best values of cst-rcc construction

SL. NO.	TYPE OF REACTOR	TYPE OF CONSTRUCTION	VOF (M3) RT *(D)*	VOF1 (M3) RT1 *(D)*	ROGP (M3) VSP *(%)*	VSD (%) HOD *(M)*	AMTEMP (oC) TEMPRQ *(oC)*	ROI (%) Y *(YEAR)*	VOD (M3)	COST OF CONSTRUCTION	BENEFIT COST RATIO	PAY-BACK PERIOD (YEARS)
1	CST	RCC	3.00 20.00	0.00 0.00	0.30 6.00	75.00 0.00	30.00 35.00	0.12 20.00	60.00	83241.29	1.33	5.62
2	CST	RCC	3.00 20.00	0.00 0.00	0.30 5.00	65.00 0.00	15.00 35.00	0.12 20.00	60.00	89282.72	0.82	9.07
3	CST	RCC	3.00 20.00	0.00 0.00	0.30 5.00	65.00 0.00	5.00 35.00	0.12 20.00	60.00	95624.32	0.77	9.71
4	CST	RCC	3.00 20.00	0.00 0.00	0.30 5.00	55.00 0.00	15.00 35.00	0.12 20.00	60.00	87850.49	0.67	11.18
5	CST	RCC	3.00 20.00	0.00 0.00	0.30 5.00	55.00 0.00	5.00 35.00	0.12 20.00	60.00	94162.87	0.62	11.98

TABLE : 1.29
BESTOUT2.TXT
The best five (if exists) values for cst - brick construction

SL. NO.	TYPE OF REAC-TOR	TYPE OF CONSTR-UCTION	VOF (M3)	VOF1 (M3)	ROGP (M3)	VSD (%)	AMTEMP (oC)	ROI (%)	VOD (M3)	COST OF CONSTR-UCTION	BENEFIT COST RATIO	PAY-BACK PERIOD (YEARS)
			RT *(D)*	RT1 *(D)*	VSP *(%)*	HOD *(M)*	TEMPRQ *(oC)*	Y *(YEAR)*				
1	CST	BRICK	3.00	0.00	0.30	75.00	30.00	0.12	60.00	180500.82	0.53	14.09
			20.00	0.00	6.00	0.00	35.00	20.00				
2	CST	BRICK	3.00	0.00	0.30	75.00	15.00	0.12	60.00	180500.82	0.50	14.94
			20.00	0.00	6.00	0.00	35.00	20.00				
3	CST	BRICK	3.00	0.00	0.30	75.00	5.00	0.12	60.00	18050082	0.48	15.51
			20.00	0.00	6.00	0.00	35.00	20.00				
4	CST	BRICK	3.00	0.00	0.30	65.00	30.00	0.12	60.00	178773.78	0.44	16.88
			20.00	0.00	6.00	0.00	35.00	20.00				
5	CST	BRICK	3.00	0.00	0.30	75.00	30.00	0.12	60.00	178342.02	0.42	17.77
			20.00	0.00	5.00	0.00	35.00	20.00				

TABLE : 1.30
BESTOUT3.TXT
The best five (if exists) values of af-rcc construction

SL. NO.	TYPE OF REAC-TOR	TYPE OF CONSTR-UCTION	VOF (M3)	VOF1 (M3)	ROGP (M3)	VSD (%)	AMTEMP (oC)	ROI (%)	VOD (M3)	COST OF CONSTR-UCTION	BENEFIT COST RATIO	PAY-BACK PERIOD (YEARS)
			RT *(D)*	RT1 *(D)*	VSP *(%)*	HOD *(M)*	TEMPRQ *(oC)*	Y *(YEAR)*				
1	AF	RCC	3.00	0.00	0.30	75.00	30.00	0.12	9.00	35841.40	3.64	2.05
			3.00	0.00	6.00	4.00	35.00	20.00				
2	AF	RCC	3.00	0.00	0.30	75.00	15.00	0.12	9.00	41130.62	3.17	2.36
			3.00	0.00	6.00	4.00	35.00	20.00				
3	AF	RCC	3.00	0.00	0.30	65.00	30.00	0.12	9.00	35841.40	3.12	2.39
			3.00	0.00	6.00	4.00	35.00	20.00				
4	AF	RCC	3.00	0.00	0.30	75.00	30.00	0.12	9.00	35841.40	2.99	2.49
			3.00	0.00	5.00	4.00	35.00	20.00				
5	AF	RCC	3.00	0.00	0.30	75.00	5.00	0.12	9.00	44656.77	2.92	2.56
			3.00	0.00	6.00	4.00	35.00	20.00				

TABLE : 1.31
BESTOUT4.TXT
The best five values of AF-brick construction

SL. NO.	TYPE OF REACTOR	TYPE OF CONSTRUCTION	VOF (M3)	VOF1 (M3)	ROGP (M3)	VSD (%)	AMTEMP (oC)	ROI (%)	VOD (M3)	COST OF CONSTRUCTION	BENEFIT COST RATIO	PAY-BACK PERIOD (YEARS)
			RT *(D)*	RT1 *(D)*	VSP *(%)*	HOD *(M)*	TEMPRQ *(oC)*	Y *(YEAR)*				
1	AF	BRICK	3.00	0.00	0.30	75.00	30.00	0.12	9.00	48352.18	2.69	2.77
			3.00	0.00	6.00	4.00	35.00	20.00				
2	AF	BRICK	3.00	0.00	0.30	75.00	15.00	0.12	9.00	53972.20	2.41	3.09
			3.00	0.00	6.00	4.00	35.00	20.00				
3	AF	BRICK	3.00	0.00	0.30	75.00	5.00	0.12	9.00	57718.89	2.26	3.31
			3.00	0.00	6.00	4.00	35.00	20.00				
4	AF	BRICK	3.00	0.00	0.30	65.00	30.00	0.12	9.00	48352.18	2.31	3.23
			3.00	0.00	6.00	4.00	35.00	20.00				
5	AF	BRICK	3.00	0.00	0.30	65.00	15.00	0.12	9.00	53972.20	2.07	3.60
			3.00	0.00	6.00	4.00	35.00	20.00				

TABLE : 1.32
BESTOUT5.TXT
The best five (if exists) values of uasb-rcc construction

SL. NO.	TYPE OF REAC-TOR	TYPE OF CONSTR-UCTION	VOF (M3)	VOF1 (M3)	ROGP (M3)	VSD (%)	AMTEMP (oC)	ROI (%)	VOD (M3)	COST OF CONSTR-UCTION	BENEFIT COST RATIO	PAY-BACK PERIOD (YEARS)
			RT *(D)*	RT1 *(D)*	VSP *(%)*	HOD *(M)*	TEMPRQ *(oC)*	Y *(YEAR)*				
1	UASB	RCC	3.00	0.00	0.30	75.00	30.00	0.12	4.00	21983.98	5.99	1.25
			18.00	0.00	6.00	6.00	35.00	20.00				
2	UASB	RCC	3.00	0.00	0.30	65.00	30.00	0.12	4.00	21983.98	5.16	1.45
			18.00	0.00	5.00	6.00	35.00	20.00				
3	UASB	RCC	3.00	0.00	0.30	75.00	15.00	0.12	4.00	25902.11	5.09	1.47
			18.00	0.00	5.00	6.00	35.00	20.00				
4	UASB	RCC	3.00	0.00	0.30	75.00	30.00	0.12	4.00	21983.98	4.95	1.51
			18.00	0.00	5.00	6.00	35.00	20.00				
5	UASB	RCC	3.00	0.00	0.30	75.00	5.00	0.12	4.00	28514.19	4.62	1.62
			18.00	0.00	4.00	6.00	35.00	20.00				

TABLE : 1.33
BESTOUT6.TXT
The best five (if exists) values of uasb-brick construction

SL. NO.	TYPE OF REACTOR	TYPE OF CONSTRUCTION	VOF (M3) / RT *(D)*	VOF1 (M3) / RT1 *(D)*	ROGP (M3) / VSP *(%)*	VSD (%) / HOD *(M)*	AMTEMP (oC) / TEMPRQ *(oC)*	ROI (%) / Y *(YEAR)*	VOD (M3)	COST OF CONSTRUCTION	BENEFIT COST RATIO	PAY-BACK PERIOD (YEARS)
1	UASB	BRICK	3.00	0.00	0.30	75.00	30.00	0.12	4.00	74604.22	1.68	4.44
			18.00	0.00	6.00	6.00	35.00	20.00				
2	UASB	BRICK	3.00	0.00	0.30	75.00	15.00	0.12	4.00	79895.98	1.57	4.75
			18.00	0.00	6.00	6.00	35.00	20.00				
3	UASB	BRICK	3.00	0.00	0.30	75.00	5.00	0.12	4.00	83423.82	1.51	4.98
			18.00	0.00	6.00	6.00	35.00	20.00				
4	UASB	BRICK	3.00	0.00	0.30	65.00	30.00	0.12	4.00	74604.22	1.44	5.20
			18.00	0.00	6.00	6.00	35.00	20.00				
5	UASB	BRICK	3.00	0.00	0.30	75.00	30.00	0.12	4.00	74604.22	1.38	5.43
			18.00	0.00	5.00	6.00	35.00	20.00				

TABLE : 1.34

BESTOUT7.TXT

The best five values of diphasic digester - rcc construction

SL. NO.	TYPE OF REAC-TOR	TYPE OF CONSTR-UCTION	VOF (M3) / RT *(D)*	VOF1 (M3) / RT1 *(D)*	ROGP (M3) / VSP *(%)*	VSD (%) / HOD *(M)*	AMTEMP (oC) / TEMPRQ *(oC)*	ROI (%) / Y *(YEAR)*	VOD (M3)	COST OF CONSTR-UCTION	BENEFIT COST RATIO	PAY-BACK PERIOD (YEARS)
1	DIPHASE	RCC	3.00	3.00	0.30	75.00	30.00	0.12	6.00	454794.80	0.30	24.61
				6.00	6.00	0.00	35.00	20.00				
2	DIPHASE	RCC	3.00	3.00	0.30	65.00	30.00	0.12	6.00	402305.29	0.30	25.12
			2.00	6.00	6.00	0.00	35.00	20.00				
3	DIPHASE	RCC	3.00	3.00	0.30	75.00	30.00	0.12	6.00	389182.91	0.30	25.27
			2.00	6.00	5.00	0.00	35.00	20.00				
4	DIPHASE	RCC	3.00	3.00	0.30	75.00	15.00	0.12	6.00	474551.31	0.29	25.68
			2.00	6.00	6.00	0.00	35.00	20.00				
5	DIPHASE	RCC	3.00	3.00	0.30	55.00	30.00	0.12	6.00	349815.77	0.29	25.82
			2.00	6.00	6.00	0.00	35.00	20.00				

TABLE : 1.35
BESTOUT8.TXT
The best five values of diphasic digester - brick construction

SL. NO.	TYPE OF REACTOR	TYPE OF CONSTRUCTION	VOF (M3) / RT (D)	VOF1 (M3) / RT1 (D)	ROGP (M3) / VSP (%)	VSD (%) / HOD (M)	AMTEMP (oC) / TEMPRQ (oC)	ROI (%) / Y (YEAR)	VOD (M3)	COST OF CONSTRUCTION	BENEFIT COST RATIO	PAYBACK PERIOD (YEARS)
1	DIPHASE	BRICK	3.00	3.00	0.30	75.00	30.00	0.12	6.00	157359.94	0.88	8.52
			2.00	6.00	6.00	0.00	35.00	20.00				
2	DIPHASE	BRICK	3.00	3.00	0.30	75.00	15.00	0.12	6.00	168844.02	0.82	9.14
			2.00	6.00	6.00	0.00	35.00	20.00				
3	DIPHASE	BRICK	3.00	3.00	0.30	75.00	5.00	0.12	6.00	176500.08	0.78	9.55
			2.00	6.00	6.00	0.00	35.00	20.00				
4	DIPHASE	BRICK	3.00	3.00	0.30	65.00	30.00	0.12	6.00	154399.89	0.77	9.64
			2.00	6.00	6.00	0.00	35.00	20.00				
5	DIPHASE	BRICK	3.00	3.00	0.30	75.00	30.00	0.12	6.00	153659.87	0.75	9.98
			2.00	6.00	5.00	0.00	35.00	20.00				

TABLE : 1.36
BSTOFALL.TXT
The best ten (if exists) values of all type of reactors

SL. NO.	TYPE OF REAC-TOR	TYPE OF CONSTR-UCTION	VOF (M3) / RT *(D)*	VOF1 (M3) / RT1 *(D)*	ROGP (M3) / VSP *(%)*	VSD (%) / HOD *(M)*	AMTEMP (oC) / TEMPRQ *(oC)*	ROI (%) / Y *(YEAR)*	VOD (M3)	COST OF CONSTR-UCTION	BENEFIT COST RATIO	PAY-BACK PERIOD (YEARS)
1	UASB	RCC	3.00	0.00	0.30	75.00	30.00	0.12	4.00	21983.98	5.99	1.25
			18.00	0.00	6.00	6.00	35.00	20.00				
2	UASB	RCC	3.00	0.00	0.30	65.00	30.00	0.12	4.00	21983.98	5.16	1.45
			18.00	0.00	6.00	4.00	35.00	20.00				
3	UASB	RCC	3.00	0.00	0.30	75.00	15.00	0.12	4.00	25902.11	5.09	1.47
			18.00	0.00	6.00	6.00	35.00	20.00				
4	UASB	RCC	3.00	0.00	0.30	75.00	30.00	0.12	4.00	21983.98	4.95	1.51
			18.00	0.00	5.00	6.00	35.00	20.00				
5	UASB	RCC	3.00	0.00	0.30	75.00	5.00	0.12	4.00	28514.19	4.62	1.62
			18.00	0.00	6.00	6.00	35.00	20.00				
6	UASB	RCC	3.00	0.00	0.30	65.00	15.00	0.12	4.00	25902.11	4.88	1.71
			18.00	0.00	6.00	6.00	35.00	20.00				
7	UASB	RCC	3.00	0.00	0.30	55.00	30.00	0.12	4.00	21983.98	4.32	1.73
			18.00	0.00	6.00	6.00	35.00	20.00				
8	UASB	RCC	3.00	0.00	0.30	65.00	30.00	0.12	4.00	21983.98	4.25	1.76
			18.00	0.00	5.00	6.00	35.00	20.00				
9	UASB	RCC	3.00	0.00	0.30	75.00	15.00	0.12	4.00	25902.11	4.20	1.78
			18.00	0.00	5.00	6.00	35.00	20.00				
10	UASB	RCC	3.00	0.00	0.30	65.00	5.00	0.12	4.00	28514.19	3.98	1.88
			18.00	0.00	6.00	6.00	35.00	20.00				

REFERENCES

1. Abbasi, S. A., (1987). *Renewable energy from aquatic biomass. In:* proceedings of the 1986 International congress on Renewable Energy Sources, CSIC, Madrid, 60.

2. Abbasi, S. A., and Abbasi N., (1996). *Alternative Energy: Development and Management,* International Book Distributors,Dehradun, Vii+222 pages

3. Abbasi, S. A., and Nipaney, P. C., (1984). Generation of biogas from Salvinia molesta (Mitchell) on a commercial biogas digester. Environmental Technology Letters, 5, 75-80.

4. Abbasi,S. A. and Nipaney, P. C., (1984). *The catalytic effect of copper (II), Zinc (II), nickel (II) on the anaerobic digestion of Salvinia molesta (Mitchell)* Energy developments: New forms, Renewables, Conservation, Fred A Curtis (edt). Pergoman Press, Oxford, 237–42.

5. Abbasi, S. A., and Nipaney P. C., (1985). *Wastewater treatment using aquatic plants. Survivability and growth of Salvinia molesta* over waters treated with zinc and the subsequent utilization of the harvested weeds Resources and Conservation 12, 47-55.

6. Abbasi, S. A., and Nipaney P. C., (1986). Infestation by aquatic weeds of the fern genus Salvinia: Its status and control, Environmental Conservation, 13, 235–41.

7. Abbasi, S. A., and Nipaney, P. C., (1993). *Modelling and Simulation of Biogas Systems Economics*, Ashish Publishing House, New Delhi, Xviii+359 Pages.

8. Abbasi, S A, and Ramasamy, E. V., (1996). Utilization of biowaste solids by extracting volatile fatty aids with subsequent conversion into methane and manure, Proceedings of Xii International conference on Solid Waste Technology and Management,Philadelphia, USA,Nov 17-19, 1996, 4c1 – 4c8

9. Abbasi,S. A., and Ramasamy, E. V., (1998). *Biotechnological Pollution Control systems: The technologies most appropriate for the third world countries* Orient Longman (Universities Press India Ltd.) Hyderabad, 120 pages (in press).

10. Abbasi, S. A., Nipaney P. C. and Soni R., (1988). *Aquatic weeds : distribution, impact and control,* Journal of Scientific and Industrial Research, 47, 650-61.

11. Abbasi, S. A., Nipaney, P. C., and Ramasamy, E. V., (1992 a). *Use of aquatic weed Salvinia (Salvinia molista, Mitchell) as full/partial feed in commercial biogas digesters*, Indian Journal of Technology, 30, 451–57.

12. Abbasi, S. A., Nipaney, P. C., and Ramasamy, E. V., (1992 b). *Studies on multi-phase anaerobic digestion of Salvinia*, Indian Journal of Technology, 30, 483–90.

13. Alvasidis, A., Wandrey, C, and Hilla, E,(1989). *Studies on reaction techniques concerning reactor control design for the anaerobic degradation of complex substrates with the example of the methanation of effluents in the fermentation industry*. Bioprocess Engineering, 4, 63–74.

14. Anderson, G. K., Saw, C. B., and Fernandas, M I A P, (1986). *Application of porous membranes for biomass retention in biological wastewater processes*, Process Biochemistry, 174–82.

15. Anderson, G. K., Saw, C. B. and Fernandes, M I A P (1986). *Applications of porous membrances for biomass retention in biological wastewater processes*. Process Biochemistry, 174–82.

16. Andreoni, V., Bonfanti, P., Daffonchio, D., Sorlini, C., and Villa, M., (1993). *Anaerobic digestion of olive mill effluent - micro-biological and processing aspects*. Journal of Environmental science and health. Part A: Environmental science and Engineering, 28 (9), 2041–59.

17. Aruldhas, (1981). *Data Matrix for Public Health Engineers*: Table 17. Assistant Engineer's Association, TWAD Board, Madras.

18. Barbose, R. A., and Sant Anna Jr., G. L., (1989). *Treatment of raw domestic sewage in an UASB reactor*. Water research, 23, (12),1483–90.

19. Barry, M., and Colleran, E., (1982). *Anaerobic digestion of Silage effluent using upflow fixed bed reactor*. Agricultural Wastes, 4., 231–39 Biomass Energy Institute Inc.Biogas Production from Animal Manure Biomass Energy Institute Inc., 304–870 Cambridge Street Winnipeg,Manitokb a,21p.

20. Barry, M., (1983). *Anaerobic digestion of agricultural wastes*. Ph.D. Thesis, National University of Ireland, Dublin.

21. Barry,M. and Colleran,E (1982). *Anaerobic digestion of silage effluent using upflow fixed bed reactor*. Agricultural Wastes, 4, 231–39.Biomass Energy Institute Inc.Biogas Production from Animal Manure.

22. Barry, M., (1983). *Anaerobic digestion of agricultural wastes*. PhD Thesis, National University of Ireland, Dubliln.

23. Bell, B. A., Jeris, J., Welday, J. M., and Carrol, R., (1980). *Anaerobic fluidized bed treatment of thermal sludge conditioning decant liquor*, proceedings of the U.S. Department of Energy Workshop/Seminar on Anaerobic Filters. 9-10 January Howey–In–The–Hills. FL U.S. Department of Energy. Argon National Laborotories. Chicago. IL.pp. 171-78.

24. Bhatti, Z. I., Furukawa, K. and Fujita, M., (1993). *Treatment performance and microbial structure of a granular consortium handling methanolic waste*. Journal of fermentation and Bioengineering, 76 (3).

25. Bhoyar, R. V., and Bhide, A. D., (1996). *Studies on solid state digestion of municipal solid waste: a diphasic approach*. Energy Environment monitor, 12, (2), 71-78.

26. Dogte, J. J., Breure, A. M., Vanandel, J. G., and Lettinga, G.,(1993). *Anaerobic treatment of domestic wastewater in small scale UASB reactors*. Water science technology, 27 (9), 75-82.

27. Boller, M. A., (1993). *Upflow anaerobic filteration of a sugar containing waste water*. Water science and Technology. 28 (2) 125-34.

28. Borja, R., and Banks, C. J., (1994). *Treatment of palm oil mill effluent by upflow anaerobic filteration*. Journal of Chemical Technology and Biotechnology, 61 (2), 103-09.

29. Borja, R., and Banks, C. J., (1995). *Response of an Anaerobic fluidized bed reactor treating icecream wastewater to organic, hydraulic, temperature and pH shocks*, Journal of Biotechnology, 39, (3), 251-59.

30. Borja R., Banks, C. J., and Wang Z., (1995). *Performance of a hybrid anaerobic reactor, combining a sludge blanket and a filter, treating slaughter house wastewater*, Applied microbiology and Biotechnology, 43, (2), 351-57.

31. Borja R., Martin, A., Duran, M. M., Laque, M. and Alanso, V. (1993). *Kinetic study of anaerobic digestion of brewary wastewater*, Process Biochemistry, 29, (8), 645-50.

32. Calzada J.F., de Arriola, M.C., Castaneda, H.O., Godoy, J.E., and Ralz, C, (1984). *Methane from coffee pulp juice: experiments using polyurethane foam reactors*. Biotechnology Letters, 6, 385-88.

33. Calzada J. F,de Arriola, M. C., astaneda, H. O., Godoy, J. E. and Ralz, C. (1984). Methane form coffee pulp juice: experiments using polyurethane foam reactors. Biotechnology Letters, 6,385-388.

34. Chavadej, S., (1980). *Anaerobic filter for biogas production*. Reg. J. Energy Heat Mass Transfer, 2, 31-44.

35. Chua, H., and Fung, J. P. C., (1995). *Hydrodynamics in packed bed of anaerobic fixed film reactor*. Proceedings of third International Conference on appropriate waste management technologies for developing countries, NEERI, Nagpur, Feb 25-26, 95, 13-21.

36. Chui, H. K., Fang, H. H. P., and Li, Y. Y., (1994). *Removal of formate from wastewater by anaerobic process.* Journal of Environmental Engineering, ASCE, 120, (5), 1308-21.

37. Chynoweth, D. P., and Srivastava, V. J., (1980). *Methane production from marine biomass*. International Symposium on Biogas, Microalgae, and Livestock Wastes-1980, Taipei, Taiwan, R.O.C., September 15-17, 26.

38. Clark, R. H. and Speece R. E., (1970). *The pH tolerance of anaerobic digestion*. Fifth International Water Pollution Control Conference, July-August,II-27/14.

39. Converti, A., Zilli, M., Del Borghi, M., and Ferraiodo, G., (1990). *The fluidized bed reactor in the anaerobic treatment of wine wastewater.* Bioprocess Engineering, 5, 49-90.

40. David, M. L., Cayless, S. M., and Lester, J N, (1990). *Process aiders for start-up of anaerobic fluidised bed systems.* Environmental technology, 11, 1093-1106.

41. Delgado, M., Guariola, E., and Bigeriego, M., (1992). *Methane generation from water hyacinth biomass*, Journal of Environmental Science Health, A 27, (2), 347-67.

42. Demuynck, M. Nyns, E. J. and Palz, W., (1984). *Biogas Plants in Europe*, Solar Energy R & D in the European Community, series E, 6, Energy from Biomass, D. Reidel Publishing Company, Boston.

43. Dinoupoulou, G., Santana, F. P., and Lester, J. N., (1990). *Kinetic studies of single-stage and two-phase anaerobic fluidised beds,* Environmental technology, 11, 113-22.

44. Gadre, R. V., and Godbole, S. H., (1986). *Treatment of distillery waste by upflow anaerobic filter.* Indian Journal of Environmental Health, 28, 54-59.

45. Geller. A., and Gottsching, L., (1985). *Anaerobic fermentation of Sulphite pulp mill effluents.* Water Science Technology, 17, 157-71.

46. Ghosh, S., (1982). *Novel bioconversion systems and innovative biogasification process designs.* US–China Biomass Conversion Technologies Symposium, Chengdu, Sichuan Province, China, 6p.

47. Ghosh, S., and Henry, M. P., (1982). *Application of packed-bed upflow towers in two-phase anaerobic digestion.* First International Conference on Fixed-Film Biological process, Kings Island, Ohio, April 20-23, P22.

48. Ghosh,S. Conrad,J.R. and Chynoweth,D. P, (1981). *Bioconversion of Macrocyatis pyrifera to methane.* J.Chem Tech Biotechnol., 37, 791-807.

49. Goodwin, J A S, and Stuart, J B,(1994). *Anaerobic digestion of malt whisky distillary potale using upflow anaerobic sludge blanket reactors,* Bioresource technology, 4(1), 75-81.

50. Grin, P. C., Roersma, R. E., and Lettinga, G., (1983). *Anaerobic treatment of raw sewage at lower temperatures.* BVC the European Symposium on Anaerobic Waste Water treatment, Noordwijikerhout, Netherlands, 335-47.

51. Habets, L. H. A., and Knelissen, (1985). *Application of UASB-reactor for anaerobic treatment of paper and board mill effluent.* Water Science Technology, 17, 61-75.

52 Hall, E. R., and Jovanovic, M., (1983). *Anaerobic treatment of thermal Sludge Conditioning liquor with fixed-flim and Suspended growth processes.* Proceedings of 37th Industrial Waste Conference, Purdue University, Ann Arbor Sci. Publ. Inc., 719.

53. Hall, E. R., (1982). *Biomass retention and mixing characteristics in fixed film and suspended growth anaerobic reactors.* Proceedings of the International Association of Water pollution Research Seminar on Anaerobic Treatment of Waste Water in Fixed Film Reactors. 16–18. June, 1982. Copenhagen Department of Sanitary Engineering. University of Denmark. Copenhagen.

54. Hall, E. R. and Jovanovic, M., (1983). *Anaerobic treatment of thermal sludge conditioning liquor with fixed-film and suspended growth processes.*

Proceedings of 37th Industrial Waste Conference,Purdue University, Ann Arbor Sci. Publ. Inc.,719.

55. Hanaki, K., Chatsanguthai, S., and Matsuo, T, (1994). *Characterisation of accumalated biomass in anaerobic filter treating various types of substrates*, Bioresource technology, 47, 275-82.

56. Hanson, J. F., (1983). *Produksjon au energirik biogass.* Milioteknikk, No.3., 14-20.

57. Hanson, J. F., Indergaard, M., Ostgaard, K., Baevre, O. A., Pederson, T. A., and Jensen, A., (1987). *Anaerobic digestion of Laminaria Spp. and Ascophyilum nodosum and application of end products*. Biomass, 14,1-13, 96.

58. Hanssen, J. F., (1983). Produksjon av energirik biogass. Milioteknikk,No.3.,14–20.

59. Harada, H., Uemura, S., and Momonoi, K., (1994). *Interaction between sulfate reducing bacteria and methane producing bacteria in UASB reactors fed with low strength wastes containing different levels of sulfate*. Water research, 28, (2), 355-67.

60. Hayes, T. D., Chynowelth, D. P., Reddy, K. R., and Schwegler, B., (1985). *Methane from an integrated hyacinth treatment anaerobic digestion facility*12th energy technology conference, March, 25–27, Washington D.C.

61. He, Y. L., Geng, X. L., and Yang, S. H., (1995). *Sludge granulation in a UASB reactor for the treatment of soda - anthraquinone chemical wheat-straw pulp black liquor*. Bioresource technology, 51, (2,3), 213-15.

62. Hickey, R. F., (1982). *Anaerobic fludized bed treatment of Whey: Effect of organic loading rate, temperature and substrate concentration*. Proceedings of the First International Conference on Fixed Film Biological Processes. 20-23 April 1982. Kings Island. OH. University of Pittsburgh. Pittsburgh. PA. pp. 1456-76.

63. Hills, D. J., (1980). *Biogas from a high solids combination of dairy manure and barley straw.* Transactions of the ASAE-1980. 1500-04.

64. Hilton, M. G., and Archer, D. B., (1988). *Anaerobic digestion of a sulfate rich molasses wastewater : inhibition of hydrogen sulfide production*. Biotechnology and Bioengineering, 31, 885-88.

65. Hulshoff Pol, L., and Lettinga, G., (1986). *New technologies for anaerobic wastewater treatment*. Post Conference International Seminar on Anaerobic Treatment in Tropical, Countries, Sao Paulo, Brazil.

66. Ikbal, Kida, K., Sonoda, Y, (1994). Liquefaction and gasification during anaerobic digestion of coffee waste by 2-phase methane fermentation with slurry - state liquefaction. Journal of fermentation and bioengineering, 77, (1), 85-89.

67. Jewell, W. J., Switzenbaum, M. S., and Morris, J. W., (1981). *Municipal Waste Water treatment with the anaerobic attached microbial film expanded bed process*. J. Water pollut. Control. Fed. 53: 482-90.

68. Kato, M.T., Field, J. A., Kleerebezem, R., and Letting, G., (1994). *Treatment of low strength soluble wastewaters in UASB reactors*. Journal of fermentation and Bioengineering, 77, (6), 679-86.

69. Khan, A., (1995). *Municipal waste water treatment and energy recovery*. In : Wealth from waste, Khanna S., and Mohan K. (Eds), Tata energy research institute, New Delhi, 195-213.

70. Kirby D., (1980). The design and operation of a pilot scale anaerobic for the treatment of piggery waste. Agricultural Engineering conference, 1980, Geelong.

71. Kuamr, S. (1985). *Building construction*.Standard Publishers Distribution, New Delhi, 804p.

72. Lettinga, G., and Van velson, A FM, (1974). *Anaerobic treatment of low strength wastes*, H_2O ,7, 281.

73. Lettinga, G., Van velsen, S. W. Hobma de Zeeuor, W., and Klapwikj, A., (1980). *Use of upflow sludge blanket reactor concept for biological wastewater treatment, especially for anaerobic treatment*. Biotechnology and Bioengineering, 22, 699-734.

74. Lettinga, G., Roersma, R., Grin, P., Zeenw, W. D., Hulshulf Pol, Van Velson, L., Hobma, S., and Zeeman, G., (1981). *Anaerobic treatment of sewage and low strength waste waters*. In. Anaerobic Digestion, 1981, Hughes, *et. al.*, (Eds.), Elsevier Biomedical Press, B.V., 271-91.

75. Li., A., Sutton, P. M., and Corrado, J. J., (1982). *Energy recovery from pretreatment of industrial wastes in the anaerobic fluidized bed process*, proceedings of the First International Conference on Fixed Film Biological processes. 20-23 April 1982, Kings Island. OH. University of Pittsburgh.

76. Liang, Y. M., and Hu, J., (1993). *Research on characterestics of start-up and operation of treating brewary wastewater with an anaerobic fluidised bed reactor at ambient temperatures.* Water Science and Technology, 28, (7), 187–95.

77. Lo, K. V., Liao, P. H., and White head, A. J., (1985). *Methane production at 22°C of laboratory scale fixed-film reactors receiving screened dairy manure.* Energy in Agriculture, 4, 1–13.

78. Maree, J. P., Gerber, A., Mc Laren, A. R., and Hill, E., (1987). *Biological treatment of mining effluents.* Environmental technology letters, 8, 53–64.

79. Massey, M. L., and Pohland, F. G.,(1978). *Phase seperation of anaerobic stabilisation by kinetic controls.* Journal Water Pollution Control Federation, 2204–22(1978).

80. McCarty, P. L., (1982). *One hundred years of anaerobic treatment.* In: Anaerobic Digestion 1981, Hughes, D.E..,Stafford, D.A., Wheatley,B.I., Baader, W., Lettinga, G., Nyns, E. J., Verstraete, W. and Wentworth, R.L., (Eds), Elsevier Biomedical Press, Amsterdam, 3–22

81. Monroy, O., Johnson, K. A., Wheatley, A. D., Hawkes, F, and Caine, M, (1994). *The anaerobic filteration of dairy waste : results of a pilot trial.* Bioresource technology, 50, 243–51.

82. Morris, J. W., and Jewell, W. J., (1981). *Organic particulate removal with the anaerobic attached film expanded bed process,* Proceedings of the 36th purdue Industrial Waste Conference, January 1981. Ann Arbor Science publishers Incorporated. Ann Arbor. MI. 621–630.

83. Nagaratnam S., (1973). *Fluid Mechanics,* Khanna Publishers, Delhi.

84. Ng., W. J., Chin, K. K., and Wong, K. K., (1987). *Energy yields from anaerobic digestion of palm oil mill effluent.* Biological Wastes, 19, 257–66 .

85. Nobre, C. A., and Guimaraes, M. O., (1987). *Experiments emdigestas anaerobic de esgotos urbanos.* Revista DAE, 47, 75–85.

86. Norman, J., (1982). *Treatment of a black liquor condensate form a plup mill in an anaerobic filter and an expanded bed,* proceedings of the International Association on Water pollution Research Seminar on Anaerobic Treatment of Waste Water in Fixed Film reactors, 16–18 Copenhagen. Department of Sanitary Engineering, University of Denmark. Copenhagen.

87. Palns, S. S., Loewenthal, R. E., Wentzel, M. C., and Marais, G. V. R., (1990). *Growth of biopellets on glucose in upflow anaerobic sludge bed (UASB) systems.* Water SA, 16, (3), 151–64.

88. Parkin, G. F., and Speece, R. E., (1984). *Anaerobic biological waste treatment.* Chemical engineering progrees, December 1984, 55–58.

89. Peck M. W., and Hawkes, F. R., (1987). *Anaerobic digestion of cattle slumy in an upflow anaerobic filter.* Biomass, 13, 125–33.

90. Peng, D. C., Zhang, X: W., Jim, O. T., Xiang, L. K., and Dhang, D., (1994). *Effects of seed sludge on the performance of UASB reactors for treatment of toxic wastewater.* Journal of chemical and Biotechnology, 60 (2), 171–76.

91. Pondicherry Distrilleries, (1990). Personal communication (pondicherry Distilleries), Kuruchikkupam, Pondicherry.

92. Prasertsan, P., Jung, S., Buckle, K. A., (1994). *Anaerobic filter treatment of fishery wastewater.* World journal of Microbiology and Biotechnology, 10 (1), 11–13.

93. Rajamani, S., Suthanthararajan, R., Ravindranath, E., Mulder, A., Vangroen estijin, and Lang esucerf, J. S. A., (1995). *Treatment of tannery wastewater and Biomass energy generation using UASB reactors.*, Proceedings of the third International conference on appropriate waste management technologies for developing countries, NEERI, Nagpur, Feb 25–26.

94. Ramamrutham, S.,(1983). *Design of Reinforced Concrete Structure.* Dhanpat Rai and Sons, Delhi.

95. Ramasamy, E. V., and Abbasi, S. A.,(1997). *Impact of cobalt (II) on bioges yield from Weed-Fed anaerobic digesters. Proceedings of International conference on Industrial Pollution and Control Technologies, JNTU, Nov. 17–19, Allied Publishers, Hyderabad, 328-38.*

96. Ranade, D. R., Meher, K. K., and Gadre, R. V., (1991). *Microbial pretreatment of biomass for biogas production.*, Biovigyanam 17, (2), 61–65.

97. Rico Gutierrez, J. L., Grcia Encina, P. A., and Fdz–Polanco, F., (1991). *Anaerobic treatment of cheese - production wastewater using a UASB reactor.* Bioresource technology, 37, 271–276.

98. Rintala, J. A., and Lepisto, S. S., (1992). *Anaerobic treatment of thermomechanical pulping wastewater* at 30-70° C. Water Research., 26, 1297–1305.

99. Rinzema, A., (1993). *Anaerobic digestion of long-chain fatty acids in UASB and expanded granular sludge bed reactor*, Process biochemistry, 28, (8).

100. Rinzema, A., and Schultz, C. E., (1987). *Anaerobic treatment of acid water on a semi-Technical scale.* Final report prepared for the ministry of housing, physical planning and environment. Agricultural university, dept. of water poll.control, Wageningen, The Netherlands.

101. Rozzi, A., Passino, R., and Limoni, M., (1989). *Anaerobic treatment of olive mill effluents in polyurethane foam bed reactors*. Process Biochemistry, April, 68–74.

102. Sakti Sugar Mills, Personal communication, (Sakti Sugar Mills, TTK Road, Madras) (1989).

103. Saraswat, N., and Khanna, P., (1986). *Methane recovery from water hyacinth through anaerobic activated sludge process*. Biotechnology and Bioengineering, 28, 240–46.

104. Sayed, S. K. I., Vanderspoal, H., and Truijein, G J P, (1993). *A complete treatment of slaughter house wastewater combined with sludge stabilization using two stage highrate UASB process*. Water Science Technology, 27, (9), 83–90.

105. Schellinkhout, A., Jakma, F. F. G. M., and Forero, G E, (1988). *Sewage treatment : the anaerobic way is advancing in colombia*. Proc. fifth. International Symposium on Anaerobic Digestion., Bologma, Italy, 767–70.

106. Schwitzguebel, J. P. and Peringer, P., (1987). *An assessment of seperated phase anaerobic digestion to improve methane production* In: Biochemical Engineering: A challenge for Interdisciplinary Co-operation, Chmicl, H., Hammes, W.P., and Bailey, J.E., gustav Fischer Verlag, Stuttgart, 381–384.

107. Schwitzguelbel, J. P., and Peringer , P., (1987). *An assessment of seperated-phase anaerobic digestion to improve methane production*. In : H. Chmiel, W.P. Hammes and Bailey, J.E.(Eds) , Biochemical Engineering, 381–89.

108. Sharma, S., Ramakrishna, C., Desai, J. D., and Bhatt, N. M., (1994). *Anaerobic biodegradation of a petrochemical wastewater using biomass support particals.*Microbiology and Biotechnology, 40, 768–71.

109. Shen, C. F., Kosarie, N., and Blaszczylc, R., (1993). *Properties of anaerobic granular sludge aaffected by yeast extract, cobalt and iron supplements,* Appled Microbiology and Biotechnology., 39, 132–37.

110. Shi, D. M., and Forster, C. F., (1993). *An examination of the start-up of a thermophilic upflow sludge blanket reactor treating a synthetic coffee waste.* Environmental technology,14,(10), 965–72.

111. Sorilini, C., Ranalli, G., and Merlo, S., (1990). Microbiological aspects of anaerobic digestion of swine slurry in upflow fixed-bed digesters with different packing materials. Biological wastes, 31, 231–39.

112. Soto, M., Ruiz, I., Ferreiro, M. J., Veiga, M. C., Vega, A., and Blazque Z. R., (1995). *Anaerobic treatment of domestic and Industrial (Low conc.) wastewater.* Proceedings of third International conference on appropriate waste management technologies for developing countries, NEERI, Nagpur, India, Feb 25–26.

113. Speece, R. E. and Parkin, G. F., (1984). *Nutrient requirements for anaerobic digestion.* First Symposium on Biological Advances in Processing Municipal Waste for Fuels and Chemicals – Minneapolis, 28p.

114. Stafford, D. A., Wheatley, B. I., and Hughes, D. E., (1980). Anaerobic Disgestion. Applied Sceince Publichers Ltd., London, 528 p.

115. Sutton, P. M., and Huss, D. A., (1984). *Anaerobic fluidized bed biological treatment : pilot to full-scale demonstration.* Water pollution control federation conference, New orleans, Louisiana, 1984.

116. Switzenbaun, M. S., (1980). *Anaerobic expanded bed treatment of wastewater.* Proceedings of the U.S. Department of Energy Workshop/ Seminar on Anaerobic Filters. 9–10 January 1980. Howey–In–The–Hills. FLU.S. Department of Energy. Argon National Laboratories. Chicago. IL. 115–27.

117. Szendrey, L. M., (1983). *Start-up and operation of the Bacardi Corporation anaerobic filter.* Third Int. Conf. Anaerobic Digestion, Boston, 13p.

118. Tanemura, K., Kida, K., Ikbal, Matsumoto, J., and Sonoda, Y., (1994). *Anaerobic treatment of wastewater with high salt content from a pickled-plum manufacturing process.,·* Journal of fermentation and Bioengineering., 77 (2), 188–193.

119. Toldra, F., Flors, A., Lequerica, J. L., and Valles, S., (1986). *Fluidizid bed biomethanation of acetic acid.* Applied microbiology and biotechnology, 23, 336–41.

120. Uemura, S., and Harade, H., (1993). The effect of temperature on the performance of thermophilic UASB reactors., Environmental technology., 14, (9), 897–900.

121. Van den berg, L., and Kennedy, K. J., (1982). *A Comparison of advanced anaerobic reactors.* BVC Third International Symposium on Anaerobic Digestion, Boston.

122. Van Velsen, A F M (1979). *Anaerobic digestion of piggery waste,* Netherlands Journal of Agricultural Sciences, 27, 142–152.

123. Van den Berg, L. and Kennedy, K. J., (1983). *A Comparision of advanced anaerobic reactors.* BVC Third International Symposium on Anaerokbic Digestion, Boston.

124. Van Groenestijn, J. W., Mulder, A., Rajamani, S., Langerweri, J S A, *and* Kortekaas, S, (1995). *Treatment of tannary waste water using an upflow anaerobic sludge blanket (UASB) process.* Proceeding of third International conference on appropriate waste management technologies for developing countries, NEERI, Nagpur, India, 45–46.

125. Veiga, M. C., Mendez, R., and Lema, J. M., (1994). *Anaerobic filter and DSFF reactors in anaerobic treatment of tuna processing wastewater.*, Water science and technology, 30, (12), 425–32.

126. Vieria, S. M. M., Crvalho, J. L., Barijan, F. P. O., and Rech, C. M., (1994). *Application of the UASB technology for sewage treatment in a small community at Sumare*, Sao Paulo state Water Science and Technology, 30 (12), 203–10.

127. Visser, A., Beeksma, I., Vander Zee, F., Stams, A J M, and Lettinga, G, (1993). *Anaerobic degradation of volatile fatty acids at different sulphate concentrations.* Appled Microbiology and Biotechnology, 40, 549–56.

128. Wilkie, A., and colleran, E., (1984). *Start-up of anaerobic filters containing different Support materials using pig Slurry Supernatant.* Biotechnology Letters, 6, 735–40.

129. Wilkie, A., and colleran, E., (1986). *The development of the anaerobic fixed-bed reactors and its application to the treatment of agriculture and industrial wastes.* In:International Biosystems, III. D.L. Wise (ed.), CRC press, Boca Raton, Florida, p. 53.

130. Wilkie, A. and Colleran, E., (1984). *Start-up of anaerobic filters containing different support materials using pig slumy supernatant.* Biotechnology Letters, 6, 735–40.

131. Wilkie, A. and Colleran, E., (1986). *The development of the anaerobic fixed-bed reactors and its application to the treatment of agricultural and industrial wastes*. In: International Biosystems, iII. D.L. Wise (ed.), CRC Press, Boca Raton, Florida, p. 53.

132. Yoda, M., Shein, S. W., Watanable, A., Watanabe, M., Kitagawa, M., and Miyaji, Y., (1987). *Anaerobic fluidized bed treatment with a steady-state biofilm*. Water science technology, 19, 287–98.

Chapter 2

WATER QUALITY MODELLING : APPROACHES TO WATER POLLUTION FORECASTING AND MANAGEMENT

The pollution of surface water bodies - such as rivers, ponds, lakes, reservoirs–is among the environmental problems of greatest concern because water pollution effects the health of the humans, other animals, vegetation, soil, and even air.

In order to control water pollution we need to know how the water quality in any water body changes with the type of pollutants that come into that water body, the amount of water that is present in the water body, and ambiant factors such as aeration and temperature. But we can't keep experimentally monitoring these factors all round the year. Such an effort would be prohibitively costly. If the water quality of a water body can be modelled, we can then use the model–after due validation and caliberation–to forecast the behaviour of the water body with the help of limited monitoring.

The science of water quality modelling attempts to achieve this objective.

In this chapter we have introduced the concept of water quality modelling and have described the systematic progression of the models beginning with the first-ever model presented by Streeter and Phelps in 1925.

The limitations of the various models and the attempts to overcome are described, leading to one of the most widely used computer-automated tool in water quality modelling: QUAL 2E.

INTRODUCTION

The ultimate objective of all the water quality monitoring, modelling, and treatment exercises is to ensure that water is either not allowed to get polluted at all or if it does get polluted, to eliminate or cushion its adverse impacts.

Water quality modelling enables us to translate a real-life water course into a table-top model – be it physical, mathematical, analog or any other type of model. Once the model has been developed and validated it then becomes possible to 'tinker' with it, manipulate it, or 'play with it' in a manner one can't with the actual water-course (Figure 1). We can do a very large number of assessments and develop innumerable scenarios with a model–by varying the flow, varying the stream water quality, varying the types and quantities of waste inputs, varying the ambient conditions etc – and then seeing what happens to the modelled water-course under different situations.

In order to get the same insight by experimenting with the real-life water-course one would need enormous time and money. Certain experiments (such as the behaviour of the river in spate) cannot even be organized.

The following three main issues confront a water quality expert:

i) What are the water quality standards to be stipulated for a particular water body and how is the zoning and classification of the river to be done?
ii) What are the impacts which waste water inputs would have on the water quality of the receiving body of water and, conversely, what should be the level of treatment required for the waste so as to ensure compliance with the water quality standards stipulated for the water body?

iii) Are the levels of treatment arrived at by the above investigation really attainable from techno-economic standpoints?

It could be easily appreciated that the crux of the entire problem is the estimation of the impact of the waste load on water quality. The historical development of water quality modelling began by the estimation of the impact of waste-load (in terms of biological oxygen demand, BOD) on the dissolved oxygen (DO) of receiving streams. The first DO-BOD model was developed by Streeter and Phelps (1925). BY now great improvements have occurred to this model.

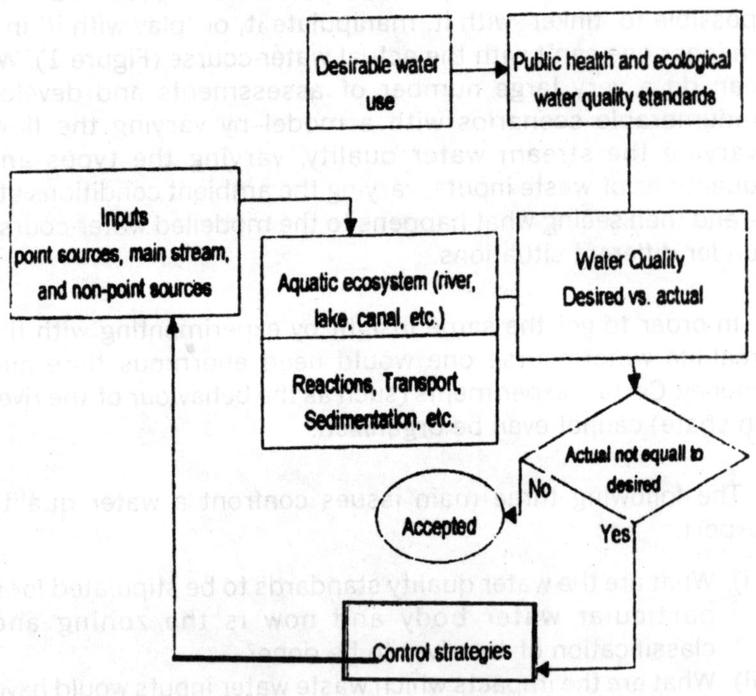

Fig. 2.1: The feedback mechanism of water quality modelling

Water quality modelling mathematically defines cause-effect relationships between the chemistry, biology, and hydrology of the river and can then be used to predict the relation between the waste loading input to water quality output. The movement and reactions of waste materials through an aquatic system is a result of hydrodynamics transport and biological and chemical reactions by the biota, suspended materials, plantgrowths and bottom sediments. These processes occurring in the system can be expressed mathematically. Figure 2.1 presents an example of a relatively simple water quality management system with a sequence of input/output relationships, water quality objective comparisons, and a feedback loop for waste effluent and/or stream modification, calibration, validation and final acceptability.

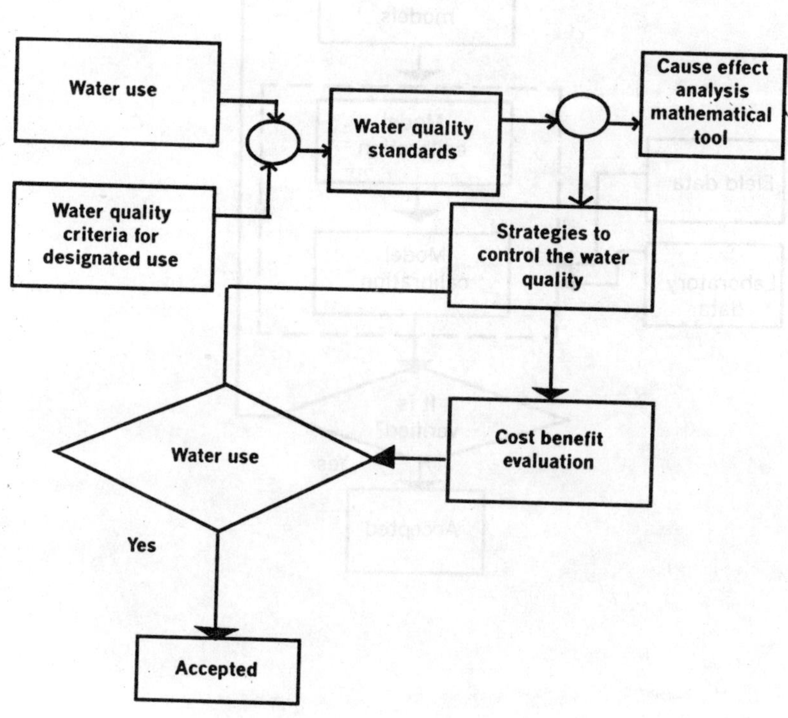

Fig. 2.2 Main steps in water quality assessment

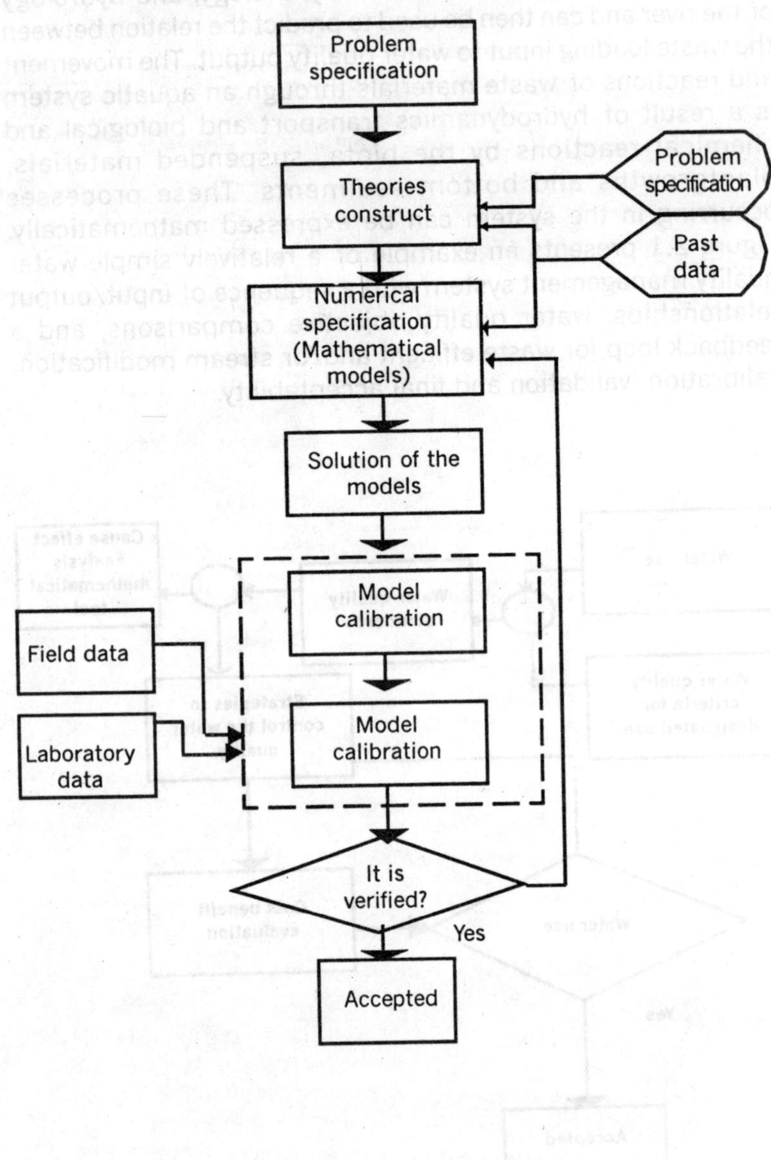

Fig. 2.3: Principal Components of Modelling Framework

TYPES OF APPROACHES TO WATER QUALITY MODELLING

In general then, the role of the water quality engineer and scientists is to analyze water quality problems by dividing the problem into its principal components. Figure 2.1 shows such principal components as:

1. Inputs, that is, the discharge of residue into the environment from man's and nature's activities.
2. The reactions and physical transport, that is, the chemical and biological transformations, and water movement that result in different levels of water quality at different locations in time in the aquatic ecosystem.
3. The output, that is, the resulting concentration of a substance, such as dissolved oxygen or nutrients, at a particular location in the water body, during a particular time of the year or day.

The inputs are discharged into an ecological system such as a river, lake, estuary, or oceanic region. As a result of chemical, biological, and physical phenomena (such as bacterial biodegradation, chemical hydrolysis, physical sedimentation) these inputs result in a specific concentration of the substance in the given water body. Concurrently, through various mechanisms of public hearings, legislation, and evaluation, a desirable water use is being considered or has been established for the particular region of the water body under study. Such a desirable water use is translated into public health and/or ecological standards, and such standards are then compared to the concentration of the substance resulting from the discharge of the residue. This desired versus actual comparison may result in the need for an environmental engineering control, if the actual or forecasted concentration is not equal to that desired. Environmental engineering controls are then instituted on the inputs to provide the necessary reduction to reach the desired concentration. The presentation of various environmental engineering control alternatives to reach the same objective, forms a central role in the decision-making process of water quality management. This is made more specific in the application of the principles of waste load allocation (Figures 2.2 and 2.3).

The ultimate objective of all the water quality monitoring, modelling, and treatment exercises is to ensure that water is neither not allowed to get polluted at all or if it does get polluted, to eliminate or cautions its adverse impacts.

Water quality *modelling* enables us to translate a real-life water course into a table-top model–be it physical, mathematical, analog or any other type of model. Once the model has been developed and validated it then becomes possible to 'tinker' it, manipulate it or 'play with it' in a manner one can't with the actual water-course (Figure 2.3). We can do a very large number of assessments and develop innumerable scenarios with a model - by varying the flow, varying the stream water quality, varying the types and quantities of waste inputs, varying the ambient conditions etc.,–and then seeing what happens to the modelled water-course under different situations.

The order to get the same insights by experimenting with the real-life water-course one would need enormous time and money. Certain experiments (such as the behaviour of the river in spate) cannot even be organized.

The following three main issues confront a water quality expert:

i) What are the water quality standards to be stipulated for a particular water body and how is the zoning and classification of the river to be done?
ii) What are the impacts which waste water inputs would have on the water quality of the receiving body of water and conversely, what should be the level of treatment required for the waste so as to ensure compliance with the stipulated water quality standards stipulated for the water body?
iii) Are the levels of treatment arrived at by the above investigation really attainable from techno-economic standpoints?

It could be easily appreciated that the crux of the entire problem is the estimation of the impact of the waste load on water quality. The historical development of water quality modelling began by the estimation of the impact of waste-load (in terms of BOD) on the dissolved oxygen of receiving streams. The first DO-BOD model was developed by Streeter and Phelps (1925).

APPROACHES TO WATER QUALITY MODELLING

Water quality modelling mathematically defines cause-effect relationships between the chemistry, biology, and hydrology of the river and can then be used to predict the relation between the waste loading input to water quality output. The movement and reactions of waste materials through an aquatic system is a result of hydrodynamics transport and biological and chemical reactions by the biota, suspended materials, plant growths and bottom sediments. These processes occurring in the system can be expressed mathematically. Figures 2.1-2.3 presents an example of a relatively simple water quality management system with a sequence of input/output relationships, water quality objective comparisons, and a feedback loop for waste effluent and/or stream modification, calibiration, validation and final acceptability.

Types of approaches to water quality modelling

A surface water-course can be considered as a single system or made up of a large number of sub-systems. The various system parameters defining the water-course can be flow, volume, turbulence, temperature, oxygenation rate, etc.

A water-course can be modelled with varying degrees of sophistication depending on one of the following assumptions:

a) *Fixed parameter steady state input* : in this system the waste load input, system parameters and water quality output remain constant with time.
b) *Fixed parameter dynamic input* : here, the waste load input and water quality output are time variable.
c) *Variable parameter steady state input* : here the output varies.
d) *Variable parameter dynamic input*: both load and system parameters vary and hence also output.

In general, the degree of complexity of problem solution and hence the sophistication of the model, increases as one proceeds down the list. The extent of sophistication necessary would depend on the ultimate task we want the model to perform. For example, for a number of capital improvements and, long-term water quality management problems, where time varying detail is not important, the approach of the first system would be adequate. On the other hand, if one were investigating the case of a relatively constant waste load but had to incorporate variable uncontrolled stream flow, it would be necessary to use the approach defined in the third system.

In a single system a functional relation exists between an input (the discharge of material) and a single output (concentration). The further action of material in the stream on some other variable leads to coupled systems phenomena, which is discussed later. A deterministic relation for a single system can be derived using mass balance equation. A river can be divided into number of reaches or sub-systems so that the system parameters can be considered constant for each reach. Hence, it can be assumed that the water quality variables have a gradient only along the length of the river. Further 'plug flow' is assumed; in other words it is assumed that there is no mixing due to diffusion of material.

Let V represent an elemental volume of the stream of cross-sectional area A and elemental slice of thickness x. Also, let $S(M/L^3)$ be the total mass entering the upstream face in time t, Q be the river flow.

The overall mass balance equation of material can be written as :

$$V\,S = QSt \cdot (Q+Q)(S+\underline{S}\,x)\,t \pm K\,V\,S\,t \pm \text{(Sources and Sinks)} \qquad (1)$$

where K (1/T) represents a source sink coefficient of S assuming a first order reaction, which is often used for natural waters. Mathematically, the system remains linear under this first-order kinetics so principle of superposition applies. For higher order nonlinear type reactions, simulation type of analysis has to be made.

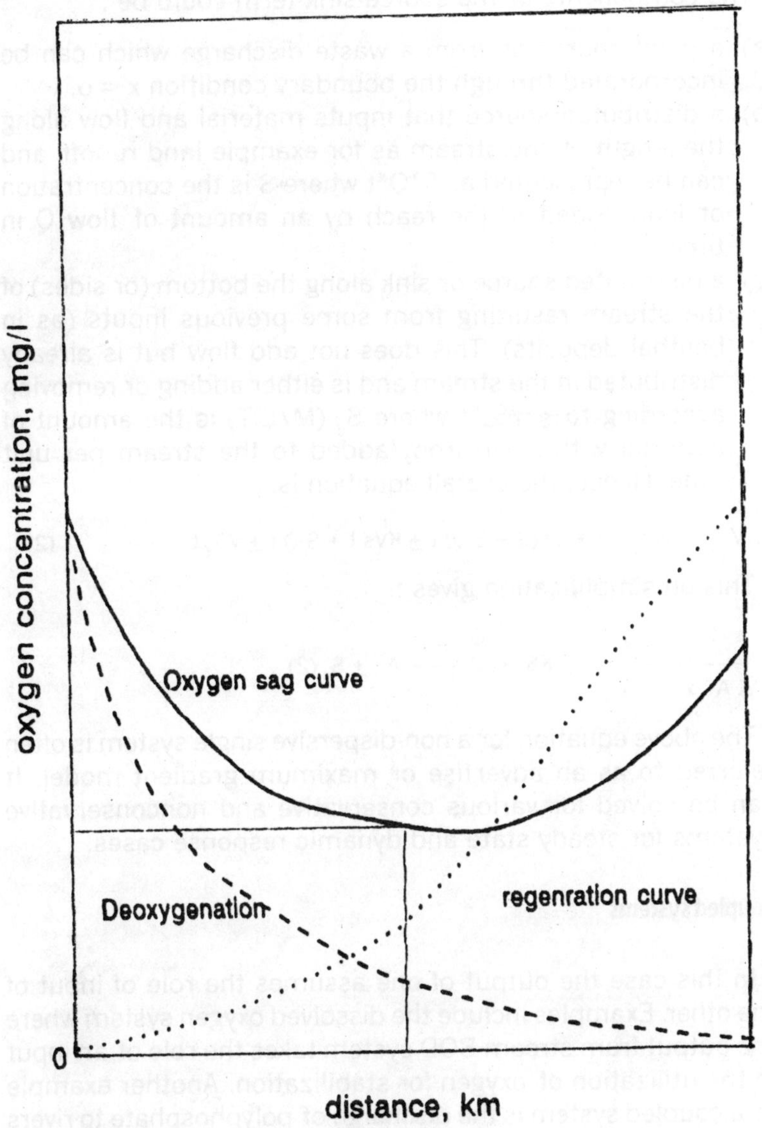

Fig. 2.4: Oxygen-Sag curve

The components of the source-sink term could be :

a) a point source as from a waste discharge which can be incorporated through the boundary condition x = o.
b) a distributed source that inputs material and flow along the length of the stream as for example land runoff, and can be represented as S*Q*t where S is the concentration of load added to the reach by an amount of flow Q in time t.
c) a distributed source or sink along the bottom (or sides) of the stream resulting from some previous inputs (as in benthal deposits). This does not add flow but is already distributed in the stream and is either adding or removing according to $\pm VS_d{}^*t$ where S_d (M/L^3T) is the amount of material withdrawn from/added to the stream per unit time. Hence, the overall equation is:

$$V S = Qs\,t \cdot (Q + Q)(S \pm S\,Q)\,t \pm KVs\,t + S\,Q\,t \pm VS_d\,t \quad (2)$$

This on simplification gives :

$$\frac{s \cdot 1}{t A} \; \underset{x}{=} \; \frac{(QS)}{x} \pm KS + \overset{Q}{S} \; \frac{}{} \; A \pm S_d \; (2)$$

The above equation for a non-dispersive single system is often referred to as an advertise or maximum gradient model. It can be solved for various conservative and nonconservative systems for steady state and dynamic response cases.

Coupled systems

In this case the output of one assumes the role of input of the other. Examples include the dissolved oxygen system where the output from stream BOD system takes the role of an input in the utilization of oxygen for stabilization. Another example of a coupled system is the discharge of polyphosphate to rivers leading first to the distribution of polyphosphate in the stream and then subsequently forcing the orthophosphate systems by virtue of hydrolysis of the complex phosphates.

The procedure for constructing a mathematical model of coupled streams is identical to single systems and the same assumptions can be retained. One of the sources (sinks) in coupled systems will be the output from a preceding system. Thus for two variables S_1 and S_2 the following set of equations would be obtained.

$$\frac{s_1}{t} = \cdot\cdot \frac{1}{A} \frac{(QS_1)}{x} \cdot k_{11} S_1 + S_1 \left(\frac{Q}{x} A\right) \pm Sd_1 \tag{3}$$

$$\frac{s_2}{t} = \cdot\cdot \frac{1}{A} \frac{(QS_2)}{x} \cdot k_{22} S_2 + S_2 \left(\frac{Q}{x} A\right) + K_{12} S_1 \pm Sd_2 \tag{4}$$

with boundary conditions

$$S_1 (o, t) = S_{o1} (t)$$

$$S_2 (o, t) = S_{o2} (t)$$

In the text section this model is applied to the DO–BOD coupled system.

THE DISSOLVED OXYGEN (DO) - BIOCHEMICAL OXYGEN DEMAND (BOD) MODELS

Water quality models can be applied to a number of water pollutants including chlorides, nitrogen and phosphorous, coliform bacteria distribution, pesticides, and dissolved oxygen-biochemical oxygen demand studies. Because of the great significance of dissolved oxygen for water quality, modelling of the relationship between dissolved oxygen (DO) and biochemical oxygen demand (BOD) has received considerable attention. Because organic wastes, reflected in BOD concentrations, account for approximately 90 percent of water pollutants in India, DO-BOD modelling is of special significance. Of particular importance for water quality management is that the model enables a calculation of the assimilation capacity of the maximum BOD loading, that can be discharged into an aquatic system without degrading water quality.

The DO–BOD model determines the relationship of DO concentrations with distance downstream resulting from a given input of BOD. This general relationship is derived from the mass balance equation for oxygen :

c c —— = · U — ± S t (5)

where, c = concentration of dissolved oxygen (ML^{-3})

t = time at a stationery point (T)
u = velocity of flow in the x direction (LT^{-1})
x = distance downstream (L) and
s = sources and sinks of oxygen

Thus, the sources and sinks of dissolved oxygen in the system must be mathematically expressed.

Sources and sinks of dissolved oxygen
The sources of dissolved oxygen are as follows :

a) Quantity of incoming or tributary flow

Oxygen will be supplied from that present in the upstream water and in the tributary flow. This is considered as an initial condition in the equation. The dissolved oxygen in waste discharge should also be considered if the waste flow is large relative to stream flow.

b) Reaeration

Water saturated with respect to dissolved oxygen concentration undergoes atmospheric reaeration.This is primarily related to the degree of turbulence and natural mixing in the water body (high in rapids, low in impounded areas). The oxygen transfer from air to water can be defined as :

$$\frac{dc}{dt} = k_2 (c^* - c) \quad (6)$$

where, c^* = oxygen saturation concentration (ML^{-3})
and k_2 = reaeration coefficient (T^{-1})
k_2 has been found to be a function of turbulence as

$$k_2 = \frac{D_m U^{1/2}}{H^{3/2}} \quad (7)$$

where, D_m = molecular diffusion coefficient for oxygen in water (L^2T^{-1})

and H = average depth (L)

The molecular diffusion coefficient is temperature-dependent. Expressed in units of square centimeters per second, it is:

$$D_m = 2.037 \times 10^{-5} (1.037)^{T-20} \quad (8)$$

where, T = temperature, °C.

c) Photosynthesis

The photosynthetic source of dissolved oxygen depends upon many factors, such as sunlight, temperature, and concentration of phytoplankton and nutrients. This exhibits a diurnal variation with zero at night and during the day varies with sunlight intensity. In the application of the model, it is convenient to combine photosynthesis and respiration in a constant rate term, as described below :

The sinks of dissolved oxygen are as follows:

a) Biochemical oxygen demand

BOD can be the most significant oxygen sink in streams and estuaries. One considering BOD, distinction should be made between removal of BOD through biologic activity and removal by physical factors. When a portion of the BOD exerting material is in particulate form, sedimentation may occur, especially when flow velocities are sufficiently low. Furthermore, when the ratio of bottom area to flow volume is relatively large, bottom slimes may result in the removal of a significant portion of BOD exerting materials through adsorption. Effectively, BOD removed through sedimentation and/or adsorption is no longer available to exert a demand on the dissolved oxygen in the water from which it was removed.

The time rate change in DO concentration caused by the removal of BOD through biologic activity can be expressed as

$$\frac{dc}{dt} = -k_1 L \quad (9)$$

where, L = first stage BOD (ML^{-3})
and k_1 = deoxygenation constant (T^{-1})

The time rate of change in the concentration of BOD, however is given by

$$\frac{dL}{dt} = -k_1 L - k_3 L \qquad (10)$$

where, k_3 = rate constant for BOD removal by sedimentation and/or adsorption (T^{-1})

b) Bottom deposits and runoff

Organic material once it settles out along the bottom of a stream undergoes decomposition, resulting in organic materials in the form of organic acids and reduced gases being released to the water above. These materials then become part of the BOD of the following stream.

The rate at which BOD is added to a stream from bottom deposits is generally expressed in a term that includes also the rate at which BOD is added to the stream by surface runoff:

$$\frac{dL}{dt} = L_a \qquad (11)$$

where, L_a = rate of addition of BOD by local runoff and/or by resuspension of organic from bottom sludge deposits ($ML^{-3}T^{1}$)

c) Respiration

The respiration of both microscopic and macroscopic aquatic plants may be significant sink for dissolved oxygen. Respiration is continuous and independent of the diurnal photosynthetic pattern. Hence, respiration may be expressed as a constant rate term. In many cases, however, photosynthesis, respiration and other sources and sinks not explicitly included in the weight balance for dissolved oxygen are combined in a constant rate term :

$$\frac{dc}{dt} = -s_R \qquad (12)$$

where, s_R = rate at which dissolved oxygen changes as a result of photosynthesis, respiration and other sources and sinks not explicitly included in the weight balance for oxygen ($ML^{-3}T^{-1}$).

A REVIEW OF DO-BOD MODELS

The first-ever DO-BOD model was formulated by Streeter and Phelps in 1925. This was also the first-ever water quality model and in spite of its limitations is used even today.

In development of the DO model, steady-state conditions are frequently assumed. Under steady-state conditions, there is no change in loading with time at the BOD source and hence;

$- \frac{c}{t} = 0$. Thus, equation (5) becomes

$$U \frac{D}{x} \pm S = 0$$

where the dissolved oxygen concentration is expressed in terms of the dissolved oxygen saturation deficit, D, where,

$$D = c^{*} \cdot c \qquad (13)$$

In the original development of the model by Streeter and Phelps in 1925, only reaeration and deoxygenation were considered as the source and sink of oxygen, respectively. Assuming the

$$k_3 = S_R = L_a = 0,$$

$$U \frac{D}{x} = \cdot k_2 D + K_1 L_{(x)}$$

where, $L_{(x)}$ = first-stage BOD at distance x (ML^{-3})

Dividing throughout by velocity and integrating the equation yields the following expression :

$$D_{(x)} = \frac{k_1 L_o}{(k_2 - k_1)} (e^{-k_1 x/u} - e^{-k_2 x/u}) + Do\, e^{-k_2 x/u} \quad (15)$$

where, $D_{(x)}$ = dissolved oxygen deficit at distance x (ML^{-3})

D_o = initial dissolved oxygen deficit (ML^{-3})

and L_o = initial first-stage BOD (ML^{-3})

This expression describes the oxygen sag curve and is shown graphically in Figure 2.4.

The oxygen sag curve provides a visual representation of the spatial distribution of the dissolved oxygen concentration in a stream. Of particular interest is the low point along the curve located at a critical distance downstream, x_c. The deficit at this point, the critical deficit, D_c, is the maximum deficit that will occur under the given conditions of waste loading and stream phenomena.

By differentiating the oxygen sag curve and equating the derivate equal to O, an expression is obtained for the critical time of flow,

$$x_c = \frac{U}{(k_2 - k_1)} \ln \left[\left(\frac{k_1 L_o - k_2 D_o + k_1 Do}{k_1 L_o} \right) \frac{k_2}{k_1} \right] \quad (16)$$

at the critical deficit

$$D_c = \frac{k_1}{k_2} L_o e^{-k^1 X_{c/u}}$$

Thus, the dissolved oxygen model can predict the critical oxygen deficit and the critical distance that will result from a given BOD loading. The maximum permissible BOD loading in order to maintain dissolved oxygen concentration above a specified level can also be determined from these equations.

Since development of the Streeter-Phelps model, additional work has been done to include other sources and sinks of dissolved oxygen. Including photosynthesis and respiration, sedimentation and adsorption, and bottom deposits and runoff as sources and sinks of dissolved oxygen in addition to reaeration and deoxygenation, and assuming steady-state (D/t= 0; L/t=0),

then

$$U \frac{D}{X} = - k_2 D + k_1 L_{(x)} + S_{R(x)} \tag{18}$$

$$\text{and } U \frac{L}{X} = - k_1 L(x) - k_3 L_{(x)} + L_{a(x)} \tag{19}$$

By substituting L into equation (18) and integrating, the oxygen sag curve is given by:

$$D_{(x)} = D_o e^{-k_2 x/u} + \frac{k_1}{k_2 - (k_1 + k_3)} (L_o - \frac{L_a}{k_1 + k}) \times (e^{-(k_1 + k_3) x/u} - e^{-k_2 x/u}) + [\frac{S_R}{k_2} + \frac{k_1 L_a}{k_2 (k_1 + k_3)}] (1 - e^{-k_2 x/u}) \tag{20}$$

By differentiating and equating the derivative equal to O, the critical distance is:

$$x_c = \frac{U}{k_2 - (k_1 + k_3)} \ln [\frac{K_2}{k_1 + k_3} + \frac{k_2 - (k_1 + k_3)}{(k_1 + k_3) L_o - L_a} \times (\frac{L_a}{k_1 + k_3} - \frac{k_2 D_o - S_R}{k_1})] \tag{21}$$

As elaborated earlier, the major mechanisms affecting the DO and BOD in a stream could be summarized as follows :

i) Dispersion and dilution
ii) Decay of BOD by biochemical action and subsequent depletion of in-stream DO levels.
iii) Deposition of suspended portion of the BOD on the river bed and the formation of benthal deposits.

iv) Scour of benthal deposits and resuspension and leaching of intermediate products of anaerobic decomposition therefrom to the water above them.

v) Entratinment of atmospheric air into the water as a result of surface turbulence and its dissolution in water.

vi) Photosynthesis and respiration of aquatic plants, weeds and algae which affect the DO concentration.

Quantification of the above mechanisms has been done to varying extents in the work done so far. Streeter and Phelps had assumed that the balance between DO and BOD is dependent on two processes only viz. biochemical decay and reaeration from atmosphere.

No significant improvement in the detail of the Streeter and Phelps equation was made till the mid 1960s when Dobbins (1964) presented the following expressions for the source and sink terms :

$$\text{BOD} : -K_1X_1 - K_3X_1 + L_A \tag{22}$$

$$\text{DO} : K_2(C_s - X_2) - K_1X_1 + D_B \tag{23}$$

The additional parameters introduced are as defined below :

K_3 = rate constant for BOD removal by sedimentation and/or adsorption on benthal layers, day[1].

L_A = rate of addition of BOD to the reach of the river by local run off etc. (mg/1)/day.

D_B = Net rate of additions to the reach by the combined action of decomposition of bottom mud deposits and the photo-synthesis/respiration of aquatic plants (mg/1)/day

Dobbins presented his model as a second order linear partial differential equation by including the effect of longitudinal mixing. He defined a dimensionless ratio R given by

$$R = \left[\frac{(U^2)^2}{(2KD_L)}\right] + \frac{2U^2}{2KD_L} - \frac{U^2}{2KV_L} \tag{24}$$

Here, U = velocity of the stream, ft/day

K = $K_1 + K_3$ or K_2 depending on which equation (DO or BOD) is being analysed

D_L=rate of dispersion, ft^2/day

for $U^2/2KD_L$ exceeding 100, R was found to be greater than 0.995 indicating that the error involved in neglecting longitudinal dispersion is negligible.

Dobbins and Dresnack (1968), proposed a numerical solution to the equations developed by Dobbins. Temporally and spatially varying inputs were considered and the Marching and Jury procedures were used for solution. The term D_B was also modified to the following form :

$$BOD : \cdot K_1X_1 \cdot K_3X_1 + L_A \quad (25)$$

$$DO : K_2 (C_s \cdot X_2) \cdot K_1X_1 \cdot D_B \cdot R + P \quad (26)$$

D_B=rate of removal of DO by decomposition of bottom deposits, (mg/1)/day

R =rate of removal of DO by plant respiration, (mg/R)/ day

P= rate of addition of DO, (mg/1)/day by plant photosynthesis

Working on lines similar to Dobbins, Camp (1976) presented an alternative set of equations which were :

$$BOD : K_1X_1 \cdot K_3X_1 + P1 \quad (27)$$

$$DO : K_2 (Cs \cdot X_2) \cdot K_1Y_1 + P2 \quad (28)$$

where P2 = rate of addition of DO by algae through photosynthesis, (mg/l)/day.

P1=rate of addition of BOD to the following stream by resuspension of bottom mud.

The significance of the studies by Camp and Dobbins is evident when one considers the numerous applications their models have found in subsequent contributions to DO-BOD studies. Frankel and Hansen presented a discussion on the work of Camp and Dobbins. The following factors are recommended by them for inclusion into DO-BOD equations :

i) The variation of K_1, particularly at the onset of the nitrification stage.
ii) Change in channel configuration and its effect on characteristics of surface turbulence and consequently, the rate of oxygen transfer from the atmosphere.
iii) The effects of suspended and dissolved solids on the rate of diffusion from the surface into the main body of the stream.
iv) Diurnal variations in DO due to photosynthesis.

The equations developed by Frankel and Hansen were :

$$BOD : -(K_{12} + K_{11})X_1 - K_3X_1 + P \tag{29}$$

$$DO : K_2 (Cs - X_2) - (K_{12} + K_{11}) X_1 - D_B + P_m \text{Cos} (wt - O) \tag{30}$$

where K_{12} and K_{11} are the first order reaction rates for carbonaceious and nitrogenous oxygen demand respectively (day^{-1}); K_3 the rate constant for BOD removal by sedimentation and/or absorption on the river bed, (day^{-1}); P_m is the amplitude of diurnal oscillations in the rate of addition of DO by photosynthetic and respiratory activity of stream biota.

w = frequency of oscillations (2 radians/day)

O = Phase lag (radians)

Frankel and Hansen modified Camp's formulation of benthal oxygen demand. Camp's equation was based on the assumption that the rate of BOD escaping from the benthos to the overlying water equals the rate of anaerobic decomposition in the benthos.

viz :

$$Hp = \frac{dL_d}{dt} = -K_4L_d \tag{31}$$

where H = depth of water, ft.
P = scour rate, (mg/1)/day per ft^3 of water
L_d = BOD of benthos, (mg/1)/ft^2
K_4 = anaerobic fermentation rate constant, day^1.

An alternative equation was suggested by Frankel and Hansen which obviated the assumption made by Camp :

$$\frac{1}{H} * \frac{dL_d}{dt} = -K_4 \frac{L_d}{H} - p \qquad (32)$$

Towards the late 1970's the second generation models were freely used and more complex descriptions for diurnal DO variation were developed. O'Connor and Di Toro (1970) derived a Fourier series representation of the oxygen released due to photosynthesis, which was based on the assumption that the photosynthetic rate P(x,t) varies as the sunlight intensity during the day :

$$P(x,t) = P^m [\frac{2T}{} + 2^{n-1} a_n \cos (2^n (t - \frac{T}{2})] \qquad (33)$$

$$\text{where } a_n = \frac{2 \ /T \ \cos (n \ T)}{(\ /T)^2 - (2n \ )^2} \qquad (34)$$

and T = period of sunlight, day

p_m = maximum photosynthetic rate, (mg/1)/day

The photosynthetic rate would actually depend on the following factors in addition to sunlight:

1. Temperature
2. Mass of algae
3. Nutrient content of water

The claims of O'Connor and Ditoro were that the above factors can be incorporated in P_m. However, no experimental procedure has been delineated by them for estimation of this term.

Looking back over the contributions that have been made since the study of Streeter and Phelps, it can be seen that through more and more realistic interpretations and representations of DO-BOD interactions have been put forward, the basic approach is still the same which is a tribute the pioneers of water quality modelling! Significant departures are found only rarely; in this light, Thomann's work conducted during 1970s and 1980s assumes special significance. He was the first to introduce the systems analysis approach to the modelling and control of water quality. Linear control systems theory was adopted by him for water quality application. Thomann was also a pioneer in the use of time series analysis of water quality data for forecasting purposes.

As the models evolved and increased in complexity, more and more parameters were introduced in the basic equations. There is still a considerable amount of ambiguity in the field evaluation of these parameters. The state of art as regard field evaluation of DO-BOD system parameters is presented subsequently.

SOME PRACTICAL PROBLEMS IN FIELD ESTIMATION OF MODEL PARAMETERS

It goes without saying that the problems and hurdles of field estimation of model parameters are commensurate in magnitude and complexity with the degree of sophistication of the model. While increasingly complex DO-BOD models are being formulated by research workers, the aspect of parameters estimation is unfortunately receiving scant attention. Some observation on the major parameters hitherto introduced by research workers as regard their measurement, would be in order at the stage.

a) First order Decay Constant for BOD

The evaluation of the biochemical decay rate, is carried out by collecting samples at each of a number of locations along the stream and analysing them for BOD. The oxygen uptake rate is then determined by using any one of several available methods such as the Method of Moments. Such estimates of the BOD decay coefficient are referred to as 'bottle estimates' and may be drastically different from the true values of the co-efficient in the river. These differences arise due to biological extraction of BOD on the river bottom and are not accounted for in the bottle reaction rates has reported an in-stream estimate of K_1 for the Clinton River as being about ten times the bottle rate. The lack of reproducibility of the BOD test results for low BOD values further decreases the reliability of the value of the parameters as obtained experimentally.

b) Re-aeration Rate Constant

The determination of the reaeration rate constant is however, relatively more reliable. This coefficient has been the subject of much study and investigation for the past few decades. Studies in this area have ranged from the theoretical investigations of O'Connor and Dobbins (1958) to the empirical studies of Churchill et al (1962) and Ownes et al (1964). These studies have covered a wide range of river situations.

c) Oxygen Uptake By Benthal Deposits

In-situ measurements of this parameter involve careful submersion of a bell-jar device on the bottom and then measuring the oxygen uptake with time. Laboratory measurements are conducted by extracting samples (preferably undisturbed) from the river bed and placing each sample in a large vessel with oxygenated water.

The DO reduction over time gives the benthal demand as weight of O_2 per unit area of the river bed per unit time. The common drawback of these procedures is the requirement of extensive sampling and measurements as the nature and thickness of benthal deposits could vary substantially over the river bed.

d) Net Photosynthetic Rate

The contribution of DO by photosynthesis of aquatic plants and its utilization by respiration can be determined by a variety of methods. These methods include :

i. light and dark bottles
ii. photosynthetic chambers
iii. evaluation of diurnal DO measurements, etc.

Unfortunately, however, no direct method of obtaining estimates due to algae above, independent of bacterial respiration, has been developed so far. This indeed is a serious limitation of net photosynthetic rate estimations.

LACUNAS IN CONVENTIONAL APPROACH

It is evident from the above mentioned that the direct approach to the measurement of parameters involves extensive field investigations and experimentation and, to crown it all, there is no guarantee as regards reliability of the values obtained therefrom. The conventional modelling and validation procedure could be summed up as in Figure 2.3.

Two lacunas in the approach are evident

1. In case of mismatch between predicted and measured output variables, the error could originated either in the model structure itself or estimates of parameters or even possibly, the measurement of in-stream BOD values. The exact source of error therefore is difficult to trace.
2. The measurement of parameters and of in-stream BOD often proves to be a futile exercise as described earlier in this paper.

The state variable approach delineated here below obviates the above mentioned limitation and reduces drastically the requirement of field data.

Prior to proceeding with the analysis, the state space representation is introduced.

QUAL2E AND QUAL2E–UNCAS

QUAL2E is a comprehensive and versatile stream water quality model. It can simulate up to 15 water quality constituents in any combination desired by the user. Constituents which can be simulated are :

1. Dissolved Oxygen
2. Biochemical Oxygen Demand
3. Temperature
4. Algae as Chlorophyll a
5. Organic Nitrogen as N
6. Ammonia as N
7. Nitrite as N
8. Nitrate as N
9. Organic Phosphorus as P
10. Dissolved Phosphorus as P
11. Coliforms
12. Arbitrary Nonconservative Constituent
13. Three Conservative Constituents

The model is applicable to dendritic streams that are well mixed. It assumes that the major transport mechanisms, advection and dispersion, are significant only along the main direction of flow (longitudinal axis of the stream or canal). It allows for multiple waste discharges, withdrawals, tributary flows, and incremental inflow and outflow. It also has the capability to compute required dilution flows for flow augmentation to meet any prespecified dissolved oxygen level.

Hydraulically, QUAL2E is limited to the simulation of time periods during which both the stream flow in river basins and input waste loads are essentially constant. QUAL2E can operate either as a steady-state or as a dynamic model, making it a very helpful water quality planning tool. When operated as a steady-state model, it can be used to study the impact of waste loads (magnitude, quality and location) on instream water quality and also can be used in conjunction with a field sampling program to identify the magnitude and quality characteristics of nonpoint source waste loads. By operating the model dynamically, the user can study the effects of diurnal variations in meteorological data on water quality (primarily dissolved oxygen and temperature) and also can study diurnal dissolved oxygen variations due to algal growth and respiration. However, the effects of dynamic forcing functions, such as headwater flows or point loads, cannot be modeled in QUAL2E.

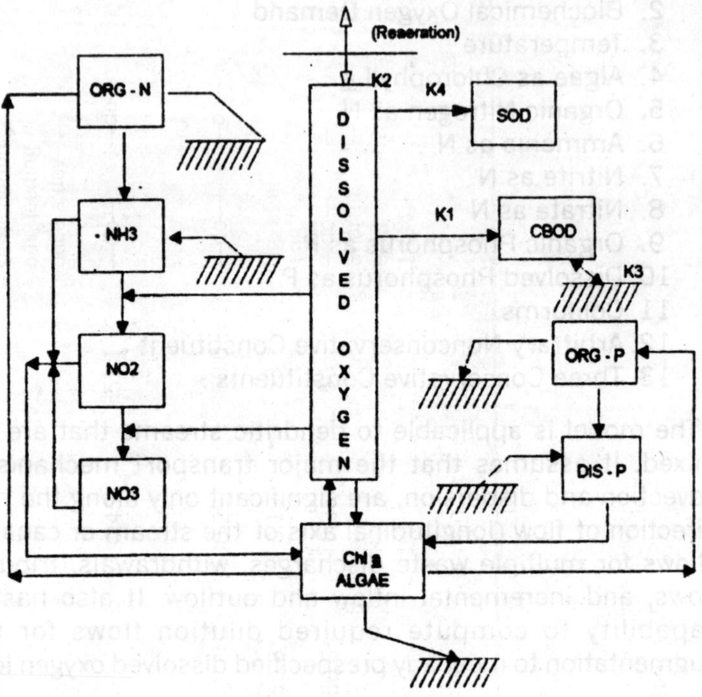

Fig. 2.5: DO simulation with QUAL2E

QUAL2E-UNCAS is a recent enhancement to QUAL2E which allows the modeler to perform uncertainty analysis on the steady state water quality simulations. Three uncertainty options are available : sensitivity analysis, first order error analysis, and Mmonte Carlo simulations. With this capability, the user can assess the effect of model sensitivities and of uncertain input data on model forecasts. Quantifications of the uncertainty in model forecasts will allow assessment of the risk (probability) of a water quality variable being above or below an acceptable level. The uncertainty methodologies provide the means whereby variance estimates and uncertainty prediction can become as much a part of water quality modelling as estimating expected values is today. An evaluation of the input factors that contribute most to the level of uncertainty will lead modelers in the direction of most efficient data gathering and research. In this manner the modeler can assess the risk of imprecise forecasts, and recommend measures for reducing the magnitude of that imprecision.

Brief history of QUAL series of models

The original QUAL-II model was an extension of the stream water quality model QUAL-I developed by F.D. Masch and Associates and the Texas Water Development Board (1971) and the Texas Water Development Board (1970). In 1972, Water Resources Engineers, Ins. (WRE) under contract to the U.S. Environmental Protection Agency, modified and extended QUAL-I to produce the first version of QUAL-II. Over the next 3 years, several different versions of the model evolved in response to specific user needs. In March 1976, the Southeast Michigan Council of Governments (SEMCOG) contracted with WRE to make further modifications and to combine thc best features of the existing versions of QUAL-II into a single model. The significant modifications made in the SEMCOG version by WRE (Roesner et al., 1981a and b) were :

- Option of English or metric units on input data
- Option for English or metric output-choice is independent of input units
- Option to specify channel hydraulic properties in terms of trapezoidal channels or stage-discharge and velocity-discharge curves
- Option to use Tsivoglou's computational method for stream reaeration
- Improvement in output display routines
- Improvement in steady-state temperature computation routines

The SEMCOG version of QUAL-II was later reviewed, documented, and revised (NCASI, 1982). The revised SEMCOG version has since been maintained and supported by the EPA Centre for Water Quality Modelling (CWQM). In 1983, EPA, through the CWQM, contracted with NCASI to continue the process of modifying QUAL-II to reflect state-of-the-art water quality modelling. Extensive use of QUAL-II/SEMCOG had uncovered difficulties that required corrections in the algal-nutrient-light interactions. In addition, a number of modifications to the program input and output had been suggested by users. The enhanced QUAL-II model was renamed QUAL2E (Brown and Barnwell, 1985) and incorporated improvements in eight areas.

1. Algal, nitrogen, phosphorus, dissolved oxygen interactions
   Organic nitrogen state variable
   Organic phosphorus state variable
   Nitrification inhibition at low DO
   Algal preference factor for NH_3
2. Algal growth rate
   Growth rate dependent upon both NH_3 and NO_3 concentrations
   Algal self-shading
   Three light functions for growth rate attenuation
   Three growth rate attenuation options
   Four diurnal averaging options for light
3. Temperature
   Link to algal growth via solar radiation
   Default temperature correction factors
4. Dissolved Oxygen (DO)
   16th Edition *Standard Methods* DO saturation function
   Traditional SOD units (g/m^2-day or g/ft^2-day)
   Dam reaeration option
5. Arbitrary non-conservative constituent
   First order decay
   Removal (settling) term
   Benthal source term
6. Hydraulics
   Input factor for longitudinal dispersion
   Test for negative flow (i.e. withdrawal greater than flow)
   Capability for incremental outflow along reach
7. Downstream boundary
   Options for specifying downstream boundary water quality constituent concentrations
8. Input/output modifications
   Detailed summary of hydraulic calculations
   New coding forms

Local climatological data echo printed
Enhanced steady-state convergence
Five part final summary including components of DO deficit and plot of DO and BOD.

Enhancements to QUAL2E

Since the first release of QUAL2E in 1985, enhancements to the model have continued. The modifications, listed below, are designed to improve the computational efficiency of the code, as well as to assist the user in model calibration and verification. The reach variable climatology modifications were added in response to applications of QUAL2E to the river network in Madrid, Spain. In that system, large changes in elevation presented difficulties in calibrating QUAL2E for temperature and dissolved oxygen. The major addition to the current release of QUAL2E is the uncertainty analysis capability. Inclusion of this feature resulted from a project which investigated various methodologies for incorporating uncertainty analysis as an integral part of the water quality modeling process. The QUAL2E model was chosen for this application because it is a general purpose computer code, widely used by consultants and state regulatory agencies in waste load allocation and other planning activities.

Enhancements to QUAL2E in the current release include:

1. Option for reach variable climatology input for steady state temperature simulation.
2. Option for including observed dissolved oxygen data on the line printer plots of predicted dissolved oxygen concentrations.
3. Changing the steady state convergence criteria for algal, nitrification, and dissolved oxygen simulations from an absolute error to a relative error.
4. Updating the formulation for estimating reaeration effects of water flowing over a dam.

Capabilities of the uncertainty analysis model, QUAL2E-UNCAS, include the following :

1. Sensitivity analysis–with an option for factorially designed combinations of input variable perturbations.
2. First order error analysis–with output consisting of a normalized sensitivity coefficient matrix, and a components of variance matrix.
3. Monte carlo simulation–with summary statistics and frequency distributions of the output variables.

QUAL2E COMPUTER MODEL

Prototype Representation

QUAL2E permits simulation of any branching, one-dimensional stream system. The first step in modelling a system is to subdivide the stream system into reaches which are stretches of stream that have uniform hydraulic characteristics. Each reach is then divided into computational elements of equal length. Thus, all reaches must consist of an integer number of computational elements.

There are seven different types of computational elements :

1. Headwater element
2. Standard element
3. Element just upstream from a junction
4. Junction element
5. Last element in system
6. Input element
7. Withdrawal element

Headwater elements begin every tributary as well as the main river system and as such, they must always be the first element in a headwater reach. A standard element is one that does not qualify as one of the remaining six element types. Because incremental flow is permitted in all element types the only input permitted in a standard element is incremental flow. A type 3 element is used to designate an element on the main steam just upstream of a junction. A junction element (type 4) has a simulated tributary entering it. Element type 5 identifies the last computational element in the river system; there should be only one type 5 element. Element types 6 and

7 represent inputs (waste loads and unsimulated tributaries) and water withdrawals, respectively. River reaches, which are aggregates of computational elements, are the basis of most data input. Hydraulic data, reaction rate coefficients, initial conditions, and incremental flows data are constant for all computational elements within a reach.

Model limitations

QUAL2E has been designed to be a relatively general program; however, certain dimensional limitations have been imposed during program development. These limitations are :

Reaches	:	a maximum of 25
Computational elements	:	no more than 20 per reach or a total of 250
Headwater elements	:	2 maximum of 7
Junction elements	:	a maximum of 6
Input and withdrawal elements	:	a maximum of 25

QUAL2E incorporates features of ANSI FORTRAN 77 that allow these limitations to be easily changed.

Model Structure and Subroutines

QUAL2E is structured as one main program supported by 51 different subroutines. Figure 2.5 illustrates the functional relationships between the main program and the subroutines (specifically for DO simulation). New state variables can be added or modifications to existing relationships can be made with a minimum of model restructuring through the simple addition of appropriate subroutines.

The structural framework of QUAL2E has been modified from prior versions of QUAL-II. The large MAIN program and subroutine INDATA have been divided into smaller groups of subroutines, each with a more narrowly defined task. The new subroutines in QUAL2E include the algal light functions (GROW/ LIGHT), the steady state algal output summary (WRPT1), the organic nitrogen and phosphorus state variables (NH2S,

PORG), and the line printer plot routine (PRPLOT). This reorganization of QUAL2E into smaller programmatic units is the first step in adapting the model to micro and minicomputers that have limited memory.

QUAL2E Version 3.0 retains this modular program structure.

Program language and Operating Requirements

QUAL2E is written in ANSI FORTRAN 77 and is compatible with mainframe and personal computer systems that support this language. QUAL2E typically requires 256 k bytes of memory and uses a single system input device (cards or disk file) and the system's line printer (or disk file) as the output device.

If the system's normal FORTRAN input device unit is not unit 1 or the output unit is not unit 7, then the variables 'NI' and 'NJ' in the main program (files Q2E3PO or Q2U3PO) should be changed to reflect the system's I/O unit identifiers.

REFERENCES

1. American public Health Association, Inc., 1985. Standard methods for the examination of water and waste water, APHA (19th edition), New York.

2. Brown, L.C., and Barnwell, T.O., Jr. 1985. Computer program documentation for the enhanced stream water quality model Qual2E, EA, 600-3-85/065. US, Environmental Protection Agency, Environmental Research Laboratory, Athens, GA.

3. Churchill, M.A., Elmore, H.L., and Buckingham, R.A., 1962. The prediction of stream reaeration rates, *International J. Air and Water Pollution*, (6), 45-504.

4. Dobbains, W.E., 1964. BOD and oxygen relationships in streams – *Journal of the American society of Civil Engineers (Sanitary Engg Division)* 90 53-78.

5. Frank, D. Mash and Associates, and Texas Water Development Board., 1971. Simulation of water quality in streams and canals, theory and description of the Qual-I, Mathematical modelling system, Report 128.

6. JRB Associates., 1983. Users manual for Vermount Qual II intel prepared for U.S. Environmental Protection Agency, Washington D.C.

7. O'Connor, D.J. and Dobbins, W.E., 1958. Mechanism of reaeration in natural streams, *Transations of the American Society of Civil Engineers*, 123, 541–54.

8. Owens, M. Edwards, R.W. and Gibbs, J.W., 1964. Some reaeration studies in streams, *International J. of Air and Water Pollution*, 8 (8/9), 469–86.

9. Roesner, L.A., Giguere, P.R., and Evenson, D.E., 1981a. Computer program documentation for stream quality modelling (Qual-II), EPA-600/9-82114, US Environmental Protection Agency, Athens, GA.

10. Roesner, L.A., Giguere, P.R., and Evenson, D.E., 1981b. User manual for stream quality model (Qual-II), EPA – 600/9-81-015, US Environmental Protection Agency, Athens, GA.

11. Streeter, H.W., and Phelps, E.B., 1925. Study of the pollution an natural purification of the Ohio river, US Public Health Service, Washinton D.C., Buletin number 146.

12. Thommann, R.V., 1972. System Analysis and Water Quality Management, McGraw-Hill New York.

13. The Enhanced Stream Water Quality Models Qual-2E and Qual-2E Uncas., 1987. Documentation and User Model, U.S. Environmental Protection Agency, Athens.

14. Water Resources Engineers, Inc., 1972. Upper Mississippi River Basin Model Project, Sponsored by the Environmental Protection Agency, Environmental Protection Agency, Athens, GA.

Chapter 3

MODELLING OF BUCKINGHAM CANAL WATER QUALITY

In this chapter, which is a follow-up of Chapter 2, we present a case study of the modelling of the water quality of a canal situated in a petrochemical industrial complex. The canal receives wastewaters from Madras Refineries Limited (MRL), and Madras Fertilizers Limited (MFL).

The canal—well known Buckingham Canal which passes through Chennai (Madras), India—has been modelled using the software QUAL2E-UNCAS. After testing and validation of the model, simulations have been carried out.

The exercise enables forecasting the impacts of different seasons, base flows, and waste water inputs on the water quality of the Buckingham Canal. It also enables development of water management strategies.

STUDY AREA: THE BUCKINGHAM CANAL

The Buckingham Canal is a coastal channel running through the backwaters, depressions, and low-lying areas for 419 km between the towns of Peddagangam and Marakkanam at the east coast of the Indian peninsula. The canal runs close to the coast and is not more than one to two km away from the sea at any point along its length.

The canal was completed in 1897 and was the main means of communication between the rich deltaic areas of Coramandal coast and Madras during the later part of the 19th century and the early 20th century, when rail and road transport facilities were not well developed. Subsequently as the other modes of transport came up the traffic in the Buckingham canal gradually declined.

Now, due to improper maintenance, effect of cyclonic storms, and high level of siltation hastened by pollution, the canal is in disuse at several places. It now exists in bits and pieces serving small stretches, notably in the city of Chennai (Madras).

One of the still useful portions of the Buckingham canal is running through the Manali industrial area, at Chennai (Madras), for a length of ~10 km (latitude 80.33°, longitude 13.13°). In this region (Map 1) the canal carries run-off water which is occasionally used for irrigation, laundry, and other domestic purposes by economically weak population living near its banks. They also occasionally fish in the canal waters.

We have modelled the water quality of the canal and have carried out simulations to forecast the impact of waste inflows on the water quality.

MODELLING OF THE CANAL

The computer-automated tool: The model has been developed using the computer-automated tool QUAL 2E-UNCAS. QUAL 2E-UNCAS has been developed for the Environmental Protection Agency (EPA) of The United States of America and is the latest version (USEPA 1987) of a series of such tools beginning with QUAL I developed by F.D.Masch and Associates and the Texas Water Development Board (1970, 1971). QUAL 2E-UNCAS can handle upto 15 water quality constituents including DO, BOD, temperature, algae as chlorophyll a, various forms of N, P, coliforms, upto three conservative, and an arbitery non-conservative, constituents. It can be used to model dendritic streams that are well-mixed.

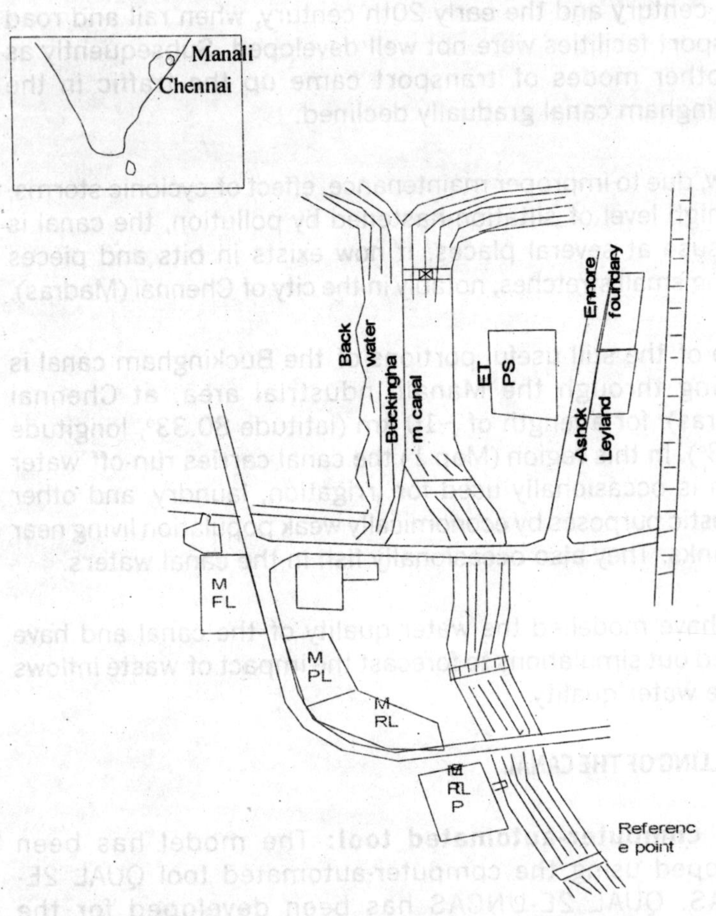

Map. 3.1. The study area

As the Buckingham canal is shallow (average water levels 1.2 m in summer and 2.0 m during rains) and turbulant, the assumption of well-mixed stream is justified in this case. QUAL 2E-UNCAS further assumes that the major transport mechanisms, advection and dispersion, are significant only along the main direction of flow (longitudinal axis of the stream or canal). It allows for multiple waste discharges, withdrawals, tributary flows, and incremental inflow and outflow.

Hydraulically QUAL 2E-UNCAS can be operated in steady state and dynamic modes. In the former it can be used to study the impact of waste loads on instream water quality and to identify the contributions to it of nonpoint source waste loads. In the dynamic mode the user can study the effects of diurnal variations in meteorological data on water quality (primarily dissolved oxygen and temperature) and diurnal dissolved oxygen variations due to algal growth and respiration.

The UNCAS facility allows the modeler to perform uncertainty analysis on the steady state water quality simulations. Three uncertainty options are available: sensitivity analysis, first order error analysis, and Monte Carlo simulations.

Hydrological, including water quality data: The requisite data on canal dimensions, flow, and water quality was generated by us, using standard procedures (Linsely-1975, APHA-1995). Similar set of data earlier generated by an independent agency (MRL 1989) was used to validate the model after we had caliberated it with our data.

Model validation: Typical results validating the model are presented in Figures 3.1and 3.2. The observed values of temperature, BOD, TDS (total dissolved solids) and SS (suspended solids) match well with the predicted values all the way. Only in case of chloride, the observed values are 50% higher in summer than the predicted value after about 4.6 km from the reference point. This may be due to the tidal influence from the sea.

The model was validated for the canal water quality for three seasons which occur in Chennai-summer (April–September), monsoon (October-December) and past monsoon (January–March). The city does not experience much of a winter.

During summer the match between observed value and predicted values is closest for temperature, TDS, SS and BOD. Only for chloride, there is a mismatch after about 4.6 km. from the reference point, as mentioned above. We attribute this to tidal influence. Such an influence is noticed only during summer because the flow in the canal is lowest during summer allowing the tidal influence to penetrate deeper upstream than in other seasons.

During monsoon and post monsoon seasons, too, the trend is similar. The predicted values match with the observed ones all the way upto the point about 4.6 km. downstream wate discharge point. Beyond this range the water quality is influenced by tidal inflows.

SIMULATIONS

The stretch of Buckingham canal studied by us has four sources of pollutants: the main canal stream, and wastewater inflows at two points from Madras Refineries Limited (MRL), Madras Fertilizers Limited (MFL) and 'non-point' load or incremental flow.

We have carried out Monte-carlo simulations on the impact of changes in the volume and characteristics of the water, in one or more of the four sources of pollution, on the canal water quality.

Impact of the main canal streams: As flow in the canal increases, its temperature remains more or less constant but the concentrations of TDS, SS, Chloride and BOD decrease. An increase in the temperature has little effect on TDS, SS and Cl, but causes BOD to decrease, due perhaps, to the increase in microbial activity hastening the rate of BOD decay.

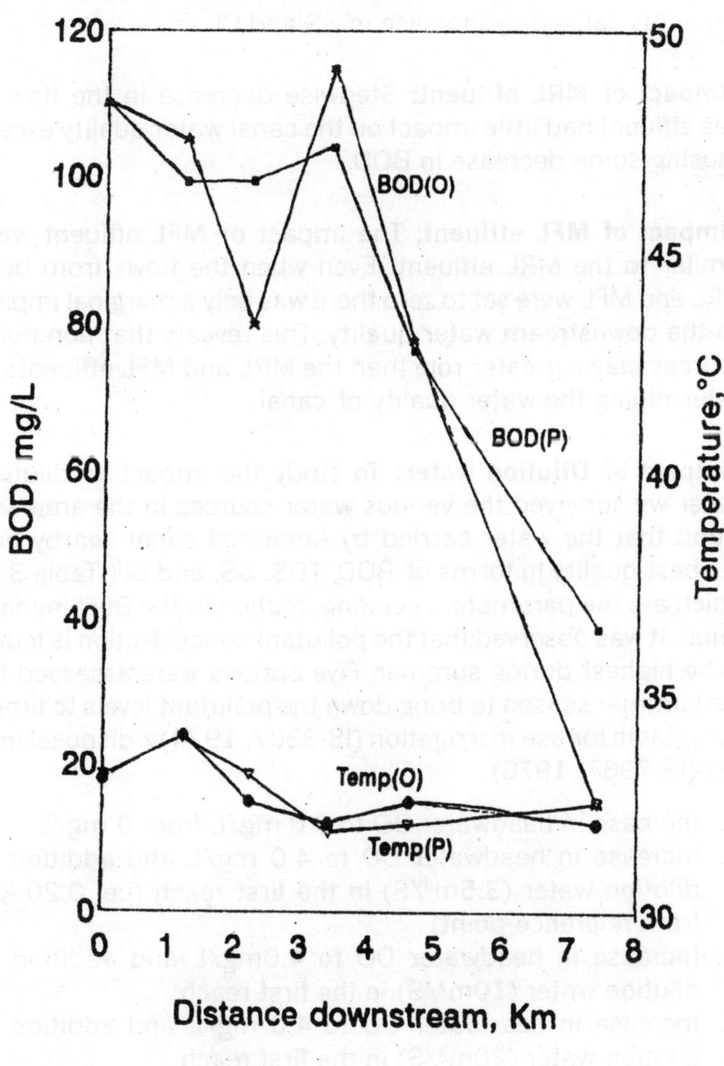

Fig. 3.1: Validation curves for Temperature and BOD in summer season

Step increase in TDS concentration has little impact on temperature and BOD but causes Cl and SS to increase. Similar impact is caused by increase in SS and Cl.

Impact of MRL effluent: Stepwise decrease in the flow of this effluent had little impact on the canal water quality except causing some decrease in BOD.

Impact of MFL effluent: The impact of MFL effluent were similar to the MRL effluent. Even when the flows from both MRL and MFL were set to zero there was only a marginal impact on the downstream water quality. This reveals that non point sources play a greater role than the MRL and MFL effluents in determining the water quality of canal.

Impact of Dilution water: To study the impact of dilution water we surveyed the various water sources in the area and found that the water carried by Amulavoil canal nearby had the best quality in terms of BOD, TDS, SS, and Cl (Table 3.1) which are the parameters needing dilution in the Buckingham canal. It was observed that the pollutant concentration is found to be highest during summer. Five options were assessed for the summer season to bring down the pollutant levels to limits acceptable for use in irrigation (IS-3307, 1974) / disposal into sea (IS-7967, 1976).

I. Increase in headwater DO to 8.0 mg/L from 0 mg/L
II. Increase in headwater DO to 4.0 mg/L and addition of dilution water (3.5m^3/S) in the first reach (i.e. 0.20 km from reference point)
III. Increase in headwater DO to 4.0mg/L and addition of dilution water (10m^3/S) in the first reach.
IV. Increase in headwater DO to 4.0 mg/L and addition of dilution water (20m^3/S) in the first reach.
V. Increase in headwater DO to 4.0 mg/L and addition of dilution water (3.5m^3/S, 1.5m^3/S, and 2.0m^3/S) at the first, the second and the third reach respectively (i.e. 0.20km, 1.80km and 4.80km from the reference point).

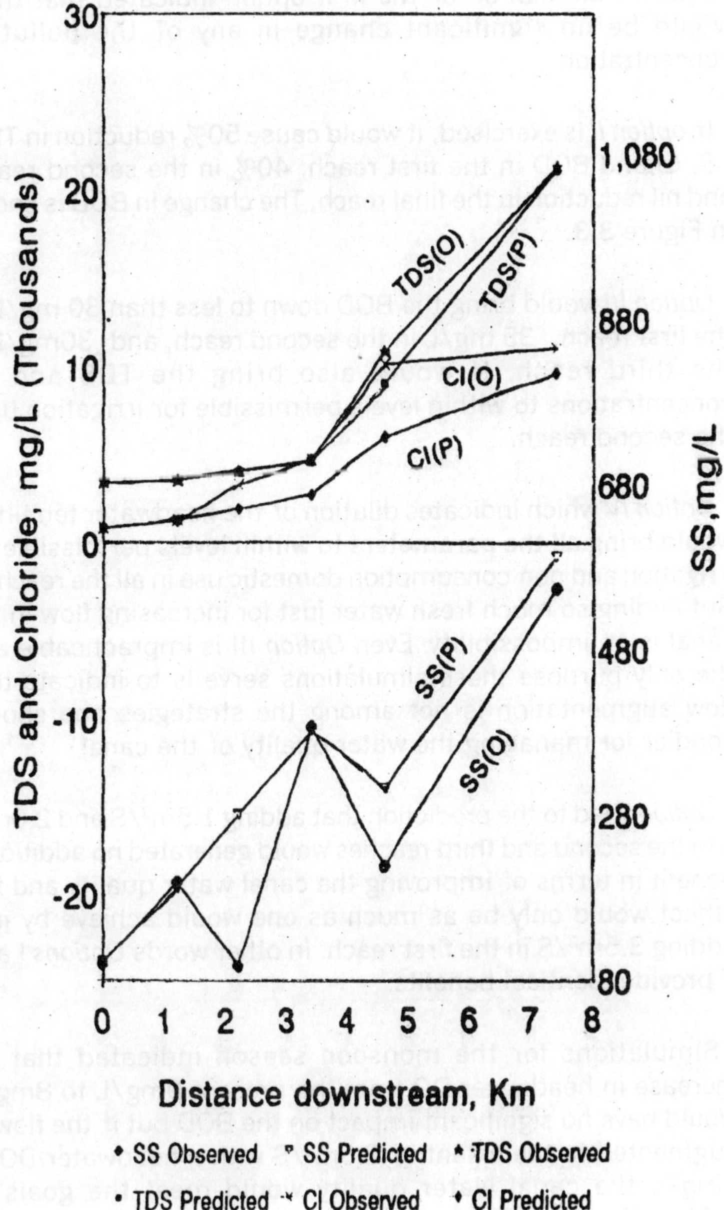

Fig. 3.2: Validation curves for TDS, SS and chloride in summer season

Option I Simulation of the first option indicated that there would be no significant change in any of the pollutant concentration.

If *option II* is exercised, it would cause 50% reduction in TDS, SS, Cl,and BOD in the first reach, 40% in the second reach, and nil reduction in the final reach. The change in BOD is shown in Figure 3.3.

Option III would bring the BOD down to less than 30 mg/L in the first reach, _35 mg/L in the second reach, and _30mg/L in the third reach. It would also bring the TDS and SS concentrations to within levels permissible for irrigation upto the second reach.

Option IV which indicates dilution of the headwater ten times would bring all the parameters to within levels permissible for irrigation and non-consumption domestic use in all the reaches. But finding so much fresh water just for increasing flow in the canal is an impossibility. Even *Option* III is impracticable and the only purpose these simulations serve is to indicate that flow augmentation is not among the strategies one should condier for managing the water quality of the canal.

Option V led to the prediction that adding 1.5m^3/S and 2.0m^3/S in the second and third reaches would generated no additional benefit in terms of improving the canal water quality and the efffect would only be as much as one would achieve by just adding 3.5m^3/S in the first reach. In other words *Options* I and V provide identical benefits.

Simulations for the monsoon season indicated that an increase in headwater DO from the present 0mg/L to 8mg/L would have no significant impact on the BOD but if the flow is augmented to the extent of 2.5m^3/S with a headwater DO of 4mg/L the canal water quality would meet the goals of irrigation and non-consumption domestic use. The reduction in BOD is given in Figure 3.3.

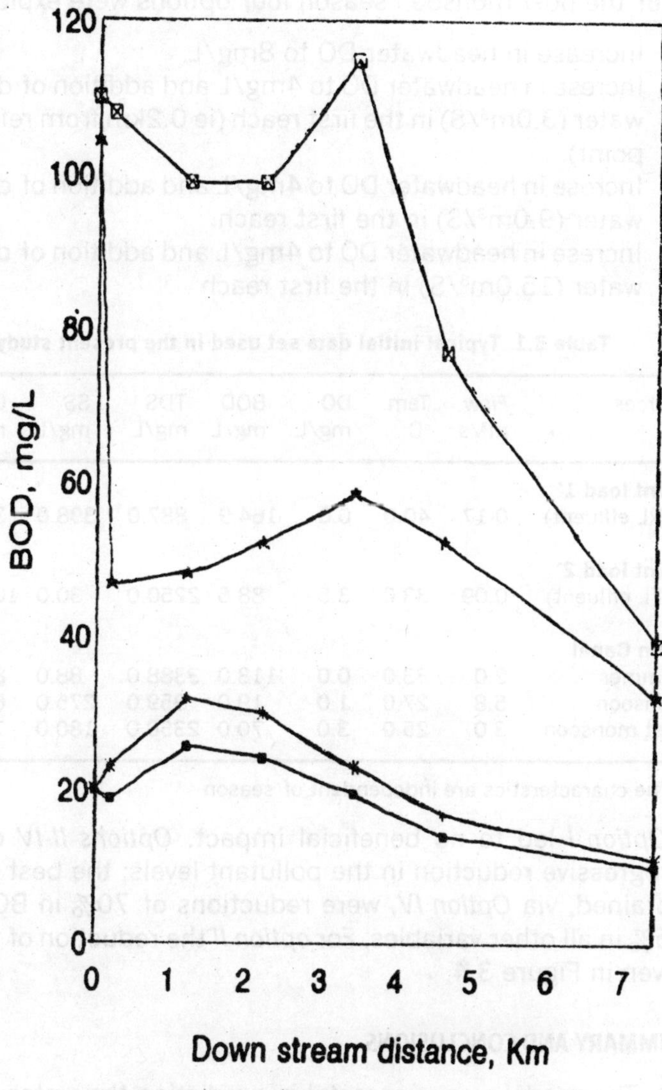

Fig. 3.3: Impact of addition of fresh water stream (3.5 m3/s and 2.5 m3s) and increase in head water DO (4 mg/L) on BOD of the canal during summer and monsoon seasons.

For the post-monsoon season four options were explored.

i. Increase in headwater DO to 8mg/L
ii. Increse in headwater DO to 4mg/L and addition of dilution water (3.0m³/S) in the first reach (ie 0.2km from reference point).
iii. Increse in headwater DO to 4mg/L and addition of dilution water (9.0m³/S) in the first reach.
iv. Increse in headwater DO to 4mg/L and addition of dilution water (15.0m³/S) in the first reach

Table 3.1. Typical initial data set used in the present study

Sources	*Flow* m^3/s	Tem. °C	DO mg/L	BOD mg/L	TDS mg/L	SS mg/L	Cl_2 mg/L
Point load 1*							
(MRL effluent)	0.17	40.0	0.0	164.9	887.0	398.0	356.0
Point load 2*							
(MFL effluent)	0.09	33.0	3.5	88.6	2250.0	30.0	1011.0
Main Canal							
Summer	2.0	33.0	0.0	113.0	3388.0	88.0	810.0
Monsoon	5.8	27.0	1.0	19.0	959.0	275.0	661.0
Post monsoon	3.0	25.0	3.0	70.0	2350.0	180.0	700.0

* The characterstics are independent of season

Option I led to no beneficial impact. *Options II-IV* caused progressive reduction in the pollutant levels; the best results obtained, via *Option IV*, were reductions of 70% in BOD and 75% in all other variables. For *option II* the reduction of BOD is given in Figure 3.4.

SUMMARY AND CONCLUSIONS

1) The model was successful in predicting the water quality of Buckingham canal at 90-95% confidence levels across the 7.80 km stretch studied by us, throughout the year. The only exception was the mismatch between the observed and the predicted values of chloride in the third reach, about 4.6 km from the reference point (headwater)–this was due to the impact of tidal inflow which was particularly

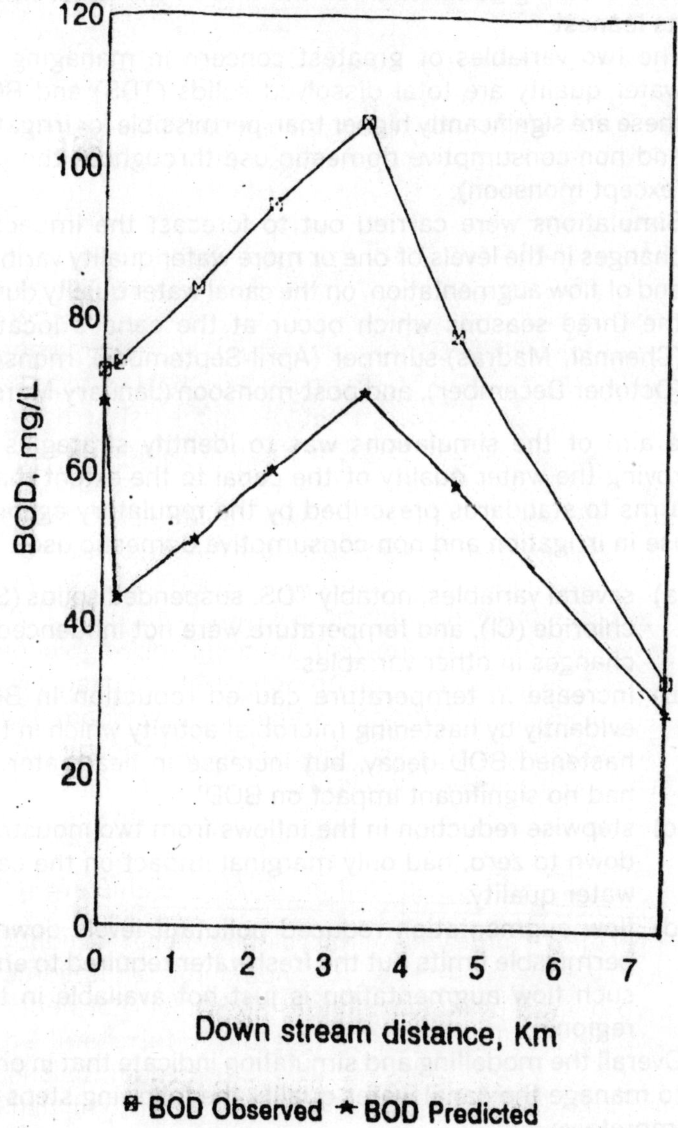

Fig. 3.4: Impact of addition of fresh water stream (3.0 m³/s) and increase in head water DO (4 mg/L) on BOD of the canal during post monsoon seasons.

severe during summer when the flow in the canal was at its leanest.

2) The two variables of greatest concern in managing the water quality are total dissolved solids (TDS) and BOD; these are significantly higher than permissible for irrigation and non-consumptive domestic use throughout the year (except monsoon).
3) Simulations were carried out to forecast the impact of changes in the levels of one or more water quality varibles, and of flow augmentation, on the canal water quality during the three seasons which occur at the canal's location (Chennai, Madras)-summer (April-September), monsoon (October-December), and post-monsoon (January-March).

The aim of the simulations was to identify strategies for improving the water quality of the canal to the extent that it confirms to standards prescribed by the regulatory agencies for use in irrigation and non-consumptive domestic use.

a) several variables, notably TDS, suspended solids (SS), chloride (Cl), and temperature were not influenced by changes in other variables;
b) increase in temperature caused reduction in BOD, evidently by hastening microbial activity which in turn hastened BOD decay, but increase in headwater DO had no significant impact on BOD'
c) stepwise reduction in the inflows from two industries, down to zero, had only marginal impact on the canal water quality.
d) flow augmentation reduced pollutant levels down to permissble limits but the freshwater required to effect such flow augmentation is just not available in that region.

4) Overall the modelling and simulation indicate that in order to manage the canal water quality the following steps are impretaive.
   a) sources of BOD and TDS upstream must be identified and steps taken to reduce or eliminate their contributions;
   b) the BOD and TDS levels of the effluents from the two industries (MRL and MFL) which join the canal in the

study area should be reduced drastically by implementing appropriate treatment at the industries before disposal.

REFERENCES

1. Environmental Impact Assessment of Lube plant expansion, *MRL final report*, Madras, 1989.

2. IS-3307–1974. Tolerence limits for industrial effluents discharged on land for irrigation, *ISI*, 1974.

3. I.S. 1967–1976, Criteria for controlling pollution of marine coastal areas caused by discharge of effluents, sewage and waste, *ISI*, 1976.

4. Linsley, R.K., Jr. M. A.Kohler and J.L.H.Pauthus, Hydrology for Engineers 2d ed., Mc Graw Hill, New York 1975.

5. Standard methods for the examination of water and wastewater, *American Public Health Association*, New York, 19th Edition, 1995.

6. The Enhanced Stream Water Quality Models QUAL-2E and QUAL-2E UNCAS, Documentation and user model, *U.S.Environmental protection agency*, Athens, 1987.

Chapter 4

MODELLING AND SIMULATION OF HEAVY GAS DISPERSION ON THE BASIS OF MODIFICATIONS IN PLUME PATH THEORY

When concern for environmental pollution began to grow in the late 1960s, the greatest attention was paid initially to water pollution. But gradually the impact of air pollution was also felt with increasing intensity.

In this chapter we discuss the problem of dispersion of 'heavy gases' and present a model with which we can forecast the manner in which a 'plume' or a cloud of heavy gas will move in atmosphere after the gas is released from any source (such as chimney of an industry or valve of a storage vessel). The model takes into account the variations in temperature, density, and specific heat during the movement of heavy gas plume.

The model has been tested for three hazardous gases - chlorine, natural gas and liquefied petroleum gas. The results have been compared with the recently generated experimental data as also with the outputs of other models. A good agreement is observed qualitatively as well as quantitatively.

A study has also been carried out to simulate the effect of the wind speed, density of the gas, and venting speed on dispersion. Based on the simulation study a set of empirical equations have been developed. The equations have been validated by theoretical as well as experimental studies.

INTRODUCTION

Modelling of the dispersion of the 'dense' gases - gases with density higher than air–has been assuming ever greater importance as many of the hazardous gases (chlorine, hydrogen fluoride, liquefied petroleum gas) are denser than air. Numerous air pollution models, which were developed for lighter-than-air or light-as-air gases, have not been successful with dense gases; accentuating the need for mathematical models appropriate for dense gas dispersion. In recent years, a few mathematical models have been proposed for the study of heavy gas dispersion–notably by Ooms *et al.* (1974), Ooms and Duijm (1983), Colenbrander (1980), Eidsvik (1980), Ermak and Chan (1985), Van Ulden (1983,1987), Langlo and Schatzmann (1991) and Deaves (1983,1992).

One of the first attempts to model the heavy gas dispersion was made by Ooms *et al.*. (1974). They proposed an analytical model based on using the conventional transport phenomena and the plume path theory (Ooms, 1972). Later Colenbrander (1980) proposed slab (continuous release box) model which assumes normal distribution of concentration within the slab. Around the same time Eidsvik (1980) proposed a refined box model with equations modified to estimate vertical entrainment of air. Van Ulden (1983), Van Ulden and Holtslag (1985), and Van Ulden (1987) used K-theory with atmospheric scaling parameters to model heavy gas dispersion. Ermak and Chan (1985) proposed a model based on the turbulence dissipation and boundary layer parameters. In subsequent years, Langlo and Schatzmann (1991) modelled the heavy gas dispersion using Langrangian approach based on similarity theory. Deaves (1992) modelled the atmospheric turbulence and analysed the way it affects dense gas dispersion. He also used K-theory and employed more extensive meteorological data than the other models did : wind profile, turbulence and boundary layer profiles.

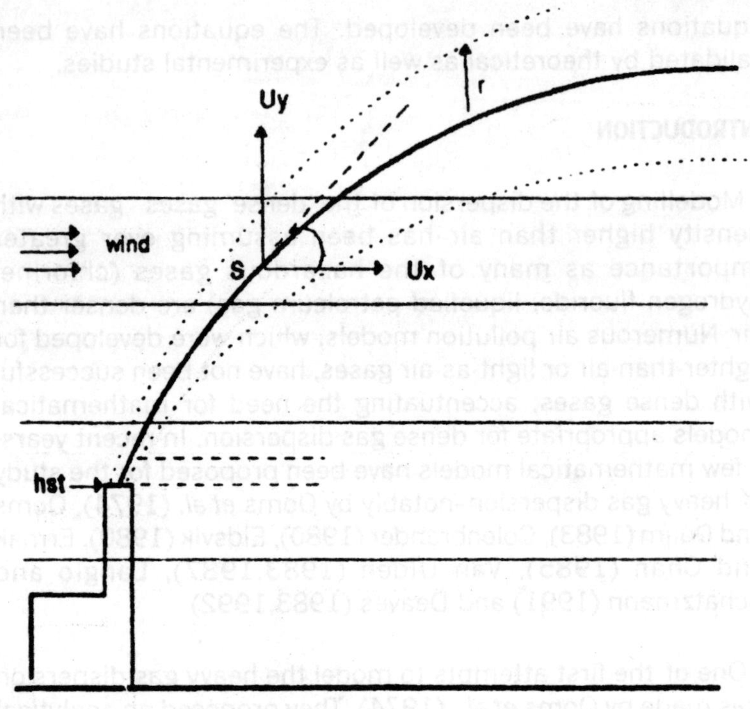

Fig. 4.1. The plume coordinates as used in the present study

Surprisingly, in the present age of computer-based application software, very few such models have been proposed. They include finite difference models which use K-theory : MARIAH (Taft *et al.* 1983), SMART (Tran and Liu, 1981), SIGMET (England *et al.* 1978; Havens, 1982) and MERCUR-GL (Riou, 1986), or box model such as HEGADAS (Colenbrander, 1980; Puttock, 1986). Of these the last named is the most often cited one. The output of these models are well tested with experimental values and are found to be in fairly good agreement.

Although the above mentioned models have been reasonably successful in some cases, they are limited in scope and have found applicability in only certain specific conditions only. In this paper we present a model based on plume path theory

(PPT) which, we hope, may enrich the existing repertoire of the heavy gas dispersion models presently available.

Plume path theory (PPT) was proposed by Ooms (1972) and has been used mainly to calculate the plume path of the lighter-than-air and light-as-air gases escaping from the stacks into the atmosphere at atmospheric temperature and pressure. Later this theory was extended to 'heavy' gases (of density higher than air; Ooms *et al.* 1974) with a number of assumptions. Numerous applications of the PPT have been reported for example, Rottman *et al.* (1985) and McQuaid (1986) have used the theory for analysing toxic gas dispersion; Niewstadt (1982,1992), Blewitt *et al.* (1987) and Weil (1988) have used the same for vapour cloud modelling; and Ooms and Duijm (1983) have used the theory to estimate the dispersion of heavy gases coming out of the stacks with high momentum. The theory has also been the basis of the commercial packages PLUME and HFPLUME. In spite of the obvious potential of Ooms' theory, it has not enjoyed wider applicability because the main assumptions have not yet been overcome. The authors of this paper have recently modified Ooms's theory to significantly enhance its range, accuracy, and precision. Most importantly we have enabled application of the theory to 'heavy gases' which is a much less explored domain of air quality modelling than the dispersion of the lighter-than-air or light-as-air gases. Based on the theory PPT (Ooms, 1972; Ooms *et al.* 1974; Ooms and Duijm, 1983) is a simple approach based on the fundamental principles of fluid dynamics such as Fick's law of diffusion, turbulent kinetic energy and fluid flow. It is based on the assumption that the density and the specific heat of a gas do not differ significantly from that of air. In dispersion calculations, it neglects density spread as well as buoyancy effect. However, these assumptions do not hold true for heavy gas dispersion. In the present paper we have modified PPT in an attempt to make it suitable to model heavy gas dispersion. Some empirical correlations have also been developed to show the dependency of plume variables (concentration, density, plume width, plume velocity) on atmospheric operating variables such as wind velocity, venting

Table 4.1. Redefinition and advancement in the assumption made in Oom's PPT model

Assumptions made in PPT (Ooms and cowrokers,1972 1974)	Modifications proposed by the authors
The mean flow velocity perpendicular to the main flow in the direction of the plume is negligible till a late stage of the plume movement	Secondary flow perpendicular to the plume axis is significant at much earlier stage as well
The velocity profile, density, and pollutant concentration are similar in all sections normal to the plume axis	none
Molecular transports is considered negligible in comparison with turbulent transports	Molecular transport has been treated as significant and taken into account in parameter estimation
Longitudinal turbulent transport is considered negligible compared with longitudinal convective transport	Both longitudinal as well as convective transports have been taken into account
—	Physical properties of venting gas have been considered as a function of downwind distance as well as atmospheric parameters

velocity and density difference. The applicability of the model developed by us has been demonstrated.

MATHEMATICAL REPRESENTATION

In its original form (Ooms, 1972) plume path theory takes into account only plume dimensions namely velocity and concentration. Later it was modified to take into account density and specific heat variations(Ooms and Mahiue,1974). Ooms *et al.* (1974) demonstrated its application to the modelling of venting gases heavier than air at high temperature. However, the modified PPT was based on a number of assumptions (Table 4.1). The present work is an attempt to make the existing theory more reliable by redefining the assumptions (Table 4.1) and developing empirical relations to estimate the effect of release and atmospheric parameters on dispersion.

Profiles

Let s, r, and ff be the plume co-ordinates at the plume axis as shown in Figure 4.1. These co-ordinates are related to the horizontal and vertical axis as :

$$dx/ds = \cos\phi$$
$$dz/ds = \sin\phi \qquad (1)$$

In the earlier form of PPT (Ooms and co-workers,1972; 1974; 1983) only three parameters (u,rr,c) have been considered as significant. In the present work temperature also has been taken as one of the dominant parameters. Incorporating the aspects of plume density, temperature and specific heat, the plume characteristics can be written as:

$$u(s,r,\varphi) = u_a\cos\varphi + u^*\exp(-r^2/b_s^2) \qquad (2)$$
$$\rho\rho(s,r,\varphi) = \rho_a + \rho\rho^*\exp(-r^2/\lambda^2 b_s^2) \qquad (3)$$
$$c(s,r,\varphi) = c^* * \exp(-r^2/\lambda^{2*} b_s^2) \qquad (4)$$
$$T(s,r,\varphi) = T_a + \exp(-r^2/Pr^{2*} b_s^2) \qquad (5)$$

Where $u(s,r,\varphi)$, $\rho(s,r,\varphi)$, $c(s,r,\varphi)$ represent the values of the variables at an arbitrary point in the plume; u^*, r^*, c^* denote the values of variables relative to the surroundings on the plume axis in the direction at a tangent to the plume axis.; and b represents the local characteristic width of the plume, in the present study it is equal to the radius of plume (b/2).

Plume path

As the plume moves through the atmosphere, air is entrained. Getting a precise mathematical description of this entrainment is one of the most difficult problems in the air pollution modelling (Hawthrone-1955, Abraham-1970, Briggs-1984, McQuaid-1986). Here we have tried to represent the entrainment in terms of mass flow equations keeping the following facts in mind :

a) in the vicinity of vent or release point venting velocity is higher than wind velocity,
b) at a sufficiently long distance downwind, the velocity of the plume may equal the wind velocity,
c) atmospheric turbulence is one of the most effective factors causing entrainment.

These three facts have been taken into account independently by Abraham (1970), and recently by Briggs (1984), Fay and Zemba (1985) who proposed different flow equations for each type of entrainment. The final mass flow equation will be a combination of these three entrainment modes with modified profiles of plume characteristic parameters. The mass flow equation can be written as;

$$d(\int \rho u 2\pi r dr)/ds = 2\pi b_s \rho_a \{\alpha_1 |u^*| \sin\varphi |\cos\varphi + \alpha_3 u\} \quad (6)$$

where,

$2\pi b_s \rho_a \alpha_1 |u^*|$ represents the entrainment due to the jet release of gas.

$2\pi b_s \rho_a \alpha_2 u_a |\sin\varphi| \cos\varphi$ represents the entrainment in a thermal stagnant atmosphere.

$2\pi b_s\rho_a\alpha_3 u$ represents entrainment due to atmospheric turbulence.

The values of entrainment coefficient α_1 =0.0762, α_2 =0.61, and α_3 = 1.0 have been taken from Fay and Zemba (1985,1986).

Component mass balance

The component mass balance over a cross section of plume is worked out as :

$$d(\int\rho u^2\cos\varphi 2\pi r dr)/ds = 0 \quad (7)$$

This implies that no gas is assumed to be present in the atmosphere outside the plume.

Momentum balance

In plume, the momentum occurs mainly due to

a) entrainment of air,
b) force exerted by wind.

Keeping these in view the momentum balance equation in the downwind direction can be written as :

$$d(\int\rho u^2\cos\varphi 2\pi r dr)/ds = 2\pi b_s\rho_a\{\alpha_1|u'|+\alpha_2 u_a|\sin\varphi|\cos\varphi+\alpha_3 u\}+c_d\pi b_s\rho_a u_a^2|\sin^3\varphi| \quad (6)$$

Where,

$2\pi d_a\rho_a u_a\{\alpha_1|u'|+\alpha_2 u_a|\sin\varphi|\cos\varphi+\alpha_3 u\}$ represents the increase in momentum due to inflow of air from the surrounding atmosphere.

$c_d\pi\rho_s u_a^2|\sin^3\varphi$ represents the increase in impulse due to the drag force exerted by the wind on the plume.

The momentum balance in the cross wind direction is a combination of

- density spread,
- drag force exerted by wind.

The final balance equation can be written as :

$$d(\int(\rho u^2|\sin\varphi|2\pi r dr))/ds=\int g(\rho-\rho_a)2\pi r dr\cdot c_d\pi b_s\rho_a u_a{}^2\sin^2\varphi\cos\varphi \quad (9)$$

The first term symbol $\int(\rho u^2|\sin\varphi|2\pi r dr)$ represents density spread while second term $c_d\pi b_s\rho_a u_a{}^2\sin^2\varphi\cos\varphi$ represents the impulse due to drag force exerted by wind.

Energy balance

The energy balance for the plume implies that the amount of heat emitted by the gas per unit time is conserved with respect to a chosen reference. So, for a reference temperature T the energy balance equation can be written as:

$$d(\int\rho u cp(T\cdot T_{ao})2\pi r dr)/ds=2\pi b_s\rho_a cp_a(T\cdot T_{ao})\ \{\alpha_1|u^1|+\alpha_2 u_a|\sin\varphi|\cos\varphi+\alpha_3 u\} \quad (10)$$

If it is assumed that air and vent gas obey ideal gas law, then the temperatures (plume and air temperature) can be expressed as

$$T = M_w P/(R\rho) \text{ and } T_a = M_{wa}P/(R\rho_a) \quad (11)$$

Where, M_w and M_{wa} represents molecular weight of plume and air respectively at any point in the plume. As molecular weight and specific heat differ, unlike what was assumed in the original PPT (Ooms-1972), these variables can be expressed as:

$$M_w = M_{wo}cT / (c_oT_o) + M_{wa}(1\cdot cT/(c_oT_o)) \quad (12)$$

$$cp = \{M_{wo}cp_o cT/(c_oT_o) + M_{wa}cp_a(1\cdot cT/(c_oT_o)\} /M_w \quad (13)$$

Combining the above energy balance equations with molecular weight variation and specific heat variation, the final equation can be written as :

$$d(\int M_w cp/(M_{wa}cp_a) * u[1\cdot\rho_o/(co\rho)(Mwo\cdot 10\}]2\pi r dr)/ds=$$

$$2\pi^* b_s^*(1\cdot\rho_a/p_{ao})\{\alpha 1|\ u^1|+\alpha 2^* u_a|\ \sin\varphi|\ \cos\varphi+\alpha_3 u\}$$

SOLUTION OF THE MODEL

By substituting the similarity profile to these conservation equations, integrals can be calculated by using a suitable numerical integration technique-here we have used Siphons-1/3 technique. Sets of non-linear simultaneous equations have been solved by using Newton-Raphson method (Carnahan-1969) coupled with L-U decomposition (Carnahan-1969) technique. The model has been solved for three different gases and for different atmospheric operating conditions as presented in Table 4.2.

Experimental studies

An extensive study to measure the quality of stack emissions and the behaviour of the plume formed when ammonia is released from a pressurised storage vessel was conducted at Manali (near Madras, southern peninsula of India). The initial and boundary conditions of the release are given in Table 4.3. The study area has flat terrain and is made up of rural habitat. The study included meteorological parameters (vertical temperature profile, vertical as well as horizontal wind velocity profile), air quality (concentration profiles), and behaviour of the plumes (of heavier-than-air as well as heavy-as-air gases) under different sets of conditions influencing dispersion for a continuous as well as an instantaneous release. The following characteristics of the plume were studied measuring various parameters such as; temperature variation within the plume, concentration profiles in the cross wind and downwind directions, plume width, plume height, and the effect of the meteorological parameters on the plume behaviour. A gist of the experimental results obtained in the present study is presented in Table 4.4.

RESULTS AND DISCUSSION

The comparison of the plume path (height of plume rise) obtained by our model with the results reported by Bodurtha (1961) Moore *et al.* (1988) and our own recent experimental

Table 4.2. Ambient operating variables used in the study

	Vent characteristic			Ambient characteristics	
	Vent diameter (*m*)	Venting speed (*m/s*)	Temperature (C)	Wind speed (*m/s*)	Atmospheric stability (*based on meteorological data*)
Sulphur dioxide	0.37	50	45	5	slightly stable
	0.37	50	45	7	slightly stable
	0.37	75	45	5	slightly stable
	0.37	75	45	7	slightly stable
Ammonia	0.21	70	45	5	slightly stable
	0.21	70	45	7	slightly stable
	0.21	90	45	5	slightly stable
	0.21	90	45	5	slightly stable
Hydrogensulphide	0.25	65	25	5	slightly stable
	0.25	65	25	7	slightly stable
	0.25	95	25	5	slightly stable
	0.25	95	25	5	slightly stable
Natural gas	0.37	95	50	5	slightly stable
	0.37	95	50	7	slightly stable
	0.37	125	50	5	slightly stable
	0.37	125	50	7	slightly stable
LPG	0.25	45	40	5	slightly stable
	0.25	45	40	7	slightly stable
	0.25	63	40	5	slightly stable
	0.25	63	40	7	slightly stable
Chlorine	0.20	25	35	5	slightly stable
	0.20	25	35	7	slightly stable
	0.20	35	35	5	slightly stable
	0.20	35	35	7	slightly stable

studies are presented in Figure 4.2. Good, qualitative as well as quantitative, agreements have been observed.

Table 4.3. The initial and boundary conditions for release of ammonia from pressurized vessel through vent valve

Parameters	*Values*
Storage capacity	200 tons
Storage temperature	45 °C
Storage pressure	1625 kPa
Height of the vessel	10.5 m
Height of vent pipe	2.5 m
Vent diameter	0.2 m
Ambient temperature	27 °C
Ambient pressure	107.3 kPa
Wind speed (at 10 m)	5.5 m/s
Wind direction	North-West
Terrain of the area	Flat rural area with a roughness height of 1.2 m

The temperature and the plume velocity variations have also been compared with that the of experimental data (Table 4.4), and a fairly good agreement has been observed (Figures 4.3 and 4.4). When the concentration profile obtained by the present model (for release of ammonia) is compared with our experimental results (Figure 4.5) the predicted results are seen to lie within the confidence interval of 40-50%, a match acceptable for air pollution models.

The simulation study reveals that the plume width and the plume velocity both increase in the downwind direction, while density of the plume, temperature of the gas (above atmospheric temperature) and gas concentration all decrease downwind. We see that the trend diminishes as the distance of travel of the plume increases. This has been observed for all the three gases studied. The observations on the individual gases are summarised below.

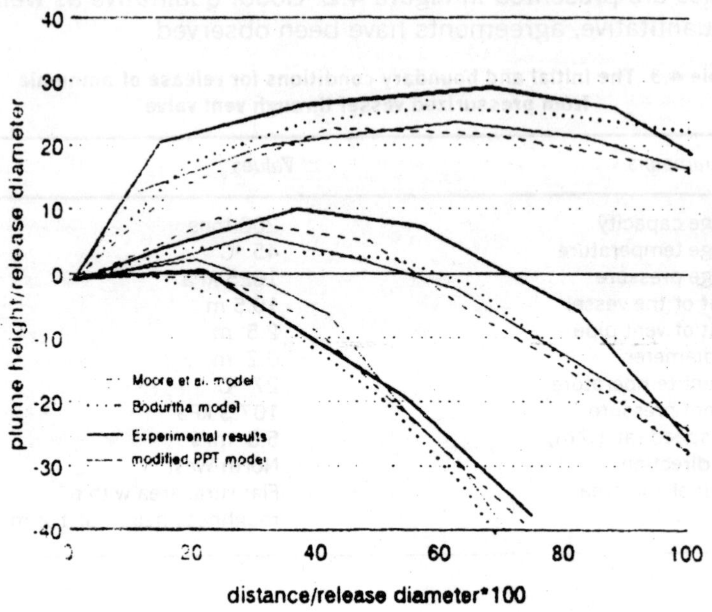

Fig. 4.2: Comparison of plume path estimated by PPT with other models and experimental results

Sulphur dioxide

Industries commonly store sulphur dioxide (SO_2) in large quantities in pressurised vessels. Possible release of SO_2 through pressure relief valve or accidental opening threatens the plant personnel as well as the surrounding population. In several industries sulphur dioxide is continuously or semi-continuously purged to atmosphere. A continuous pressurised release of SO_2 has been studied here. As the density of SO_2 is much higher (2.92 kg/m^3) than air, its plume resists entrainment of air (dilution) and travels longer distance downwind than a plume of a lighter gases would. However, due to the fast release rate (45 m/s) and consequent turbulence an initial mixing occurs (to a distance up to 600 m) but further mixing is limited and only a change of 20 % concentration occurs over the next 1500 m (Figure 4.6). It is also evident form Figure 4.6 that not only concentration but plume width also first increases very rapidly during the initial phase of dispersion but later the rate of increase slows down drastically. The plume velocity in the initial stage rises sharply upto a distance of 750 m due to the fast movement of air which dictates the flow of the plume. Subsequently the trend diminishes as the plume velocity approaches the wind velocity. For a release rate 45 m/s under the atmospheric conditions given in Table 4.2 the lethal concentration (threshold lethal value-TLV) would cover an area of 1750 meter radius.

Ammonia vapours

Ammonia is stored and processed in liquid state under refrigerated or pressurised conditions. When released it forms an aerosol of ammonia liquid carried as droplets away along with ammonia vapour. Ammonia gas has a density (0.778 kg/ m^3) lower than air but due to aerosol formation the effective density of the gas (with aerosol) becomes higher than air. Hence the dispersion of ammonia is generally treated as dispersion of a heavy gas and we have employed the same equations we had used in the sulphur dioxide study. It is seen (Figure 4.7) that ammonia disperses in a manner similar to SO_2 but the trend is much more steep. The variations in plume width and plume velocity with distance are higher compared to SO_2 . It is because droplets of ammonia liquid settle faster due to gravity leading to the formation of gradient for mixing subsequently facilitating dilution. At the initial stage of release the effective density of plume is observed 4 times higher than air but by 650 meters it falls down to 25% of the initial value. The lethal concentration of the gas at plume edge extends upto 750 m (downwind).

Table 4.4. Experimental results obtained in the present study

Down wind distance meters	Plume velocity meters	Ground level concentration (wt%)	Plume width meters	Plume temperature °C
10	0.85	0.001	0.027	42.66
50	1.55	0.002	0.065	39.69
100	2.15	0.038	0.097	35.40
200	2.75	0.045	0.137	33.48
300	3.25	0.087	0.174	30.78
400	3.65	0.098	0.202	29.43
500	4.00	0.129	0.245	28.35
600	4.35	0.157	0.312	27.81
700	4.50	0.175	0.357	27.54
1000	4.75	0.224	0.415	27.46
1300	4.85	0.254	0.447	27.41
1750	4.90	0.354	0.521	27.39

Hydrogen Sulphide

Hydrogen sulphide (H_2S) is another hazardous heavy gas studied here. An instantaneous jet release (1.54 Kg/m^3) of H_2S forms a plume whose characteristics change along downwind direction. It has been observed (Figure 4.8) that H_2S also disperses in a manner similar to sulphur dioxide and ammonia. However, at any given distance the concentration of H_2S at the plume edge is higher compared to ammonia but lower compared to sulphur dioxide. It may be due to the following reasons :

i) the density effect is higher in case of H_2S, resulting in greater resistance to mixing with air and hence dilution of the gas;

ii) a slower rate of release compared to ammonia reduces the initial turbulence and subsequently slows down the rate of dilution;

iii) a higher release rate as well as lower density effect makes H_2S dispersion faster compared to SO_2 (all other conditions remaining the same).

The plume width and plume velocity also follow the same trend as observed for concentration.

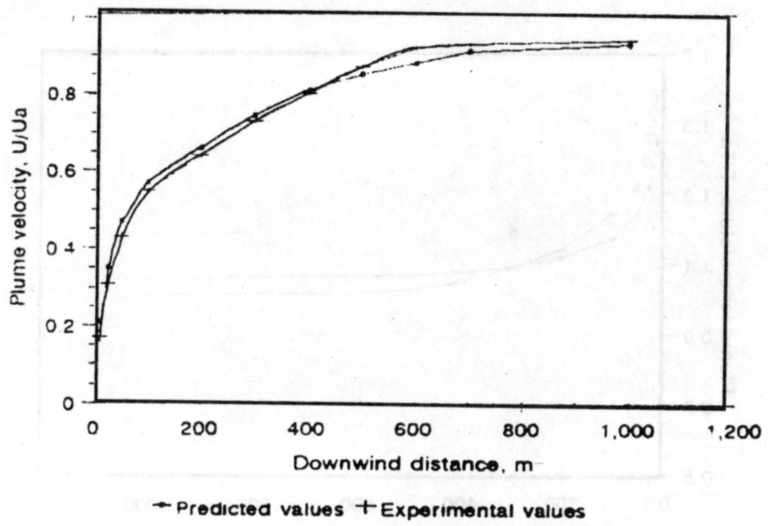

Fig. 4.3: Comparison of observed values of plume velocity with experimental values

NATURAL GAS

Figure 4.9 shows the behaviour of plume variables (dimensionless form) in downwind direction due to the change (step increase) in wind and venting speeds. An increase in the wind speed causes an increase in the plume velocity and the plume width. A similar trend is also observed for an increase in the venting speed. However, the trend is more marked in the case of increase in the venting velocity compared to the increase in wind speed. Perhaps high venting speed creates high turbulence and wake formations in the atmosphere which consequently lead to the rapid entrainment of air and swift dispersion too. Wind speed effects the downwind transportation of the plume more strongly than it does the entrainment of air and its dispersion. As the density of the gas (vapour density

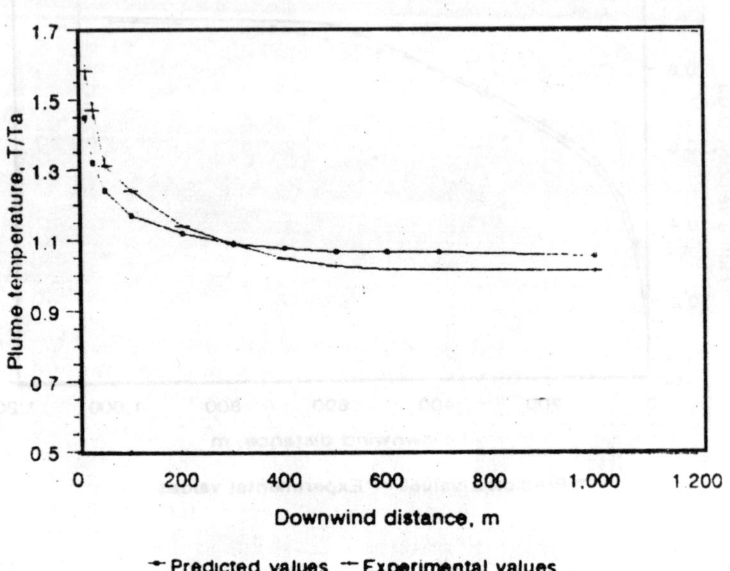

Fig. 4.4: Comparison of observed value of plume temperature with experimental values

=1.34) is not much higher than air a fast response due to a change in the controlling parameters viz, wind velocity and venting velocity, has been observed. For example, a venting velocity of 125 m/s (mass release rate 10 kg/s for 5 min) covers a smaller area under flammability limits compared to a venting velocity of 95 m/s (mass release rate 5 kg/s for 10 min) under high wind (15 m/s). This reveals that a high release (venting) velocity of this gas for shorter periods is safer than a slow release under unstable conditions for longer duration.

Liquefied petroleum gas (LPG)

The pattern of dispersion of LPG is by and large similar to the pattern of dispersion of natural gas discussed above. However a low venting speed (45 m/s) and a higher density causes slower dispersion. It is evident from Figure 4.10 that

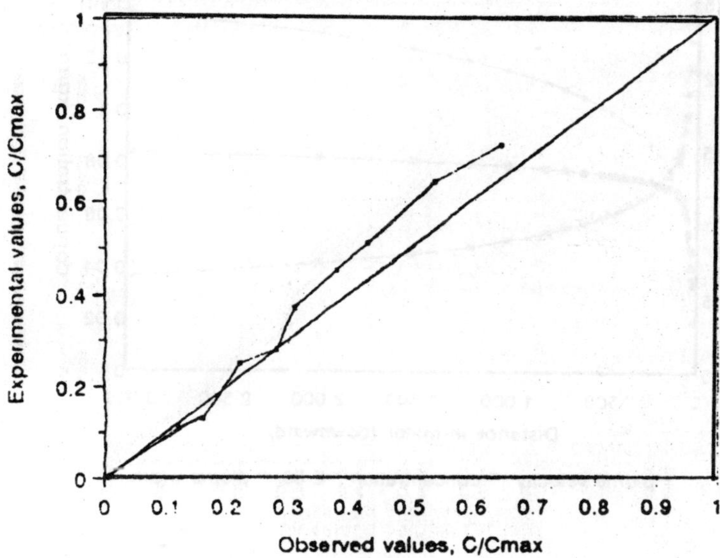

Fig. 4.5: Comparison of concentration of gas in the plume with experimental values (venting of ammonia)

at any distance along the downwind direction, plume velocity and the concentration of gas in the plume are higher while plume width is less due to an increase in the wind speed when compared with the increase in venting speed. Thus, high venting speed has stronger influence on the entrainment of air and the plume width leading to faster dispersion. But high wind speed also transports the plume to a larger distance with higher velocity.

Chlorine

Of the three hazardous gases discussed in this work, chlorine has by far the slowest rate of dispersion. Otherwise the trend observed with chlorine is broadly similar to the trends seen

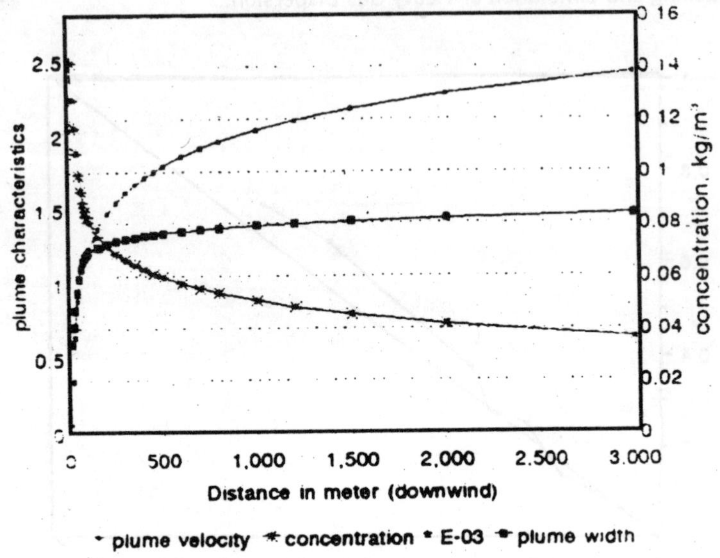

Fig. 4.6: Variation of the plume parameters along downwind distance for sulfur dioxide.

with the other two gases. As is evident from Figure 4.11 the dispersion increases with an increase in the wind speed and is faster with the higher venting speeds. For any given distance, the concentration of chlorine is higher and the plume velocity is lower compared to the other two gases. As chlorine gas is the most toxic of the three gases studied and is also the most sluggish to disperse, lethal concentration of this gas can easily build up over large areas and persist for long durations over larger areas.

PARAMETRIC EFFECT

To generalise the effect of wind speed, density and venting speed on the plume path, we have developed some empirical equations. These equations directly predict the behaviour of plume variables (gas concentration in plume, plume density, local plume width, plume velocity) with other operating variables viz : density of gas, venting speed, etc.

For this purpose a dimensionless number Q_f has been defined as:

$$Q_f = u_a/u_v*(\rho_a/\rho) \quad (15)$$

$$\rho^* = A^*\ln(s) + B$$

$$A = -0.00395848^*\log(Q_f) - 0.01925951$$

$$B = -0.16083\ ^*\log(Q_f) + 0.091579$$

$$u^* = A^*\ln(s) + B$$

$$A = 0.159874 + 0.960923^*Q_f - 1.24869^*Q_f^2 + 0.3728^*Q_f^3$$

$$B = -0.47832 + 0.0418951^*Q_f + 0.153298^*Q_f^2$$

$$c^* = A^*\ln(s) + B$$

$$A = -0.0161965 + 0.00543481 * Q_f - 0.00128631 * Q_f^2 \quad (16)$$

$$B = 0.183894 * \exp(-0.382484 * Q_f)$$

$$b_s = A^*\ln(s) + B$$

$$A = 0.0157719 + 0.104969 * Q_f - 0.0347085 * Q_f^2$$

$$B = 0.953469 * Q_f^{0.22422}$$

$$T = A * \ln(s) + B$$

$$A = 3.67477 + 7.61097 * Q_f - 2.42655 * Q_f^2$$

$$B = -8.5476 - 58.7278 * Q_f + 26.7875Q_f^2$$

The equations have been validated with experimental values. A plot representing different plume variables at Y axis with dimensionless number Q_f is shown in Figure 4.12. It can be seen that with an increase in Q_f, variables like plume width and plume velocity (shown on Y axis) increase while gas concentration and density difference both decrease.

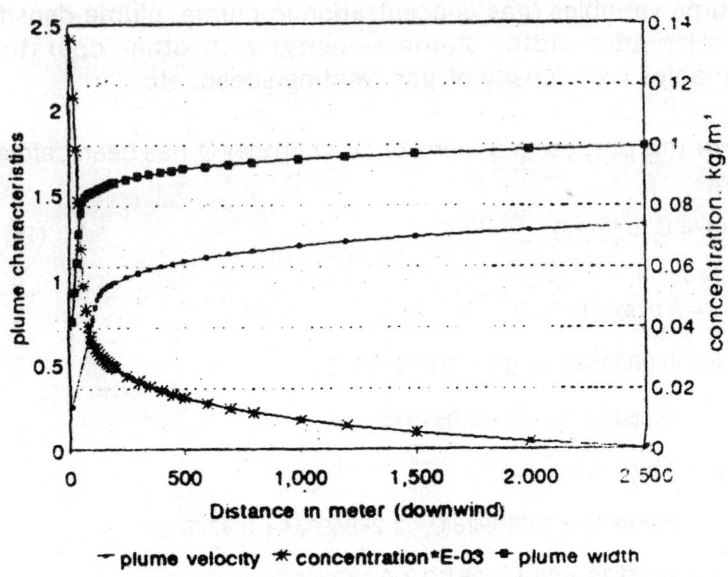

Fig. 4.7: Variation of the plume parameters along downwind distance for ammonia.

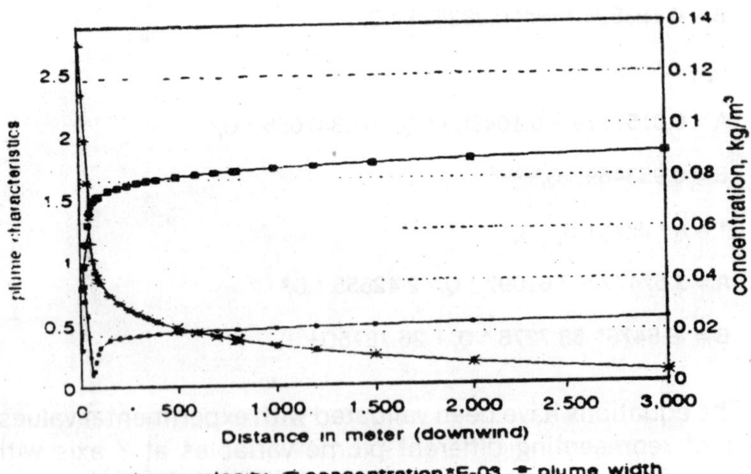

Fig. 4.8: Variation of the plume parameters along downwind distance for hydrogen sulfide.

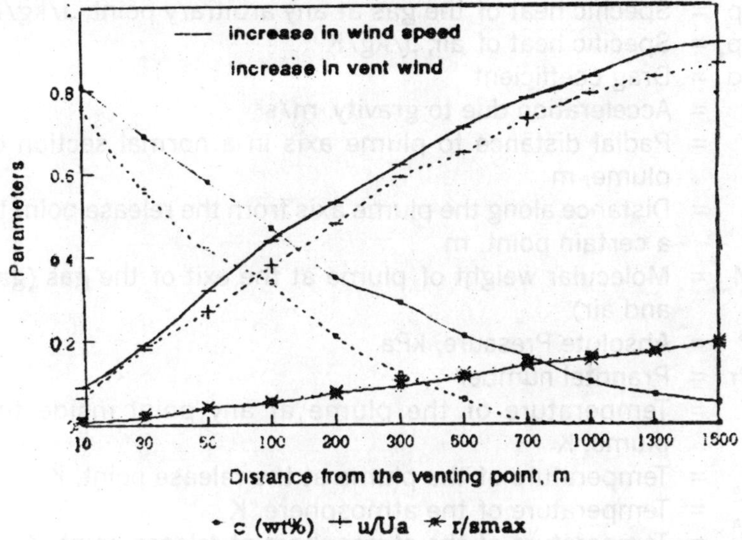

Fig. 4.9: Impact of a 40% change in wind speed (from 5 m/s to 7 m/s) as compared to the impact of 40% change in venting speed (from 95 m/s to 125 m/s) on three plume variable (Natural gas)

CONCLUSION

The plume path theory as modified by us gives satisfactory results for the dispersion of heavy gases. It also enables simulation of plume profiles along the cross section of plume as well as in the downwind directions. The empirical equations developed by us on the basis of the present model, we hope, would prove to be helpful in air-pollution studies, as they can predict responses to the different ambient operating conditions over the plume variables (concentration, density, local width, temp etc.) with relative ease and fair accuracy.

LIST OF SYMBOLS

b_s = Local characteristics width, m
c = Concentration of gas at any point inside the plume, kg/m^3
c^* = Concentration of gas on the plume axis, kg/m^3
c_o = Concentration at out let, kg/m^3
cp_o = Specific heat at outlet, J/kg/K

cp = Specific heat of the gas at any arbitrary point, J/kg/K
cp_a = Specific heat of air, J/kg/K
cd = Drag coefficient
g = Acceleration due to gravity, m/s^2
r = Radial distance to plume axis in a normal section of plume, m
s = Distance along the plume axis from the release point to a certain point, m
M_{wo} = Molecular weight of plume at the exit of the gas (gas and air)
P = Absolute Pressure, kPa
Pr = Prandtal number
T = Temperature of the plume at any point inside the plume, K
T_o = Temperature of the plume at the release point, K
T_a = Temperature of the atmosphere, K
T_{ao} = Temperature of the atmosphere at release point, K
u = Plume velocity at any point in the plume in the direction of the tangent to the plume axis, m/s
u^* = Plume velocity on the plume axis in the direction of the tangent to the plume axis, m/s
u' = Entrainment velocity due to atmospheric turbulence, m/s
u_a = Wind velocity, m/s
u_v = Venting speed, m/s
x = Cartesian coordinate (Figure 4.1)
z = Cartesian coordinate (Figure 4.1)
ρ = Plume density at any point in the plume, kg/m^3
ρ_a = Density of air, kg/m^3
ρ_{ao} = Density of air at release point, kg/m^3
ρ_g = Density of gas, kg/m^3
ρ^* = Density difference between plume and atmosphere, kg/m^3
α_1 = Entrainment coefficient of a free jet
α_2 = Entrainment coefficient due to thermal stratification
α_3 = Entrainment coefficient due to atmospheric turbulence
λ = Turbulent Schemedit number
ϕ = Angle between plume axis to horizontal component
δ = Length of transition zone, m
Q_f = Dimensional number

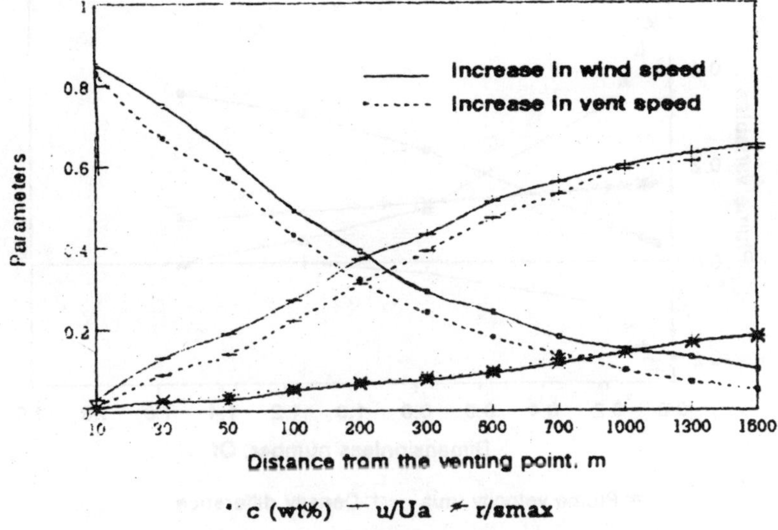

Fig. 4.10: Impact of a 40% change in wind speed (from 5 m/s to 7 m/s) as compared to the impact of 40% change in venting speed (from 95 m/s to 125 m/s) on three plume variable (Natural gas)

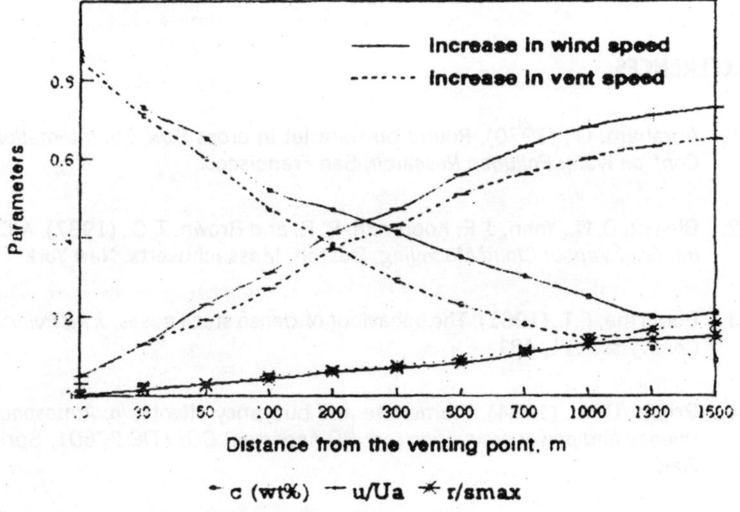

Fig. 4.11: Impact of a 40% change in wind speed (from 5 m/s to 7 m/s) as compared to the impact of 40% change in venting speed (from 25 m/s to 35 m/s) on three plume variable (Chlorine gas)

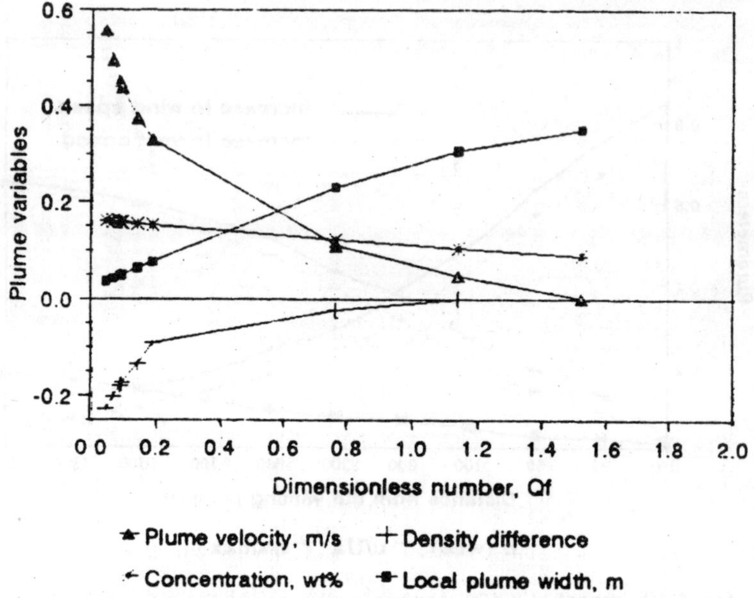

Fig. 4.12: Relationship of the dimensionless number Qf (c.f. equation 16) with different plume characteristics.

REFERENCES

1. Abraham, G., (1970). Round buoyant jet in cross flow, *5th International Conf. on Water Pollution Research*, San Francisco.

2. Blewitt, D. N., Yohn, J. F., Koopman, R. P., and Brown, T. C., (1987). *AIChE Int. Conf Vapour Cloud Modelling*, Boston, Massachusetts, New York.

3. Bodurtha, F. T., (1961). The behaviour of dense stack gases, *J. Air Pollution Contr. Asso.*,11, 431.

4. Briggs, G. A., (1984). Plume rise and buoyancy effects, *In: Atmospheric science and power production (ed: RD Andersen) DOE/TIC 27601*, Spring Field.

5. Carnahan, B., (1969). Applied numerical methods, *John Willy and Sons*, New York.

6. Colenbrander, G., (1980). A mathematical model for the transient behaviour of dense vapour clouds, *3rd Int Sym. on Loss Prevention and Promotion in the Process Industries*, Bassel.

7. Deaves, D. M., (1983). Application of advanced turbulence models in determine the structure and dispersion of heavy gas cloud, *IUTAM symposium*, Delft.

8. Deaves, D. M., (1992). Dense gas dispersion modelling, *J of Loss Prev. Process Ind*, 5(4), 219–27.

9. Eidsvik, K. J., (1980). A model for heavy gas dispersion in the atmosphere, *Atmos. Environ.*, 14, 769–77.

10. England, W. C., Teuscher, L. H., Hauser, L. E., and Freeman, B., (1978). Atmospheric dispersion of liquefied natural gas vapour clouds using SIGMET, a three-dimensional time-dependent hydrodynamics computer model, *Lecture notes heat transfer and fluid mechanics institute*, Washington state University.

11. Ermak, D. L., and Chan, S. T., (1985). A study of heavy gas effects on the atmospheric dispersion of dense gases, *15 Inter. Technical Meeting on Air Pollution Modelling and its Application*, St. Louis, USA.

12. Fay, J. A., and Zemba, S. G., (1985). Dispersion of initially compact dense gas clouds, *Atmos. Environ.*, 19 (8), 1257.

13. Fay, J. A., and Zemba, S. G.,(1986). Integral model of dense gas plume dispersion, *Atmos. Environ.*, 20(7),1347.

14. Havens, J. R., (1982). A description and computational assessment of SIGMET LNG vapour dispersion model, *J. Hazardous Materials*, 6, 181–210.

15. Hawthorne, W. R., (1955). The growth of secondary circulation in friction flow, *Proc. Camb. Phil. Soc.*, 51, 737.

16. Langlo, G. K., and Schatzmann, M., (1991). Wind tunnel modelling of heavy gas dispersion., *Atmos. Environ.*, 25A, 1189.

17. McQuaid, J., (1986). Refinement of Estimates of consequence of toxic vapour release, *Proc. IChmE Sym.*, Manchester

18. Moore, G.E., Milich, L. B., and Liu, M. K., (1988). Study of Plume behaviour using lidar and SF tracer at a flat and heavy site, *Atmos. Environ.*, 22, 1673.

19. Niewstadt, F T M, and Van Dop, H, (1982). Atmospheric turbulence and air pollution modelling, *Reidel publishing company*, Boston, USA.

20. Niewstadt, F T M, (1992). A large–eddy simulation of a line source in a convective atmospheric boundary layer–II. Dynamics of a buoyant line source., *Atmos. Environ.*, 26 A, 499.

21. Ooms, G.,(1972). A new method for the calculation of the plume path of gases emitted by a stack, *Atmos. Environ.*, 6, 899.

22. Ooms, G.,Mahiue, A. P., (1974). The plume path of vent gases, Loss prevention and safety promotion in process industries, *Elsevier science publication*, New York.

23. Ooms, G., Mahiue, A. P., and Zelis, F, (1974). The plume path theory of vent gases heavier than air, *Proceeding of the first International Loss Prevention Symposium*, Hague/Delft., The Netherlands.

24. Ooms, G., and Duijm, N. J., (1983). Dispersion of a stack plume heavier than air, *IUTAM Sym on Atmospheric Dispersion of Heavy Gases and Small Particles*, Delft, The Netherlands.

25. Puttock, J. S., (1986). Comparison of Thorny Island data with prediction of HEGABOX/HEGADAS, *2nd symposium on heavy gas dispersion trials at Thorny Island*, Sheffield.

26. Riou, Y., (1986). Comparison between MERCURE_GL code calculations, wind tunnel measurements and Thorny Island field trials, *2nd symposium on heavy gas dispersion trials at Thorny Island*, Sheffield.

27. Rottman, J. W., Simpson, J. E., Hunt, J. C. R., and Bitter, R. E., (1985). *J. Hazardous Material*, 11, 325.

28. Taft, J. R., Ryne, M. S., and Weston, D. A., (1983). MARIAH a dispersion model for evaluating realistic heavy gas spills scenarios, *Proc AGA Transmission Conf*, Seattle.

29. Tran, K. T., and Liu, C. Y., (1981). Development of a predictive dispersion modelling system for real-time emergency application, *5th symposium on turbulence diffusion and air pollution*, American met. Soc., Atlanta.

30. Van Ulden, A. P., (1983). A new bulk model for dense gas dispersion : two dimensional spread in still air, *IUTAM Sym on Atmospheric Dispersion of Heavy Gases and Small particles*, Delft.

31. Van Ulden, A. P., and Holtslag, A. A. M., (1985). Estimation of atmosphere boundary layer parameters for diffusion application, *J. Cl. App. Metrol.*, 25, 1609.

32. Van Ulden, A. P., (1987). The spreading and mixing of a dense cloud in still air, *KNMI Afdeling Fysishe meterologie*, Personal memorandum, FM-87-11.

33. Weil, J. C., (1988). Plume rise, (eds. A Venkatram, J. C. Wyngaard), *Amer. Metr. Soci.*, Boston, USA.

Chapter 5

CASE STUDY : MODELING AND SIMULATION OF HEAVY GASES RELEASED BY PETROCHEMICAL INDUSTRIES

In the preceding chapter we have discussed modelling of dispersion of heavy gases. In this chapter case studies are presented in which the model has been successfully employed to study the dispersion of several common 'heavy' gases (ethylene, propylene, isobutylene, natural gas, chlorine, and ammonia).

We also present studies to simulate the effect of venting speed (manipulated by injecting hot air with the released gas) on the plume dispersion. It is revealed that the effect of venting speed on dispersion is very pronounced and can be used to reduce the risk posed by the accidental luxury release of toxic/ flammable gases. For example, an increase of 20% in venting speed of chlorine (54.1 m/s) can reduce the distance up to which toxic concentration would occur by about 1100 meters.

INTRODUCTION

Petrochemical industries which often handle hazardous chemicals and operate reactors/storage vessels under extreme conditions of temperature and pressure are susceptible to accidents. The history of petrochemical industries is replete with several such accidents which have had catastrophic consequences (Lees, 1996). The hazardous (toxic and/or flammable) chemicals normally handled by petrochemical industries are in the form of liquids and gases (lighter-than/light-as-air, and heavier-than-air). To study the accidental release and dispersion of such chemicals one needs models that can handle various types of release scenarios. Modified plume path theory is one such model (Ooms and Duijm ,1983; Khan and Abbasi, 1997;1999).

Originally, plume path theory (PPT) was proposed by Ooms (1972) and has been used mainly to calculate the plume path of lighter-than-air and light-as-air gases escaping from stacks into the atmosphere at standard atmospheric temperature and pressure. Later this theory was extended to 'heavy' gases (of density higher than air; Ooms and Mahiue, 1974; Ooms *et al.* 1974) with a number of assumptions.

Recently, we have modified the plume path theory to improve its versatility (Khan and Abbasi, 1999). The resultant model, which has been validated with the data generated by the authors and others, overcomes several limitations which had thus far restricted the applicability of PPT. The features of the original PPT and the modifications done by us are summarized in Table 5.1. They have been detailed elsewhere (Khan and Abbasi,1997;1999).. We have subsequently developed and validated several empirical models to study the effect of various parameters on the dynamics of gaseous dispersion (Khan and Abbasi, 1997). In this chapter the authors present the applications of the aforesaid models in controlling dispersion of flammable and/or toxic gases coming out of petrochemical industries, by injecting hot air into the gaseous plume at the latter's point of exit. The models aim to handle involuntary accidental release as well as voluntary strategic release (to avert accidents).

A brief description of the methodology is presented below.

OBJECTIVES

1. To study the dispersion after accidental or voluntary release, of hazardous gases common in most of the petrochemical industries. For this purpose we have taken a real-life case

study of the storage unit of Paplani Petrochemical Limited, situated at Paplani, Gujrat, India. The release of the gases have been assumed to occur through the vent valve/ pressure relief valve of the unit around 5 to 7 meter (including height of vessel and vent) above ground level, for six different chemicals stored in different vessels. The initial set of atmospheric and operating conditions used in the present study is given in Table 5.2.

2. To study the effect of increase in venting speed (by injecting hot inert air) over the dispersion process (distance up to which flammable/lethal concentration would occur).

METHODOLOGY

To generalise the effect of wind speed, density, and venting speed on the plume path, we have developed empirical equations. These equations predict the behavior of the plume variables (gas concentration in plume, plume density, local plume width, plume velocity) as a function of such variables as density of gas, venting speed, etc.

For this purpose a dimensionless number Q_f has been defined as:

$$Q_f = w/u_v*(\rho_a/\rho_{eff}) \quad (1)$$

$\rho^* = A*\ln(s) + B$
$A = -0.00395848*\log(Q_f) - 0.01925951$
$B = -0.16083*\log(Q_f) + 0.091579$
$u^* = A*\ln(s) + B$.
$A = 0.159874 + 0.960923*Q_f - 1.24869*Q_f^2 + 0.3728*Q_f^3$
$B = -0.47832 + 0.0418951*Q_f + 0.153298*Q_f^2$
$c^* = A*\ln(s) + B$

$$A = -0.0161965 + 0.00543481*Q_f - 0.00128631*Q_f^2 \quad (2)$$

$B = 0.183894*\exp(-0.382484*Q_f)$
$b_s = A*\ln(s) + B$
$A = 0.0157719 + 0.104969*Q_f - 0.0347085*Q_f^2$
$B = 0.953469*Q_f^{0.22422}$
$T = A*\ln(s) + B$
$A = 3.67477 + 7.61097*Q_f - 2.42655*Q_f^2$
$B = -8.5476 - 58.7278*Q_f + 26.7875Q_f^2$

Table 5.1. Redefinition and advancement in the assumption made in Oom's PPT model

Assumptions made in PPT (Ooms and cowrokers, 1972 1974)	*Modifications proposed by the authors*
The mean flow velocity perpendicular to the main flow in the direction of the plume is negligible till a late stage of the plume movement	Secondary flow perpendicular to the plume axis is significant at much earlier stage as well
The velocity profile, density, and pollutant concentration are similar in all sections normal to the plume axis	none
Molecular transports is considered negligible in comparison with turbulent transports	Molecular transport has been treated as significant and taken into account in parameter estimation
Longitudinal turbulent transport is considered negligible compared with longitudinal convective transport	Both longitudinal as well as convective transports have been taken into account
—	Physical properties of venting gas have been considered as a function of downwind distance as well as atmospheric parameters

EFFECTIVE PLUME HEIGHT

The effective plume height is the height reached by the plume from the point of its exit from a stack/vent valve. The calculation of effective plume height assumes (as shown in Figure 5.1) that there is a virtual source of the plume at point A which is at a distance x_o upwind from the transition point B. This assumption allows for the mixing effect of the jet. The distance x_o is that which would be necessary to achieve the same concentration at the transition point if the mixing were done by the wind (Lees,1996). Then the rise of the plume is given as:

$$Z = 2K_7/w^*(x^{1/2}\text{-}\ x_o^{1/2}) \qquad (3)$$

Subsequently maximum rise of the plume is estimated as:

$$Z_{max} = 2^*K_7/w^*(x_{max}^{1/2}\text{-}\ x_o^{1/2}) \qquad (4)$$

Where, $x_{max} = K_8{}^*w + x_o$ and $K_8 = 200.0$; (Cude,1974), and (5)

$K_7 = (g^*Q/(2^*PP^*Dc^*tanff^*)^*ABS((rr_a\text{-}rr_{eff})/rr_a))$ (Cude,1974) (6)

The half angle of a plume (ff) from a stack is generally 5° or 6° so that tanϕ = 1. The drag coefficient Dc, as estimated by Cude (1974), is 0.4.

Since length of the jet is inversely proportional to the root of momentum flux, the transition heights of the jet in still air conditions, *ltr* and in windy conditions, *ltrw* are related as:

$$ltrw = ltr/1.2 \qquad (7)$$

Finally the maximum effective rise of the plume is estimated using the following equation,

$$hpl = H + ltrw + Z_{max} \qquad (8)$$

Briggs (1984) and Hanna *et al* (1982) have further modified the equations of maximum and effective rise of the plume taking the temperature of venting gas into consideration. They have proposed equations:

$$Z_{max} = 4(m_o/sta)^{1/4} \qquad (9)$$

$$Z_{eff} = 2.44(m_o/sta)^{1/4} \qquad (10)$$

Where,

stability is defined as, $sta = g/T(dT/dy+R)$, (11)

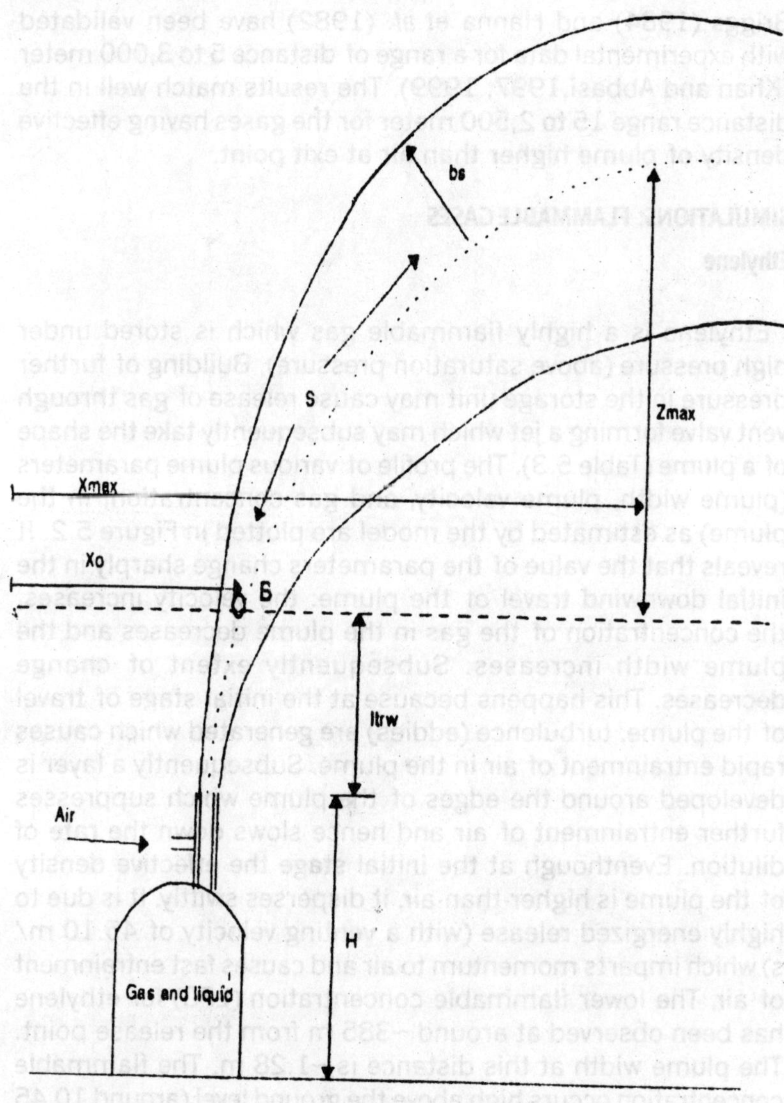

Figure 5.1. Illustrative figure showing coordinates of the plume as influnced by hot air injection

The outcome of all these equations with the modification of Briggs (1984) and Hanna *et al.* (1982) have been validated with experimental data for a range of distance 5 to 3,000 meter (Khan and Abbasi,1997; 1999). The results match well in the distance range 15 to 2,500 meter for the gases having effective density of plume higher than air at exit point.

SIMULATIONS: FLAMMABLE GASES

Ethylene

Ethylene is a highly flammable gas which is stored under high pressure (above saturation pressure). Building of further pressure in the storage unit may cause release of gas through vent valve forming a jet which may subsequently take the shape of a plume (Table 5.3). The profile of various plume parameters (plume width, plume velocity, and gas concentration, in the plume) as estimated by the model are plotted in Figure 5.2. It reveals that the value of the parameters change sharply in the initial downwind travel of the plume: the velocity increases, the concentration of the gas in the plume decreases and the plume width increases. Subsequently extent of change decreases. This happens because at the initial stage of travel of the plume, turbulence (eddies) are generated which causes rapid entrainment of air in the plume. Subsequently a layer is developed around the edges of the plume which suppresses further entrainment of air and hence slows down the rate of dilution. Eventhough at the initial stage the effective density of the plume is higher-than-air, it disperses swiftly. It is due to highly energized release (with a venting velocity of 45.10 m/s) which imparts momentum to air and causes fast entrainment of air. The lower flammable concentration (LFC) for ethylene has been observed at around ~385 m from the release point. The plume width at this distance is ~1.28 m. The flammable concentration occurs high above the ground level (around 10.45 meters).

Propylene

Propylene is another flammable gas mostly stored in large quantities under pressure. A continuous release of gas through vent valve/pressure relief valve about 7 m above the ground forms a plume. The behavior of plume variables reveals that the dispersion of propylene is sluggish compared to ethylene (Table 5.3). It is because the effective density of the propylene plume is higher than ethylene. The flammable zone has been observed over an area of 415 m radius from the release point. The release and dispersion of propylene poses considerable hazards as the flammable zone envelops a large area that is also close to the ground level (~7.45 meters).

Isobutylene

Isobutylene is a flammable gas stored in lesser quantities compared to ethylene and propylene. Release of vapors (gas with aerosol) through vent valve would form a plume of density higher than air, which would subsequently get diluted by entrainment of air. The behavior of dispersion parameters (Table 5.3) reveals a very slow dilution rate and hence building of flammable concentration over a large area. It is because:

i) a comparatively low venting speed would favor gravity-dominant dispersion due to the high density of the plume.
ii) relatively high density would favor formation of suppression layer around the plume edges which would inhibit entrainment of air and hence dilution.

NATURAL GAS

Natural gas is a mixture of light flammable gases (such as methane and ethane) which is stored at saturation pressure. Release of natural gas through a narrow opening would form a plume (~12 meters above the ground level). The results of the model output for this gas show a relatively fast dispersion compared to propylene and isobutylene. The lower limit of flammable concentration would be observed at a distance 90 m from point of release (at around 12.45 m above the ground level). It is because the effective density of this plume is low and the venting velocity is sufficiently high, allowing rapid entrainment of air, consequently swift dilution.

Table 5.2. The study setting

Parameters	Values
Wind velocity	3.5 m/s
Climatic conditions	Cloudy evening
Atmospheric boundary layer	45 m
Friction velocity	0.435 m/s
Atmospheric temperature	22 °C
Atmospheric pressure	1.01 atm
Terrain of the area	Urban
Height of the source	5·7 m
Atmospheric stability	Slightly stable
Normal venting speed	45.15 m/s

SIMULATIONS: TOXIC GASES

Chlorine

Of the two toxic heavier-than-air gases studied here, chlorine (Cl_2) has the slowest rate of dispersion. Otherwise the trend observed with chlorine is similar to the trends seen with the other five gases. For any given distance, the rate of change in concentration and plume velocity as well as plume width are lower compared to the other toxic gas ammonia (Table 5.3). As chlorine is the most toxic of the gases studied here and is also the most sluggish to disperse, lethal concentrations of this gas would easily build-up over large areas and persist for long durations.

Ammonia

Ammonia is stored and processed in liquid state under refrigerated or pressurized conditions. When released, it forms an aerosol of ammonia liquid carried away as droplets along with ammonia vapor. As such ammonia gas has density (0.778 kg/m^3) lower-than-air but due to aerosol formation the effective density of the plume (released vapors with aerosol) becomes higher than air. Hence the dispersion of ammonia is generally treated as if it were a heavy gas.

The model output shows that ammonia disperses in a manner similar to chlorine but the trend is much more steep. The variations in plume width and plume velocity with distance are higher compared to Cl_2. It is because droplets of ammonia liquid settle faster due to gravity leading to the formation of gradient for mixing, subsequently facilitating dilution. At the initial stage of release the effective density of the plume is 3 times higher than air but by the time it travels 375 meters, the density drops to 21% of the initial value. The lethal concentration of the gas at the plume edge extends up to 645 m downwind at about 9.31 m above the ground.

IMPACT OF HOT AIR INJECTION : FLAMMABLE GASES

In order to contain the damage, if these hazardous gases get released (accidentally, or voluntarily to control bigger accidents such as explosions), we have studied the option of increasing the release velocity by injecting hot air as a possible means of diluting the gases and facilitate their dispersion. Such an act has the following consequences:

a) hot inert air makes the released gas move higher and faster from the exit point; the effective release velocity is a combination of the velocities of hot inert air and the hazardous gas;
b) turbulence is generated which effects the density of the plume as well as its momentum distribution;
c) the high temperature of the air jet further disturbs the momentum balance by energizing the gas; however the air temperature should not be so high that it sparks a fire;
d) thorough mixing takes place at the maximum height of jet and along the downwind direction.

We have assumed that the shape of the plume would not be disturbed substantially. We have carried out simulations within the premises mentioned above, assuming that the atmospheric stability conditions would remain the same throughout the dispersion process. The impact of 1%, 5%, 7%, 15%, and 20% increase in the effective jet velocity (*i,e* a combination of the gas and the hot air jet velocities) on the dispersion of the six aforementioned gases, has been studied. The results of the study are discussed below:

For ethylene, at an exit velocity (45.15 m/s) without any hot air injection, the minimum flammable concentration would occur at a distance of ~385 m from the point of exit while a 20% increase in the release velocity (54.1 m/s) by hot air injection would cause the plume to disperse much faster and the ethylene concentration would go below flammability limit within ~77 meters of the exit point (Figure 5.3). The variation in ethylene concentration with increase in venting speed has been depicted in Figure 5.3. It is seen that initially a rapid decrease in the concentration is observed. For example, a 20% increase in venting speed causes a 47% decrease in concentration within a short distance - 50m - of travel of the plume. However, during the subsequent dispersion, the rate of decrease becomes much slower.

The profile of other parameters such as plume width (Figure 5.4) and plume velocity shows an increasing trend with increase of venting speed. It is clear from Figure 5.4 that during the initial travel of the plume (~ 50 meters) a large increase in the plume width occurs but during subsequent travel (more than ~400 meters) this effect is much slower. Further, a high venting speed forces the gas to go higher in the atmosphere before dispersing, taking the flammable zone of the plume well above the ground level, reducing the possibility of the gas catching fire by meeting a spark.

Similar trend has been observed for propylene. When released without hot air injection, the build-up of flammable concentration would occur up to a distance of ~415 m from the exit point. This 'striking distance' would reduce to ~96 m if the venting speed is increased by 20% (Table 5.4). The plume edge would be closer (8.5 m) to the ground level than the ethylene plume (11.35 m).

Isobutylene would disperse slower compared to ethylene and propylene. Hot air injection would shift the distance up to which flammable concentration would occur from ~493m to ~115m (Table 5.4).

The results for natural gas reveal that at the initial exit velocity (45.1 m/s), the flammable concentration occurs up to observed at ~90 m (about 12.45 m above the ground). It is because the effective density of natural gas is lower than all other flammable gases studied here. An increase of 20% in venting speed (54.1 m/s) would make the gas shoot up high enough in the atmosphere to drastically reduce the probability of it catching fire.

IMPACT OF HOT AIR INJECTION : TOXIC GASES

The study of the chlorine plume reveals that at a venting speed of 45.1 m/s (without hot air injection) the lethal concentration load (based on immediate fatal concentration for 30 minutes exposure; Contini *et al*. 1991; Lees, 1996) would be observed up to a distance ~1450 m from the point of release in downwind direction. This 'striking distance' would shrink to ~380 m when the venting speed is increased by 20%. Moreover, the plume edge in the later case would be ~ 8.12 m above from the ground level. This would further reduces the probability of the gas coming in direct contact of living organisms.

Ammonia would disperse faster than chlorine. The lethal concentration (based on immediate fatal concentration for 30 minutes exposure) if ammonia is released at 45.1 m/s would spread across ~645 m downwind, this distance would reduce to ~67 m for 20% increase in the gas exit speed. The plume edge would be farther away from the ground level compared to chlorine. The possible reasons are:

a) a comparatively lower effective density of ammonia,
b) greater impact of hot air injection on the effective density of the plume and momentum distribution.

Table 5.3. Model output concerning dispersion of the gases

Chemicals	*Parameter*		*Distance along down wind direction, meters*				
			20	*50*	*100*	*300*	*500*
Flammable gases							
Ethylene	maximum plume rise	= 12.06 m					
	plume concentration*10 $^{-2@}$ kg/m^3		6.13	4.61	4.03	3.25	2.75
	plume width, m		0.870	1.051	1.171	1.263	1.343
Propylene	maximum plume rise	= 11.67 m					
	plume concentration 10 $^{-2@}$ kg/m^3		7.41	6.07	5.25	4.55	3.97
	plume width, m		0.600	0.831	1.022	1.173	1.295
Isobutylene	maximum plume rise	= 10.36 m					
	plume concentration 10 $^{-2@}$ kg/m^3		7.66	6.643	5.57	5.09	4.57
	plume width, m		0.554	0.684	0.794	0.904	0.976
Natural gas	maximum plume rise	= 12.58 m					
	plume concentration 10 $^{-2@}$ kg/m^3		6.79	5.07	3.56	1.75	0.83
	plume width, m		0.985	1.314	1.546	1.697	1.735
Toxic gases							
Chlorine	maximum plume rise	= 9.56 m					
	plume concentration 10 $^{-6\#}$ kg/m^3		5.09	4.12	3.32	2.96	2.54
	plume width, m		0.512	0.644	0.714	0.864	0.944
Ammonia	maximum plume rise	= 10.76 m					
	plume concentration 10 $^{-6\#}$ kg/m^3		4.15	2.19	1.23	0.54	0.03
	plume width, m		0.752	1.071	1.342	1.551	1.730

@ Concentration estimated at the plume axis
\# Concentration estimated at the edge of the plume

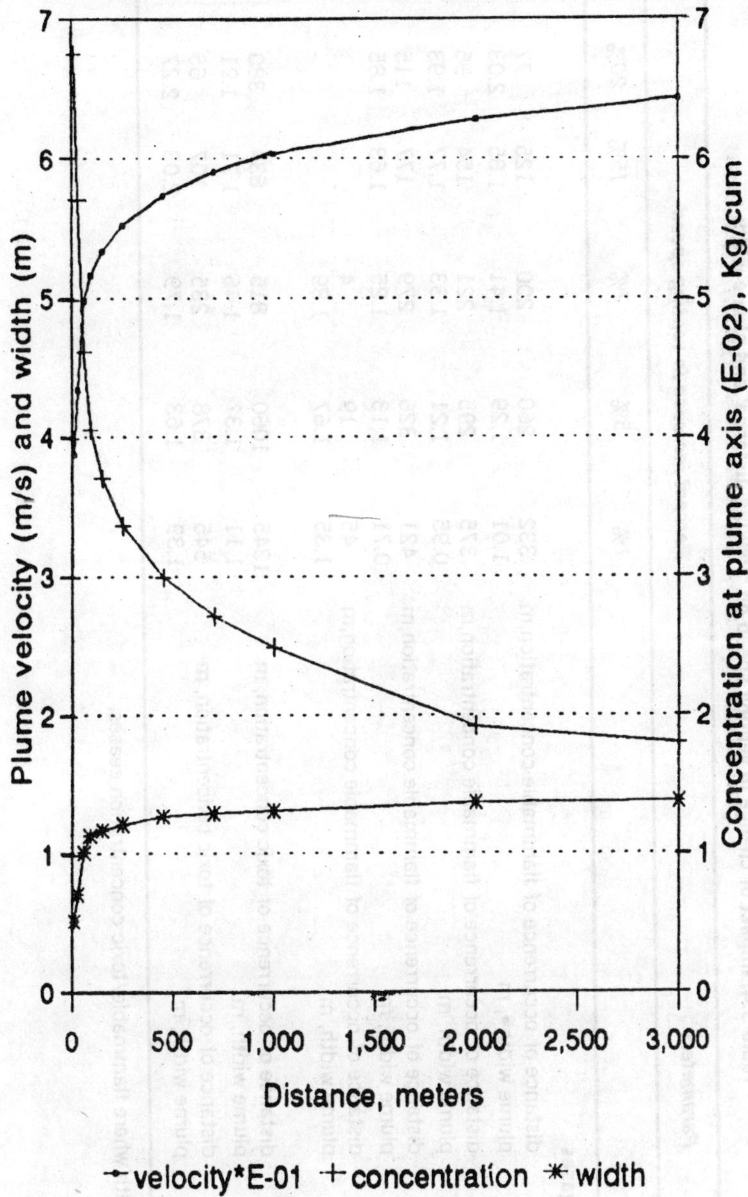

Figure 5.2: Illustrative figure showing variation of plume width and plume velocity with distance of travel (for ethylene)

Table 5.4. Impact of change in venting speed on 'striking distance' and plume width

Chemicals	*Parameters*	*Percent increase in venting speed*				
		1%	*5%*	*7%*	*15%*	*20%*
Flammable gases						
Ethylene	distance of occurrence of flammable concentration,m	332	260	200	125	77
	plume width@, m	1.01	1.29	1.41	1.85	2.03
Propylene	distance of occurrence of flammable concentration,m	375	295	221	154	96
	plume width, m	0.95	1.21	1.33	1.77	1.93
Isobutylen	distance of occurrence of flammable concentration,m	421	326	279	177	115
	plume width, m	0.71	1.13	1.25	1.63	1.85
Natural gas	distance of occurrence of flammable concentration,m	45	19	4	.	.
	plume width, m	1.35	1.67	1.89	.	.
Toxic gases						
Chlorine	distance of occurrence of toxic concentration, m	1345	1060	815	634	380
	plume width, m	1.11	1.37	1.46	1.71	1.91
Ammonia	distance of occurrence of toxic concentration, m	545	378	235	107	65
	plume width, m	1.39	1.63	1.79	2.03	2.27

@ plume width where flammable/toxic concentration ceases.

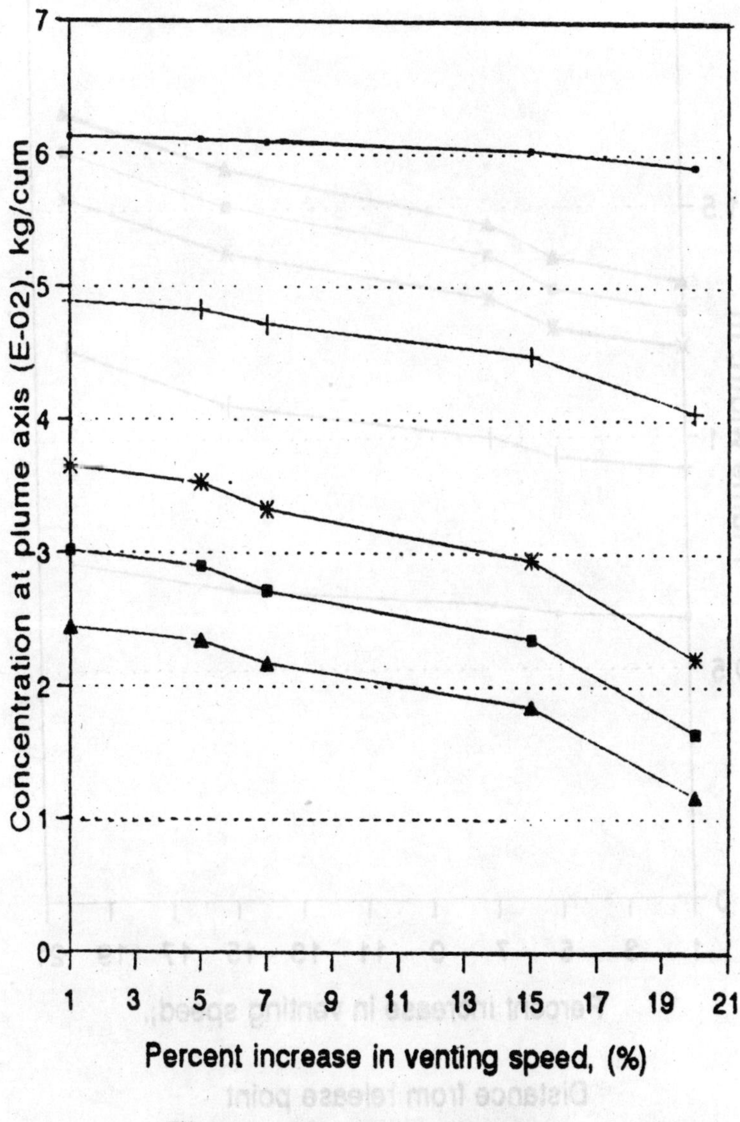

Figure 5.3: Illustrative figure showing change in concentration at plume axis with increase in venting speed (for ethylene)

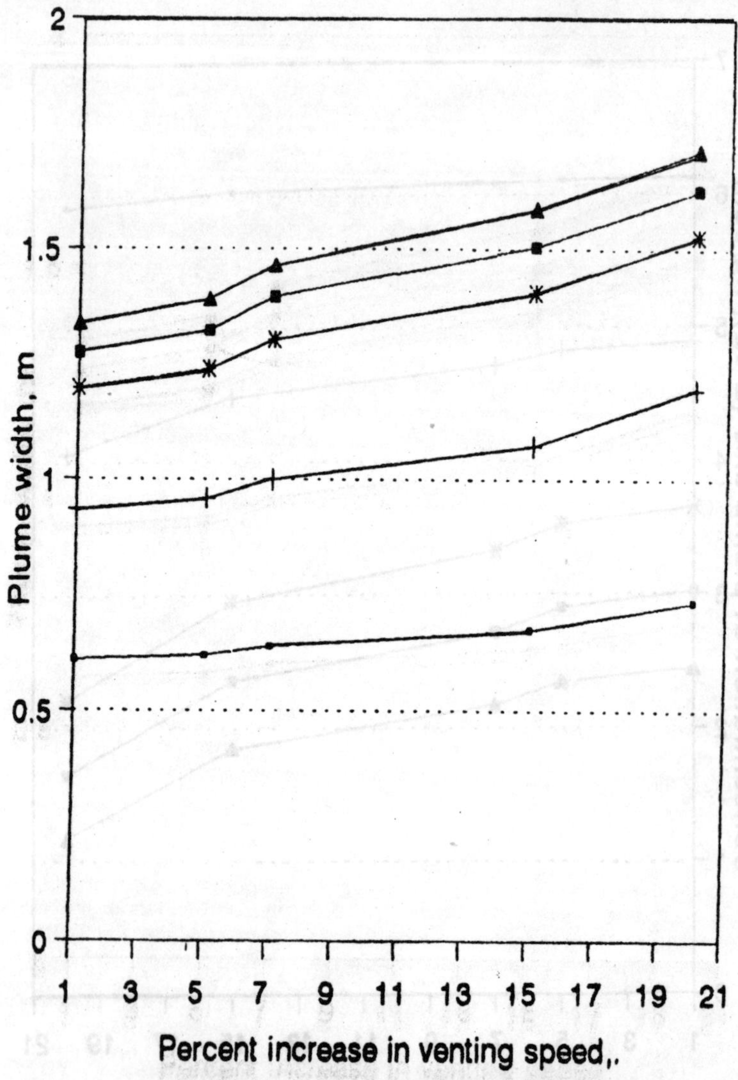

Distance from release point

• at 20 m + at 50 m ✳ at 150 m ■ at 400 m ▲ at 1000 m

Figure 5.4: Illustrative figure showing change in plume width increase in venting speed (for ethylene)

CONCLUSION

We have presented empirical equations developed by us on the basis of Modified Plume Path Theory (MPPT) reported earlier by us. The MPPT, which was duly validated by us, enables one to study the dispersion of heavier-than-air, lighter-than-air, and light-as-air gases. It is especially useful in assessing the risks posed by accidental release of hazardous gases.

The empirical equations described in this paper have been applied to possible accidental release of six hazardous gases from their storage vessels. It is seen that isobutylene release is the most hazardous as it would result in the formation of a flammable cloud covering a large distance downwind.

The effects of decreasing the concentration and increasing the exiting velocity of the gases by injecting hot inert air at the point of exit were simulated. The studies reveal that a) the plumes of hazardous gases can be raised higher (thereby reducing the probability of they catching fire); b) the rate of dilution of the hazardous gases in the plume can be made faster; and c) the 'distance of impact' (in other words 'striking distance') can be significantly reduced, by hot air injection. For example, an increase of 20% in venting speed of isobutylene by hot air injection raises the plume edge to an additional above ground and reduces its 'striking distance' by ~370 m. Similarly an increase in the venting speed of chlorine by 20% increases the plume height by 0.95m and reduces the range of lethal toxic concentration build-up by ~1100 m.

LIST OF SYMBOLS

b_s= local characteristics width, m

c^*= concentration of gas on the plume axis, kg/m^3

s= distance along the plume axis from the release point to a certain point, m

T= temperature of the plume at any point inside the plume, K

u^*= plume velocity on the plume axis in the direction of the tangent to the plume axis, m/s

u_v= venting speed, m/s

ρ_a = density of air, kg/m³
ρ_{eff} = effective density of plume/jet, kg/m³
Q_f = dimensional number
Z = vertical distance above transition point, m
Z_{max} = maximum rise of plume, m
Z_{eff} = effective rise of plume, m
x_{max} = lateral distance where maximum rise of plume occur,m
x = lateral distance along down wind direction, m
x_o = lateral distance from virtual source to transition point,m
w = wind velocity, m/s
Q = volumetric flow rate of gas, m³/s
g = gravitational acceleration , m/s²
Dc = drag coefficient
Itr = transition height in still air, m
Itrw= transition height in windy conditions, m
hpl = total effective height reached by the plume, m
H = height of the vent, m
m_o = momentum flux at exit point
R= dry adiabatic laps rate, °C/m
y = vertical distance, m
ϕ = half angle of the plume from the stack

REFERENCES

1. Briggs, G. A., (1984). *Plume rise by buoyancy effects*, Atmospheric science and power production, p.327.

2. Contini, S., Amendola, A., and Ziomas, I, (1991). *Benchmark exercise on major hazard analysis, Commission of the European Communities*, Joint Research Centre, ISPRA, Itlay, p2.

3. Cude, A. L., (1974). Dispersion of gases related to atmosphere from relief valves, *Chemical Engineer*, 290, p 629.

4. Hanna, S. R., Briggs, G. A., and Hosker, R. P. (1982). Handbook on atmospheric diffusion, *Report DOE-TIC 11223 Atmos. turbulence and diffusion lab*, Dept. of Energy, Washington, D.C.

5. Khan, F. I., and Abbasi, S. A., (1999a). Modelling of dispersion of heavy gas based on modification in Plume Path Theory, *Journal of Loss Prevention in Process Industries* (in press).

6. Khan, F. I., and Abbasi, S. A., (1997). "Application of the recently modified PPT on the dispersion of some heavy gases", *10th regional IUAPPA conference*, Gumu, Ssuyu, Istanbul, Turkey, 23-26 September.

7. Lees, F. P., (1996). *Loss prevention in process industries*, Butterworths, London.

8. Ooms, G., A. (1972). new method for the calculations of the plume path of gases emitted by stack, *Atmos Environ*, 6 , p 899.

9. Ooms, G., and Mahiue, A. P., (1974). The plume path of vent gases, *Loss prevention and safety promotion in process industries*, Elsevier, New York.

10. Ooms, G., Mahiue, A. P., and Zelis, F. (1974). The plume path theory of vent gases heavier than air, *Proceeding of the first International loss prevention symposium*, Delft, The Netherlands, 2-3 May.

11. Ooms, G. an Duijm, N. J., (1983). Dispersion of stack plume heavier than air, *Atmospheric dispersion of heavy gases and small particluates symposium*, Delft, The Netherlands.

Chapter 6

HAZDIG : A NEW COMPUTER-AUTOMATED TOOL FOR ASSESSING ACCIDENTAL RELEASE AND DISPERSION CHARACTERISTICS OF HAZARDOUS GASES

In this chapter we describe HAZDIG (HAZardous DIspersion of Gases) which is a user–friendly computer-automated tool for generating scenarios for the emissions and gaseous dispersion of hazardous chemicals. It can simulate accidental as well as normal release but has been specifically developed as a tool for studying sudden release of hazardous gases and its consequences.

HAZDIG is made-up of five main modules–data, release, scenario generation, dispersion characteristics estimation, and graphics. The features of each of these are described in this chapter.

HAZDIG incorporates latest models for estimating atmospheric stability (Van Ulden and Hostlag, 1985; Hsu, 1992; Erbink, 1995) and dispersion (Pasquill and Smith, 1983; Van Ulden, 1988; Erbink, 1993; Khan and Abbasi, 1999). The 'data module' of HAZDIG has been so designed that the data needed to run the models is easy to obtain and feed - properties of chemicals, operating conditions, ambient temperature, and a few commonly available meteorological parameters are included in the module. A database containing various proportionality constants and complex empirical data has also been built into the system. The graphics module enhances the user friendliness of the software, and enables presentation of the results in an easy-to-understand and visually appealing manner. The output of the software is so formatted that it can be directly used for reporting the results without the need of editing.

INTRODUCTION

Dispersion modelling of likely accidental releases of hazardous chemicals is an integral part of the risk assessment and management process. It forms the critical link between the hypothesised equipment failures or release scenarios in terms of loss of lives and property if accident do takes place. The dispersion analysis provides the means by which hazardous chemical vapour concentrations can be estimated, both within and beyond plant boundaries, to provide a basis for thc quantification of the risk.

The physical processes involved in the emission and dispersion of many hazardous chemicals are very complex, and often not very well understood. Much of this complexity stems from the vast array of possible release and dispersion scenarios that can exist in a given industry. Even dispersion of pollutants emitted from a well-defined and fairly steady-state source, such as a power plant stack, is difficult to accurately model. In contrast hazardous chemical releases typically are not well defined and are transient in nature. They consequently pose much greater challenges to the modeller. The release may be instantaneous or continuous from a vessel or pipe involving pressurised gases, refrigerated or pressurised liquids, or liquids at ambient pressure and temperature, resulting in vapour emissions that may or may not be heavier than air. The vapour emissions may be relatively steady-state or may vary with time if released from a pressurised vessel or from an evaporating pool of liquid. The release may involve phase changes and thermodynamic interactions with the environment, with possible rain-out of liquid droplets from the plume. Plant structures and irregular terrain may significantly affect the fate of released chemicals, which may further complicate the assessment process (Pasman *et al.* 1992; Greenberg and Crammer, 1992).

Table 6.1 List of various models in-built in™ HAZDIG

Event	Model Incorporated	References
EMISSION		
Liquid release•		
Continuous		
normal	flow through orifice	Coullson and Richardson, 1977; Pilborough, 1989
refrigerated	flow and heat transfer model	Clancey, 1974; Drake and Reid, 1975; Drivas, 1982
superheated	thermodynamic and flow model	Kletz, 1984; Leung and Epstein, 1990
Instantaneous		
normal	rupture of vessel model	Hess *et al.*, 1974; Lees, 1996; Reed and Clark, 1983
refrigerated	heat transfer and evaporation model	Clancey, 1974; Foster, 1981; Woodward and Mudan, 1991
superheated	thermodynamic and evaporation model	Kletz, 1984; Picard and Bishnoi, 1988 Greenbook, 1992; Shaw and Brisco, 1976
Gaseous release		
Instantaneous	thermodynamic model	Cox and Comer, 1980; Drivas *et al.*, 1983; Nolan *et al.*,1990
explosively	boiling liquid expanding	Prugh, 1991; Kayes, 1986; Kletz, 1984 vapor explosion Vernart *et al.*, 1993
Continuous		
sub sonic	fluid dynamic model	Greenbook, 1992; Lees, 1996; Woodward, 1990
sonic	fluid dynamic model	Coullson and Richardson, 1977; Woodward and Mudan, 1991
Two phase release		
refrigerated	thermodynamic and heat-	Fauske, 1964;1988; Fauske *et al.*, 1984; Hague and Pepe,

		1990
superheated	transfer model	Kern, 1985, Perry *et al.*, 1984; Kayes, 1986
	thermodynamic and heat-transfer model	Greenbook, 1992; Sumipathala et al., 1990 Perry *et al.*, 1984; Leung, 1990
Atmospheric Stability	boundary layer and macro-meteorological parameters based models	Van Ulden and Holtslag, 1985; Hsu, 1992 Erbink, 1995
DISPERSION		
Light gas dispersion	Pasquill, Gifford and-Turner model	Pasquill and Smith, 1983; Gifford, 1961 Turner, 1970;1985
	Robert model	Robert *et al.*, 1992; Erbink, 1993
	Sutton model	Sutton, 1948,
Heavy gas dispersion	BOX, PLUME and SLAB model	Van Ulden, 1974;1988; Deaves, 1992 Khan and Abbasi, 1997a; Predikanis *et al.*, 1994 Earmak and Chan, 1986;1988
Jet dispersion	Jet release and dispersion model	Mahieu and Oomes, 1974; Brown *et al.*, 1993 Khan and Abbasi, 1997a; Lees, 1996
CHARACTERISTICS		
Continuous release	Plume model	Mahieu and Oomes, 1974; Lees, 1996 Khan and Abbasi, 1997a; Ooms and Tennekes, 1984
Instantaneous release	Puff model	Greenberg and Cramer, 1992; Deaves, 1992 Ooms and Tennekes, 1984; Van Ulden, 1992
DAMAGE ESTIMATION		
Continuous and instantaneous	Probit model	Essienberg *et al.*, 1975; Greenbook, 1992 Clancey, 1977; Pietersen, 1990; CCPS, 1989

In order to estimate the effects of such accidental releases, a number of components of consequence analysis need to be employed. These include calculations of:

i) release rate,
ii) liquid spreading and evaporation rate,
iii) pressurised liquefied gas release,
iv) gas dispersion,
v) toxic load effects.

The depth and breadth of the problems associated with the assessment of accidental release and dispersion of hazardous chemicals has necessitated the development of a number of techniques and methodologies. There are more than 100 mathematical models of varying degrees of sophistication that attempt to address most of the physical processes that can potentially be involved in postulated accident scenarios (Lees, 1996; Khan and Abbasi, 1995;1996;1997). Many of the software based on these models are operable on micro computers : WHAZAN, HASTE, CHARM, CARE, MIDAS. A few require mainframes DENZ, DEGADIS. Most of these software require a great deal of expertise in their application to a particular problem, as well as technical background in chemical processes, thermodynamics, and turbulent diffusion theory. This limits the software's user-friendliness. Moreover most of these software are unable to handle such accidental release situations as emissions of large quantities of chemicals with great force which may effect the existing atmospheric conditions.

In this context these authors have developed a software package capable of performing consequence analysis of normal as well as accidental release of chemicals. The package can handle myriad factors effecting the dispersion: density (lighter-than/lighter-as/heavier-than-air), convection, buoyancy and momentum. The main objectives behind the development of this software are:

a) *Wider applicability*- HAZDIG incorporates larger number of models than existing packages dealing with accident

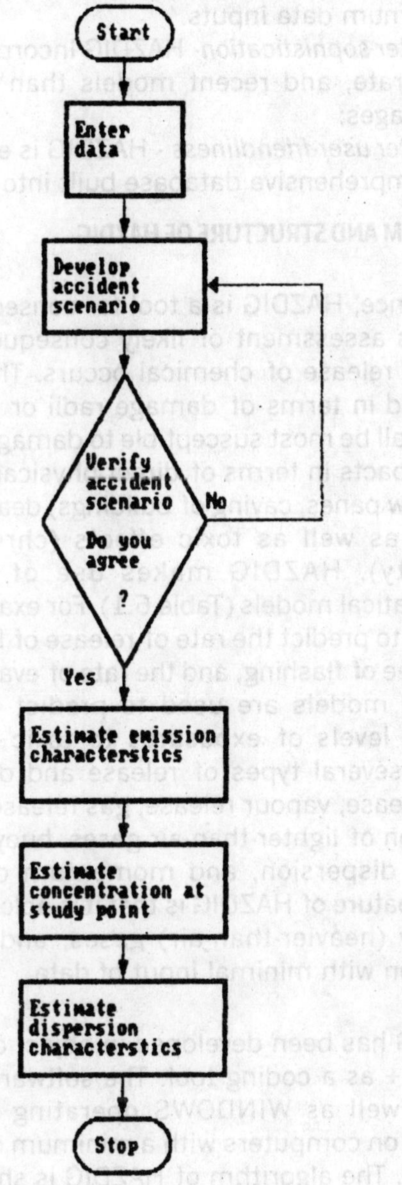

Figure 6.1: Algorithm of HAZDIG

simulation, to handle larger variety of situations with minimum data inputs.

b) *Greater sophistication*· HAZDIG incorporates more precise, accurate, and recent models than handled by existing packages;

c) *Greater user-friendliness* · HAZDIG is easier to run, and has a comprehensive database built into it.

ALGORITHM AND STRUCTURE OF HAZDIG

In essence, HAZDIG is a tool for consequence analysis, and performs assessment of likely consequences if an accident involving release of chemical occurs. The consequences are quantified in terms of damage radii or the radii of the area which shall be most susceptible to damage. HAZDIG computes likely impacts in terms of direct physical damage (shattering of window panes, caving of buildings) deaths due to blast wave injuries as well as toxic effects (chronic/acute toxicity, mortality). HAZDIG makes use of a wide variety of mathematical models (Table 6.1). For example, source models are used to predict the rate of release of hazardous materials, the degree of flashing, and the rate of evaporation. The impact intensity models are used to predict human response to different levels of exposures to toxic chemicals. HAZDIG handles several types of release and dispersion scenarios: liquid release, vapour release, gas release, two phase release, dispersion of lighter-than-air gases, buoyant dispersion, non-buoyant dispersion, and momentum driven dispersion. A special feature of HAZDIG is that it is able to handle dispersion of heavy (heavier-than-air) gases, and momentum driven dispersion with minimal input of data.

HAZDIG has been developed in object-oriented architecture using C++ as a coding tool. The software is compatible with DOS as well as WINDOWS operating environments. It is operable on computers with a minimum of 8 MB RAM and 80 MB ROM. The algorithm of HAZDIG is shown in Figure 6.1.

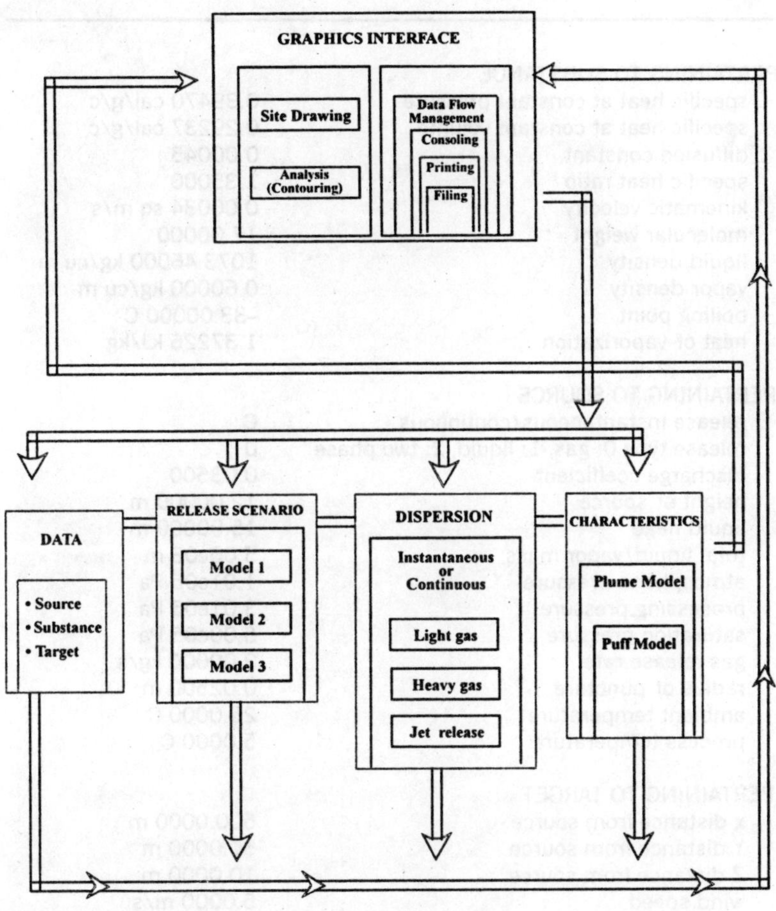

Fig. 6.2: Hierarchical structure and message flow sequence of HAZDIG.

Table 6.2. The set of input data required to run different options of HAZDIG

Parameters	Values, units
PERTAINING TO SUBSTANCE	
specific heat at constant pressure	0.39470 cal/g/c
specific heat at constant volume	0.29237 cal/g/c
diffusion constant	0.00045
specific heat ratio	1.35000
kinematic velocity	0.00034 sq m/s
molecular weight	17.00000
liquid density	1073.46000 kg/cu m
vapor density	0.60000 kg/cu m
boiling point	–33.00000 C
heat of vaporization	1.37226 kJ/kg
PERTAINING TO SOURCE	
release Instantaneous/continuous	C
release type 0: gas, 1: liquid, 2: two phase	0
discharge coefficient	0.03500
height of source	12.00000 m
liquid head	15.00000 m
total liquid/vapor mass	5.00e06 m
atmospheric pressure	1.01e05 Pa
processing pressure	3.01e05 Pa
saturation pressure	5.00e05 Pa
gas release rate	2.00000 kg/s
radius of puncture	0.02500 m
ambient temperature	25.0000 C
process temperature	5.0000 C
PERTAINING TO TARGET	
x distance from source	500.0000 m
Y distance from source	50.0000 m
Z distance from source	10.0000 m
wind speed	5.0000 m/s
duration of release	250.0000 s

DETAILS OF THE HAZDIG MODULES

As seen in Figure 6.2, HAZDIG consists of five main modules (Set of objects): data, release scenario generation, dispersion, characteristics estimation and graphics interface. Each module consists of two to three sub-modules (objects). These sub-modules depend on their parent modules as well as on other modules. A schematic diagram showing flow of information and dependency of different sub-modules to the main modules is presented in Figure 6.2.

DATA module

The main purpose of this module is to collect all relevant informations needed for the execution of other modules. The module consists of three main sub-modules derived from the main DATA module : release, dispersion and characteristics, and site specific. The dispersion and characteristic sub-module is further divided into sub-modules such as, lighter gas, heavy gas and momentum driven dispersion. The release sub-module branches into instantaneous, continuous, jet, and two-phase release.

Each sub-module needs a set of data necessary for the analysis of that particular option. This is illustrated in Table 6.2 which shows the data list related to a continuous release accident scenario. Most of the options require data pertaining to chemical properties, operating conditions, and release conditions besides synoptic meteorological data. To cater this need, the requisite data base has been built into the software.

Scenario generation module

In this module release scenarios are generated for the unit under study. It is a very important input for the subsequent steps, as the accuracy of the forecast of the type of accident is directly influenced by the realism of the release scenario. Consequently all the subsequent steps are also strongly influenced.

Each release scenario is basically a combination of different likely events that may occur in an industry. Such scenarios are generated based on the properties of chemicals handled by the industry, physical conditions under which reactions occur or reactants/products are stored, geometries and material strength of vessels and conduits, etc. External factors such as

Table 6.3. List of models available with different software to study the release and dispersion of ammonia through a large hole in the roof of a tank

Software	Models available in the package for the study of	Models usable for the study of catastrophic	Remarks
WHAZAN	1. Liquid out flow	3,4	
	2. Gas outflow		
	3. Two-phase outflow		
	4. Evaporation but not time independent		
SAFETI	1. Liquid out flow	3,4	
	2. Gas outflow		
	3. Two-phase outflow		
	4. Evaporation time dependent		
RISKIT	1. Liquid outflow	3,4	
	2. Gas outflow		
	3. Two-phase outflow		
	4. Heavy gas dispersion		
EFFECTS	1. Liquid outflow	4,5	
	2. Vapor outflow		
	3. Gas outflow		
	4. Two-phase outflow		
	5. Evaporation but not time dependent		

{Cont.{...

	1.	Liquid release		Two phase release of ammonia
	2.	Gaseous release		followed by the evaporation
MAXCRED	3.	Two-phase release	3,4,6	and heavy gas dispersion.
	4.	Evaporation but not time dependent		
	5.	Light gas dispersion		
	6.	Heavy gas dispersion		
	1.	Momentum release		
	2.	Liquid release		
	3.	Gaseous release		
	4.	Two-phase release		Two-phase release followed by
	5.	Flashing of vapor		flashing of vapor and subsequent
HAZDIG	6.	Evaporation but not time dependent	4,5,6,8,10	evaporation of remaining liquid ammonia followed by heavy gas dispersion and puff characteristics estimation.
	7.	Light gas dispersion		
	8.	Heavy gas dispersion		
	9.	Jet release and dispersion		
	10.	Plume characteristics		
	11.	Puff characteristics		

@ As listed in Contini *et al.* (1991).

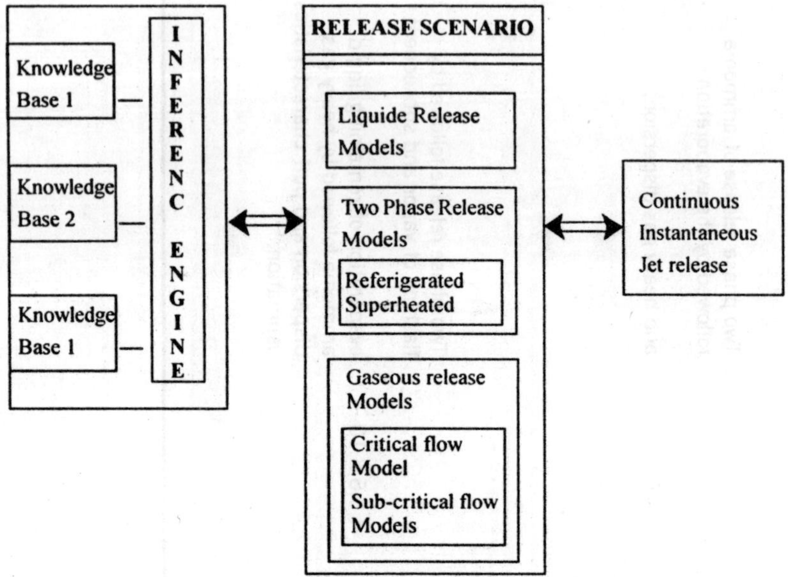

Fig. 6.3: Structure of scenario generation module

site characteristics (topography, presence of trees, ponds, rivers in the vicinity, proximity to other industries or neighbourhoods etc.) and meteorological conditions are also considered. In the available software packages such as WHAZAN (Technica,1992), EFFECTS (TNO, 1991), RISKIT (VTT, 1993), and SAVE (TNO, 1992), these aspects have been used to some extent. But the level of sophistication needs to be enhanced substantially by using advanced models of thermodynamics, heat transfer, and fluid dynamics to generate more realistic release scenarios. Furthermore, the user-friendliness of these packages have some limitations as a result of which several real-life studies conducted on the basis of these packages seem to have major lacunae. This is illustrated by the following example.

A multinational project involving 11 institutions was operated (Contini *et al.* 1991) to study the accidental release and dispersion of ammonia from a pressurised tank. The teams

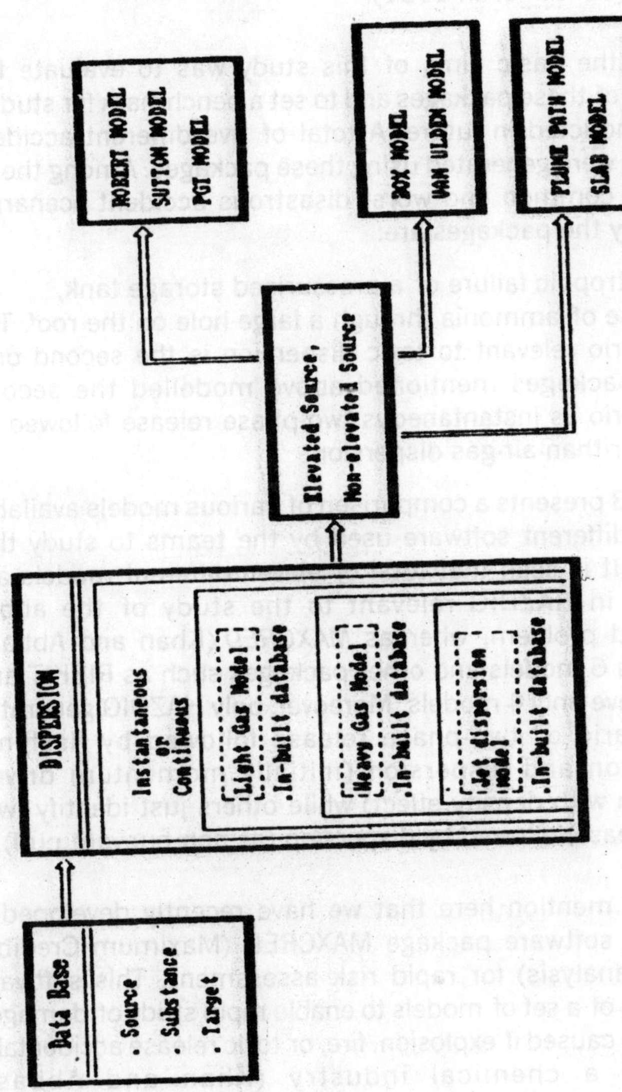

Fig. 6.4: Structure of dispersion module

utilised commercially available packages CRUNCH, DEGADIS, DENZ, HEAVY-PLUME, RISKIT, WHAZAN, SAFETI, EFFECTS, and DECARA (Contini *et al.* 1991).

One of the basic aims of this study was to evaluate the efficacies of these packages and to set a benchmark for studies to be conducted in future. A total of five different accident scenarios were generated using these packages. Among these, the most common and worst disastrous accident scenarios invoked by the packages are:

- catastrophic failure of a pressurised storage tank,
- release of ammonia through a large hole on the roof. The scenario relevant to toxic dispersion is the second one. The packages mentioned above modelled the second scenario as instantaneous two-phase release followed by denser-than-air-gas dispersion.

Table 6.3 presents a comparison of various models available with the different software used by the teams to study this scenario. It is seen that total of eleven different models are available in HAZDIG relevant to the study of the above mentioned problem, whereas *MAXCRED* (Khan and Abbasi, 1998) has 6 models and other packages such as RISKIT and SAFETI have only 4 models. Moreover, only HAZDIG generates the scenario of two-phase release followed by flashing, evaporation and dispersion (initially momentum driven dispersion with density effect) while others just identify two-phase release followed by dispersion (as non-buoyant puff).

We may mention here that we have recently developed a computer software package MAXCRED (Maximum Credible accident analysis) for rapid risk assessment. This software comprises of a set of models to enable rapid study of damages likely to be caused if explosion, fire, or toxic release accidentally occur in a chemical industry (Khan and Abbasi, 1995;1996;1997). MAXCRED has certain features, among others, for study of release pattern and dispersion of chemicals (liquid/gases). HAZDIG, on the other hand, *specialises* in the study of toxic release and dispersion occurring as a

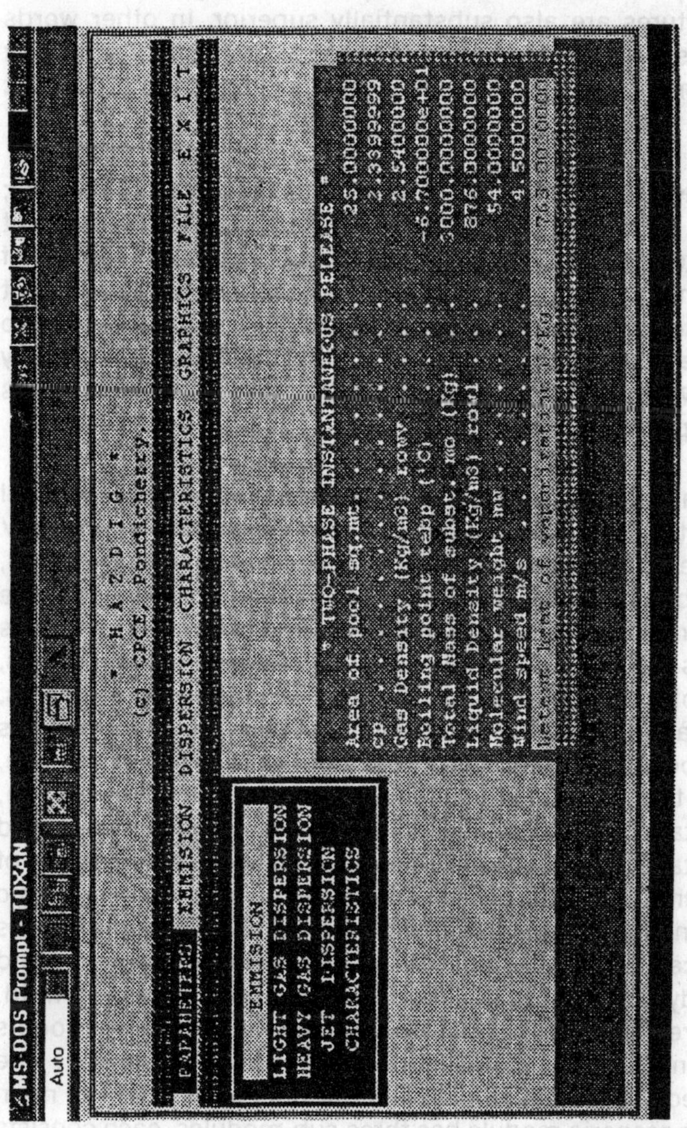

Fig. 6.5: Main menu and available options in HAZDIG

consequence of accidents in chemical process industries. It incorporates much larger number and variety of models for this purpose than MAXCRED does. Its knowledge-base and output features are also substantially superior. In other words whereas MAXCRED is a general-purpose package for rapid risk assessment, HAZDIG is a specialised package devoted to toxic release occurring in industrial accidents.

We have also come across several other reports (NEERI, 1992; CISRA, 1993; TPL, 1993) in which the existing risk assessment packages have been used in studying accidental release of toxic chemicals. In all these reports several credible accident scenarios have been left unconsidered in a manner similar to the one illustrated in the multi-institutional study quoted by us above, indicating the lack of rigor and/or user-friendliness of the packages used.

The accident scenarios are generated based on chemical properties, operating conditions, and details of the process/ storage units. Once an accident scenario has been developed, it can be processed for further verification and consequence assessment. For the same unit and same operating conditions various plausible accident scenarios can be visualised. Thus, this option helps in simulating the various likely accidents, and characterising the worst plausible ones. The option handles six type of release scenarios; continuous liquid release, instantaneous liquid release, continuous gas release, instantaneous gas release, continuous two phase release and instantaneous two phase release. A data-base consisting of constants for different thermodynamic and fluid dynamic relations has been built-in with the option. This option uses advance thermodynamic (adiabatic/isothermal expansion) and fluid dynamic (flow through orifice, uncompressed fluid, compressed fluid) models. The list of different models contained in the software is presented in Table 6.1. The architecture of this module is shown in Figure 6.3. The main release scenario module has three sub-modules, each consists of a set of functions to estimate different parameters. Brief descriptions of each of the sub-modules is presented below.

Liquid release sub-module

If a crack occurs in the liquid portion of a pressurised or refrigerated storage vessel, the discharge rate is dependent on the pressure inside the tank, the liquid head, and the size of the puncture. This sub-module works out release pattern of the chemical on the basis of storage/process conditions. It first identifies whether the release is continuous or instantaneous. It then characterises the release as normal liquid release, superheated liquid release, or refrigerated liquid release. This option calculates the release rate, quantity of the liquid flashed to vapour, rate of evaporation, and other relevant parameters.

Gaseous release sub-module

Purely gaseous releases from pressurised vessels or pipes generally occur in the form of a jet that can be characterised as critical or sub-critical flow. The occurrence of critical or "choked" flow, which has a maximum exit velocity as that of the speed of sound, is dependent on the ratio of storage pressure to atmosphere pressure.

This sub-module characterises the release of different types of gases. It identifies the release scenarios as instantaneous gaseous release (due to rupture of tanker unit, sudden opening of safety relief valve, etc.) with or without explosion, or continuous gas release. A continuous release is further characterised as sub-sonic or sonic release depending upon the property of the gas and the orifice of release.

Two-phase release sub-module

This type of release is characterised by sudden or continuous release of volatile chemical stored or processed above the normal conditions. The pattern of release depends on the properties of the concerned chemical and the storage or process conditions. This sub-module estimates the release rate and the percentage of liquid droplets going in the out flow.

Table 6.4 HAZDIG results for release and dispersion ammonia H A Z D I G (c) F I Khan & S A Abbasi, Pondicherry- 605 014, India Result of analysis of accidental release of ammonia

Parameters	Values, unit
TWO PHASE RELEASE	
BOILING LIQUID EXPANDING VAPOR EXPLOSION	
Peak over pressure	11.78460e+00 kPa
Duration of shock wave	23.54321e+00 sec
Shock wave velocity in air	53.67607e+00 m/s
Over pressure impulse	2.741312e-02 kPa/sec
PUFF RADIUS	
Total mass in the vessel	1.000000e+06 kg
Mass of vapor flashed	7.754567e+04 kg
Radius of puff	6.354876e+01 m
VAPOR GENERATION RATE	
Area of liquid pool	1.192651e+04 sq m
Evaporation ádue áto áforced ácirculation	1.145655e+03 kg/sec
Refrigerated liquid evaporation rate	7.687654e-08 kg/sec
Total evaporation (refrigerated+forced) rate	1.145655e+03 kg/sec
HEAVY GAS DISPERSION	
BOX INSTANTANEOUS MODEL	
ELEVATED SOURCE	

{cont.}.....

Front velocity of cloud	5.565452e+00 m/s
Cloud radius at time 3.6000e+03 (s)	7.756525e+02 m
Initial volume of cloud	1.000000e+06 cu m
Volume of cloud after 3.6000e+03 (s)	2.565645e+08 cu m
Initial height	5.000000e+03 m
Height of box after 3.6000e+03 (s)	2.145345e+01 m
Concentration at 5.000000e+02 m	2.644266e-06 kg/cu m
PUFF CHARACTERISTICS	
Ground level concentration of puff	2.665262e-06 kg/cu m
Ground level concentration of puff on axis	2.920399e-06 kg/cu m
Ground level concentration of puff at edge	2.920399e-07 kg/cu m
Width of puff	2.509362e+02 m
Maximum ground concentration	3.522818e-03 kg/cu m
Distance at which maximum concentration occur	5.000000e+00 m

The conditions necessary and sufficient to cause the above mentioned release scenarios (gas, liquid or two phase) have been incorporated as a knowledge-base which is a system of rules in terms of if-and-else structure. This is linked with the main release scenario module. When the user keys in information on operating conditions and properties of chemicals, the software works out appropriate release scenario which it then verifies and uses in dispersion and characteristics modules for further analysis.

The module pertaining to verification of release scenario

This feature is one of the specialities of HAZDIG. It verifies the plausibility of the proposed scenario. For example, if the scenario envisages two phase release of chemical on the basis of the characteristics of the chemical and conditions in which it is employed, this step checks whether the set of input conditions would indeed lead to the envisaged release. For example, a non-liquefied chemical may not exit in a two-phase mode, if the storage/process conditions are normal but for the same chemical there would be a high probability of two-phase release if the storage is under refrigerated or pressurised conditions. When chemical escapes from such a vessel, its mixing with the ambient air would be inefficient leading to build-up of the concentration of the chemical in the atmosphere, which subsequently may lead to severe consequences.

If HAZDIG finds that an envisaged scenario is not within the realms of probability, it modifies the scenario to the extent that it become plausible.

Dispersion module

This module consists of up-to-date mathematical models for simulating the accidents chosen as credible in the previous step. This module works out the dispersion characteristics profile, percentage of lethality, and damage radii of the release chemical. The output of this module quantifies impacts such as concentration at particular location, concentration profile, toxic dosage, damage radii of different impacts, and probabilities of causing lethality. The output of this object has been so formatted that it can be directly used in reports without having to do any editing. Moreover, using these results it is very easy to draw damage/risk contours.

The mathematical models used in this module are listed in Table 6.1. Brief explanatory notes on the phenomena simulated by these models are presented below:

This module assesses the consequences of release and dispersion of toxic gas/vapour. It simulates different types of release scenarios such as; continuous release, two-phase release, and instantaneous release. In conducting dispersion studies it takes into consideration the densities of the gases or gas-air mixtures (because of the pronounced influence density exerts on the shape of the plume) and momentum of release. The models can thus simulate dispersions of heavier-than-air, lighter-as-air and lighter-than-air gases/gas-air plumes. This module first estimates the concentration profile of the toxic gas that would develop consequent dispersion under the given meteorological conditions. It then works out the areas of toxic impact and the *extents* of toxicity that would be caused on the basis of exposure-based-toxicity data.

This module can handle the following options: heavy_gas dispersion, light–gas dispersion, momentum driven dispersion. The structure of this module is shown in Figure 6.4 and brief description of these options is presented below.

Option heavy–gas dispersion

This option estimates dispersions characteristics (concentration profile, distance travelled by the cloud/plume, and the dimensions of the cloud/plume) of the gases having effective density higher-than-air. It uses BOX model for instantaneous release (Van Ulden, 1974; Van Ulden and Hostage, 1985; Deaves, 1992; Erbink, 1993) and PLUME and

Table 6.5. HAZDIG results for release and dispersion of chlorine * H A Z D I G *
(c) F I Khan & S A Abbasi, Pondicherry- 605 014, India Result of analysis of accidental release of chlorine

Parameters	Values, unit
TWO PHASE RELEASE	
PUFF RADIUS	
Total mass in the vessel	5.000000e+05 kg
Mass of vapor flashed	2.347187e+04 kg
Radius of puff	1.308699e+01 m
LIQUID EMISSION RATE	
Area of puncture	1.963495e-03 sq m
Liquid emission rate through the puncture	4.335294e+01 kg/sec
Total time of spill	1.153324e+04 sec
Area of liquid pool after 6.000e+01 (s)	1.192651e+04 sq m
VAPOR GENERATION RATE	
Area of liquid pool	1.192651e+04 sq m
Evaporation due to forced circulation	8.915990e+02 kg/sec
Refrigerated liquid evaporation rate	2.637018e-10 kg/sec
Total evaporation (refrigerated+forced) rate	8.915990e+02 kg/sec

Heavy gass DISPERSION

PLUME CONTINUOUS MODEL

Front velocity of cloud	2.200992e+00 m/s
Cloud radius at time 3.6000e+03 (s)	2.642137e+02 m
Initial volume of cloud	2.000000e+05 cu m
Volume of cloud after 3.6000e+03 (s)	1.056855e+07 cu m
Initial height	2.546479e+03 cu m
Height of box after 3.6000e+03 (s)	4.818977e+01 m
Concentration at 5.000000e+02 (m)	5.405993e-06 kg/cu m

PLUME CHARACTERISTICS

Ground level concentration of plume	5.470422e-06 kg/cu m
Ground level concentration of plume on axis	5.994085e-06 kg/cu m
Ground level concentration of plume at edge	5.994085e-07 kg/cu m
Width of plume	2.509362e+02 m
Maximum ground level concentration	5.141720e-04 kg/cu m
Distance at which maximum concentration occur	3.500000e+01 m

SLAB model (heavy gas) for continuous release (Ermak and Chan, 1986;1988; Predikanis *et al.*1994; Langlo and Schatzmawn, 1991; Khan and Abbasi, 1996) to estimate gas concentrations and other dispersion characteristics. The results are then passed on to damage estimation options to calculate the percent likelihood of lethality and area under influence for various degrees of toxicity.

Option light gas dispersion

In this similar operations are carried out for gases having density lighter-than-air or/and lighter-as-air as in the previous option heavy_gas done for heavier-than-air gases. This option is in-built with various dispersion models: Gaussian model (instantaneous and continuous), Plume model (continuous), and Puff model (instantaneous) to estimate the dispersion characteristics for different release scenarios (Brington, 1985; Turner, 1970;1985; Robert *et al.* 1992; Erbink, 1993) . The results are then used to estimate the toxic load (concentration) at particular location, chances of lethality at that location and radii of the areas under the influence of various degrees of toxicity. Probit models proposed by Clancey (1977), Pietersen (1990), and Greenbook (1992) are used for estimating lethality.

Option momentum–driven dispersion

As long as the momentum of the escaping gas is significant the density factor does not become operative but as soon as the momentum dies down to a level where the ambient air movements could effect dispersion, the density factors take over to influence the shape of the plume.

When the gas escapes at high velocity as from a jet or a vent, the momentum effect is more prominent and lasts longer (due to higher velocity of release) than when the release velocity (venting velocity) is low.

According to Lees (1996) releases in the form of jets can be of four types : (i) turbulent momentum jet in still air, (ii) buoyant

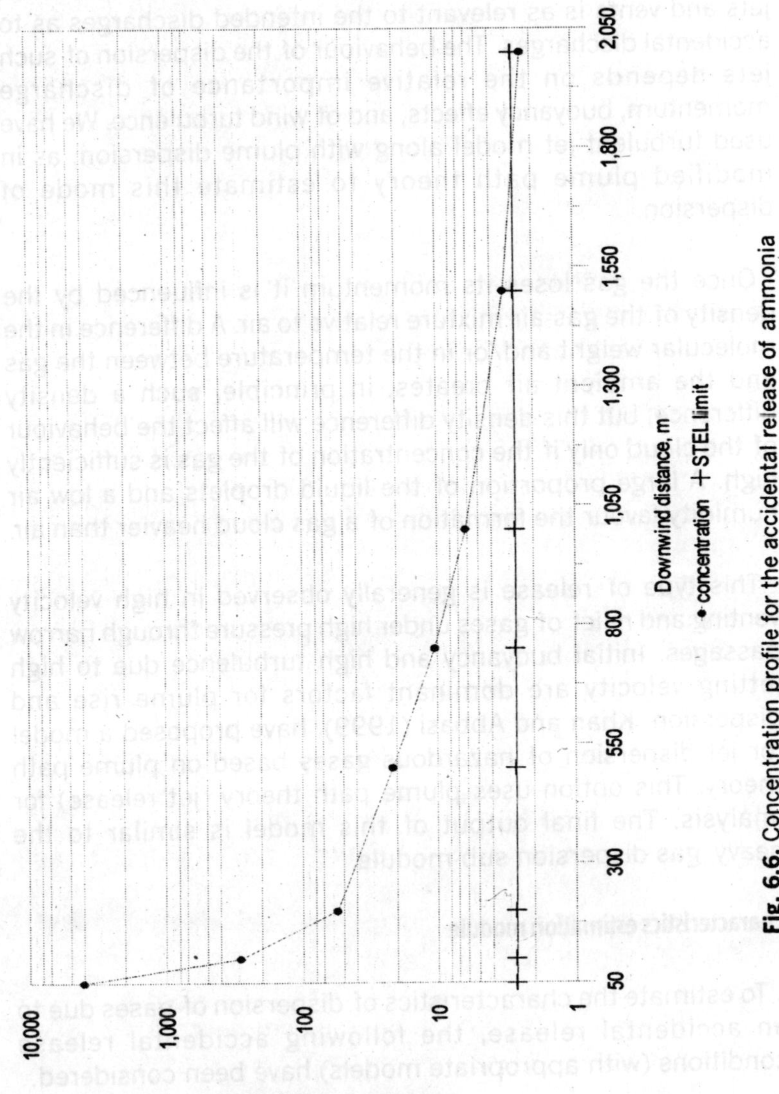

Fig. 6.6. Concentration profile for the accidental release of ammonia

plume in still air; (iii) plume dispersed by wind and (iv) jet-turbulent plume dispersed by wind. The behaviour of such jets and vents is as relevant to the intended discharges as to accidental discharges. The behaviour of the dispersion of such jets depends on the relative importance of discharge momentum, buoyancy effects, and of wind turbulence. We have used turbulent jet model along with plume dispersion, as in modified plume path theory to estimate this mode of dispersion.

Once the gas loses its momentum it is influenced by the density of the gas-air mixture relative to air. A difference in the molecular weight and/or in the temperature between the gas and the ambient air creates, in principle, such a density difference, but this density difference will affect the behaviour of the cloud only if the concentration of the gas is sufficiently high. A large proportion of the liquid droplets and a low air humidity favour the formation of a gas cloud heavier than air.

This type of release is generally observed in high velocity venting and relief of gases under high pressure through narrow passages. Initial buoyancy and high turbulence due to high jetting velocity are dominant factors for plume rise and dispersion. Khan and Abbasi (1999), have proposed a model for jet dispersion of hazardous gases based on plume path theory. This option uses plume path theory (jet release) for analysis. The final output of this model is similar to the heavy_gas dispersion sub-module.

Characteristics estimation module

To estimate the characteristics of dispersion of gases due to an accidental release, the following accidental release conditions (with appropriate models) have been considered.

a) Gaseous release.
b) Liquid release at atmospheric pressure. This condition can further be categorised as: (i) liquid with a boiling point above ambient temperature which is processed/stored at a temperature below its normal boiling point; (ii) liquid

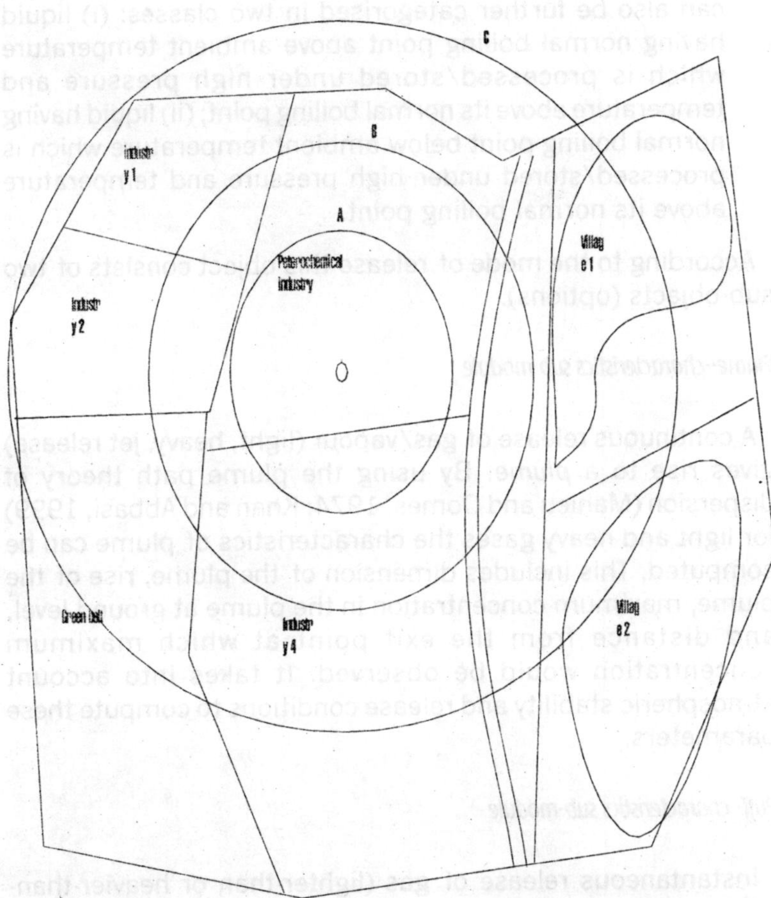

Fig. 6.7: Risk contours for accidental release of ammonia; (A) represents servere risk contour, (B) high risk contour and (C) Low risk contour

with a boiling point below ambient temperature which is processed/stored at low temperature and atmospheric pressure.

c) Two phase release (liquid under pressure). This condition can also be further categorised in two classes: (i) liquid having normal boiling point above ambient temperature which is processed/stored under high pressure and temperature above its normal boiling point; (ii) liquid having normal boiling point below ambient temperature which is processed/stored under high pressure and temperature above its normal boiling point.

According to the mode of release this object consists of two sub-objects (options).

Plume–characteristics sub-module

A continuous release of gas/vapour (light, heavy, jet release) gives rise to a *plume*. By using the plume path theory of dispersion (Mahieu and Oomes, 1974; Khan and Abbasi, 1999) for light and heavy gases the characteristics of plume can be computed. This includes dimension of the plume, rise of the plume, maximum concentration in the plume at ground level, and distance from the exit point at which maximum concentration would be observed. It takes into account atmospheric stability and release conditions to compute these parameters.

Puff–characteristics sub-module

Instantaneous release of gas (lighter-than or heavier-than-air) causes formation of vapour/gaseous cloud known as *puff*. The puff characteristics are based on the analytical modelling of a volume of chemical which disperses due to entrainment of air. The puff characteristics are estimated taking atmospheric stability parameters and release condition into account. The puff characteristics include dimension of the puff, height of the puff, volume of the puff, maximum concentration in the puff at ground level and the distance from

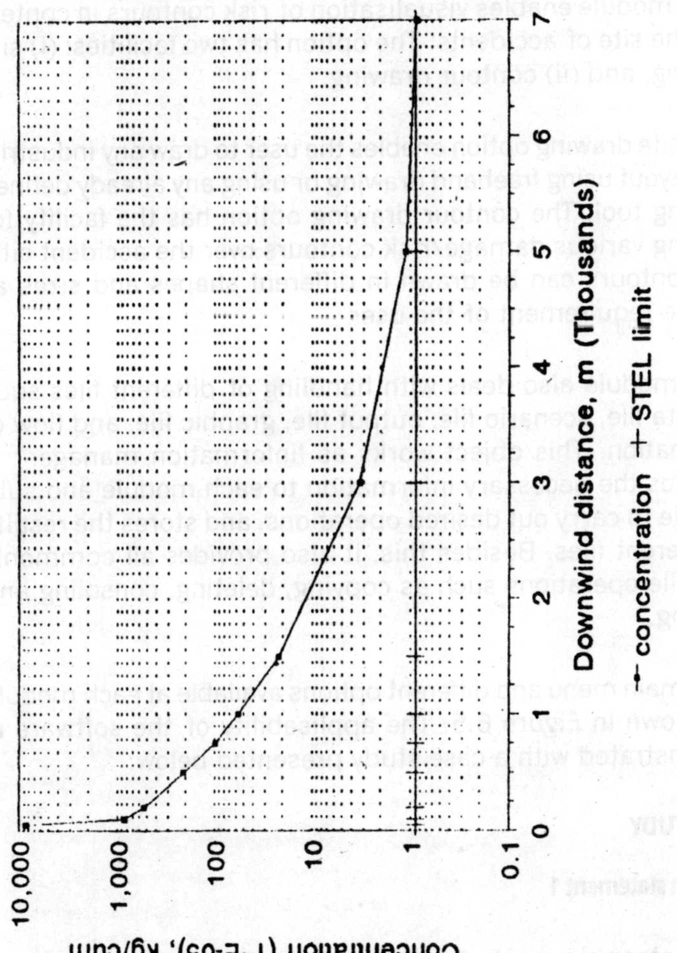

Fig. 6.8 Concentration profile for the accidental release of chlorine.

the exit point at which maximum concentration would occur (Ooms and Tennekes, 1984; Van Ulden, 1992).

Graphics-interface module

This module enables visualisation of risk contours in context with the site of accidents. The option has two facilities: (i) site drawing, and (ii) contour drawing.

The site drawing option enables the user to draw any industrial site/layout using freehand drawing or using any already defined drawing tool. The contour drawing option has the facility for drawing various damage/risk contours over the accident site. The contours can be drawn in different shapes and sizes as per the requirement of the user.

The module also deals with handling of different files such as: data file, scenario file, output file, graphic file, and flow of information. This object works as 'information manager': it provides the necessary information to each module and sub-module to carry out desired operations, and stores the results in different files. Besides this, it also provides all commonly used file operations such as copying, deleting, consoling and printing.

The main menu and different options available at each module are shown in Figure 6.5. The applicability of the software is demonstrated with a case study presented below:

CASE STUDY

Problem statement 1

Catastrophic rupture of a vessel storing ammonia at refrigerated condition. The incident occurs during early evening.

Analysis

The problem has been identified as instantaneous release (boiling liquid expanding vapour explosion-BLEVE) of ammonia under slightly stable conditions (using macro meteorological data). Eventhough ammonia vapour has density lighter-than-air, light gas dispersion model can not be used because in the present case study, ammonia is stored in refrigerated condition and an instantaneous release (as BLEVE) would lead to the formation of vapour cloud consisting of liquid ammonia droplets. Thus, the effective density of the cloud will be higher-than-air. HAZDIG also confirms this phenomenon and suggests two phase release with shock wave development and non-buoyant dispersion.

Table 6.4 shows the output of HAZDIG; it specifies accidental release scenario as-*instantaneous two phase release of gas* (flashing followed by evaporation) under sub-normal condition. It also presents the output of dispersion of heavy-gas module using BOX model. It suggests that at a distance of 500 meter from the accident site the concentration of gas would be 2.64E-06 kg/m³, which is a lethal dose considering that STEL (short term exposure limit) lethality concentration 0.027*E-09 kg/m³ (Fowcett and Wood, 1993). The concentration profile for ammonia with distance, shown in Figure 6.6, suggests that up to a distance of 1500 meters the concentration of ammonia would be above permissible limit vis a vis STEL (15 minutes exposure) for instantaneous exposure.

The puff characteristics of release and dispersion scenario can be interpreted from Table 6.4 which is an output of HAZDIG. It suggests that after 1 hour of release the puff would be around 250 meter in radius and about 10 meter high from the ground level. The maximum concentration of ammonia, 3.5*E-03 kg/m³, would occur at a distance of 5 meters from the release point. Figure 6.7 presents the damage contours of lethal dose over the area. It is clear that lethal concentration shall occur only within the industry's periphery (400 meter). However, high risk contour (representing 50% probability of lethality, estimated using probit function) would cover large area (1200 meter) which encompasses populated areas. Thus, the consequences due to release of toxic NH_3 have high risk (50% probability of lethality) to surrounding and for the populated areas near to the industrial site.

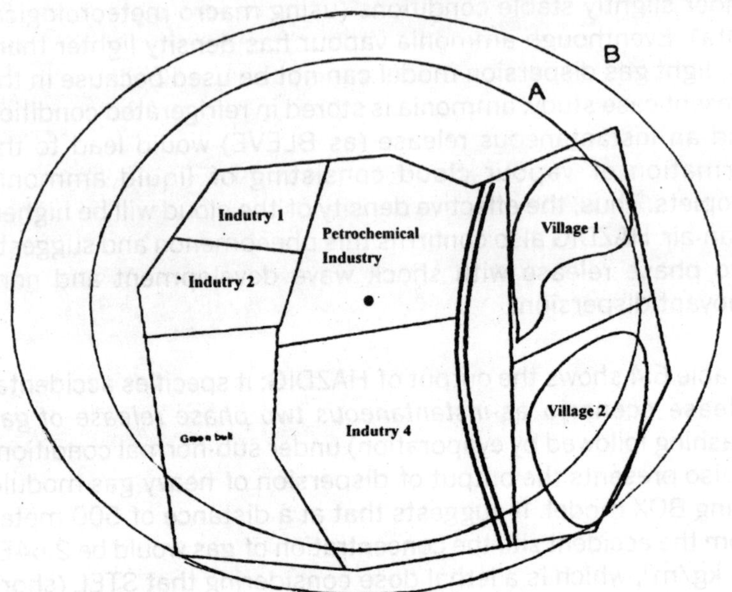

Fig. 6.9: Risk contours for the accidental release of chlorine; (A) severe risk contour, and (B) high risk contour.

Problem Statement 2

Chlorine, stored under high pressure, is accidentally released from a storage vessel due to failure of flange in the forenoon in industrial area on a clear sun-lit day.

Analysis

We have defined the problem as *continuous release of chlorine under unstable meteorological conditions*. As chlorine is a 'heavy' (denser-than-air) gas and is released at high pressure (above vapour pressure) heavy-gas continuous model is used by HAZDIG to forecast the dispersion profile.

Table 6.5 presents the output of HAZDIG for chlorine release. It provides the accident release scenario as continuous two phase release of chlorine (flashing followed by evaporation) under pressurised condition. Table 6.5 also presents the output of heavy gas module using plume path model. It suggests that at a distance of 500 meter from industrial site the concentration of gas would be 5.409E-06 kg/m³, which is above the lethal limit as per the STEL of 9.0E-09 kg/m³. The concentration profile of chlorine with distance is shown in Figure 6.8; it indicates that over a distance of 4500 meter the concentration of chlorine would be above permissible limit vis a vis STEL for instantaneous exposure (15 minutes).

The plume characteristics as per this release scenario can be interpreted from Table 6.5. The plume, after 1 hour of the accidental release would be ~250 meter in radius and shall be, ~7.5 meter above the ground level. The maximum concentration (5.147E-04 kg/m³) would occur at a distance of 35 meter from the release point. Figure 6.9 presents the damage contours over the area. It is evident from Figure 6.9 that severe risk contour (contour representing having more than 75% probability of lethality) would envelop large area (3000 meter radius) and densely populated areas are coming under the severe damage threat (severe risk).

CONCLUSION

HAZDIG is a computer software specifically developed to estimate the consequences (damage potentials and risks) due to release of toxic chemicals, accidentally or voluntarily. The modular structure of HAZDIG (developed in object oriented environment) enables swift processing of data and computation of result. It is also easy to maintain and up-grade. HAZDIG incorporates the latest available models for atmospheric stability estimation, release of chemicals (gases, liquid, two-phase, jet) and dispersion. It is capable of handling various types of release and dispersion scenarios: two phase release followed by dispersion, momentum release followed by dispersion, dispersion of heavier-than-air gases, etc. The Graphics option enables the user to draw any industrial site/ layout using freehand drawing or using any already defined drawing tool. The contour drawing option has the facility for drawing various damage/risk contours over the accident site. The contours can be drawn in different shapes and sizes as per the requirement of the user.

All-in-all HAZDIG is superior to other commercially available packages for studying accidental release of gases in following terms:

i) *Wider applicability* - HAZDIG incorporates larger number of models to handle, larger variety of situations with minimum data inputs;
ii) *Greater sophistication* - more precise, accurate, and recent models have been incorporated in HAZDIG than handled by existing packages;
iii) *Greater user-friendliness* - easy to run, has an extensive database built-in, generates output which can be directly used in reports, and enables visual comprehension of likely consequences.

Software availability: those interested in the software may contact Director, Centre for Pollution Control and Energy Technology, Pondicherry University, Pondicherry–605 014, India.

REFERENCES

1. Brington, P. W., (1985). Evaporation from a plane liquid surface into a turbulent boundary layer, *J of Fluid Mech.*, 159, 323.

2. Brown, M. J., Pal Arya, S., and Snyder, W. H., (1993). Dispersion from surface and elevated release an investigation of non-plume model, *J.Appl.Metr*, 32, 490.

3. CCPS (1989). Guidelines for chemical process quantitative risk analysis, *Center for Chemical Process Safety*, Publication number G–8, AIChE, Washington, D. C.

4. CISRA (1993). Risk assessment of Chloralkali industry, Cell for Industrial Safety and Risk Analysis, *CLRI*, Madras, India.

5. Clancey, V. J., (1974). The evaporation and dispersion of flammable liquid spillage's. *Chemical Process Hazards*, 5, 80.

6. Clancey, V. J., (1977). Dangerous clouds, their growth and explosive properties, *Chemical Process Hazards*, 6, 111.

7. Clancey, V. J., (1974). The evaporation and dispersion of flammable liquid spillage's. *Chemical Process Hazards*, 5, 80.

8. Contini, C. A., Amendal, A., and Ziomas, I., (1991). Benchmark study on major hazard study, *JRC-ISPRA*, 134, Amestradam.

9. Coullson, J. M., and Richardson, J. F., (1977). *Chemical Engineering* (SI ed), Oxford: Pergamon press, London.

10. Cox R. A., and Comer, P. J., (1980). Containment failure discharge and vapour cloud formation, *Safety promotion and loss prevention in the process industries*, Oyez, London.

11. Deaves, D. M., (1992). Dense gas dispersion modelling, *Journal of Loss Prev in Process Ind*, 5, 219.

12. Drake, E. M., and Reid, R. C., (1975). How LNG boils on soils. *Hydrocarbon Processing*, 54(5), 191.

13. Drivas, P. J., (1982). Calculation of evaporative emissions from multi-component liquid spills, *Environ. Sci. Tech.*, 16, 726.

14. Drivas, P. J., Sabnis, J. S., and Teuscher, L. H., (1983). Model simulates pipeline, storage tank failures. *Oil Gas J.*, September , 162.

15. Eissenberg, N. A., Lynch, C. J., and Breeding, R. J., (1975). Vulnerability Model: A Simulation System for Assessing Damage Resulting from Marine Spills, *Report CG-D-136-75 (Enviro control)*, Rockville M. D., USA.

16. Erbink, J. J., (1993). The advanced Gaussian model STACKS, Procd. of ERCOFTAC, *Work shop on Inter comparison of Advanced Practical Short range Atmospheric Dispersion Modelling*, Manno, Switzerland.

17. Erbink, J. J., (1995). Turbulent diffusion model from tall stacks, *Ph.D. thesis submitted to VRIJ University*, The Netherlands.

18. Ermak, D. L., and Chan, S. T., (1986). A study of heavy gas effects on the atmospheric dispersion of dense gases, *Air Pollution Modelling and its Application, (Ed: Wispleasere et al., 1986)*, Plenum Press.

19. Ermak, D. L., and Chan, S. T., (1988). Recent development on the FEM3 and SLAB atmospheric dispersion models, *Stable Stratified Flow and Dense Gas Dispersion (Ed: Puttok, J.S.)*, Clarendon Press.

20. Fauske, H. K., (1964). The discharge of saturated water through tubes. Seventh Nat. Heat transfer Conf., *Inst. Chem. Engrs*, New York, 210.

21. Fauske, H. K., (1988). Emergency relief system design for reactive and non-reactive systems: extension of the DIERS methodology. *Plant/ Operations Prog.*, 7, 153.

22. Fauske, H. K., Grolmes, M. A., and Leung, J. C., (1984). Multiphase flow considerations in sizing emergency relief systems for runaway chemical reactions. *In Multi-phase Flow and Heat Transfer III, Pt B: Applications, ed; T.N. Veziroglu and A.E. Bergles*, Elsevier, Amsterdam, 899.

23. Foster, T. C., (1981). Time required to empty a vessel. *Chem. Engr*, 88, 105.

24. Fowcett, H. H., and Wood, W. S., (1993). *Safety and accident prevention in chemical operation (third ed.)*, John willey, New York.

25. Gifford, F. A., (1961). Use of routine meteorological observation for estimating atmospheric diffusion, *Nuclear safety*, 20, 4.

26. Greenberg, H. R., and Crammer, J. J., (1992). *Risk Assessment and risk management in chemical process industries*, Van Norstand Reinhold, New York.

27. Greenbook, (1992). Methods for determining of possible damage to people and objects resulting from release of hazardous materials, Report CPR-16E, Voorburg, Warrington, UK.

28. Hague, W. J., and Pepe, W., (1990). Flow chamber simulations of aerosol formation and liquid jet break-up for pressurised releases of hydrogen fluoride. *Plant/Operations Prog.*, 9, 125.

29. Hess, K., Hoffmann, W., and Stoeckel, A., (1974). Propagation processes after the bursting of tanks filled with liquid propane - experiments and mathematical model. *Loss prevention and safety promotion*, 1, 227.

30. Hsu, S. A., (1992). An overwater stability criterion for the offshore and coastal dispersion model, *Boundary Layer Meteorology*, 60, 397.

31. Kayes, P. J., (1986). *Manual of industrial hazard assessment technique*, Technica Ltd., London.

32. Kern, D. Q., (1985). *Process heat transfer*, McGraw Hill publication, New York.

33. Khan, F. I., and Abbasi, S. A., (1995). Anatomy of industrial accidents, *Chemical Engineering World*, September, Bombay.

34. Khan F. I., and Abbasi, S. A., (1996). Accident simulation in CPI using MAXCRED, *Indian Journal of Chemical Technology*, 3, 338.

35. Khan, F. I., and Abbasi, S. A., (1997). Risk analysis of epichlorohydrin industry using computer automated tool MAXCRED, *J. of Loss Prev. Process Ind.*, 10, 11.

36. Khan, F. I., and Abbasi, S. A., (1998). MAXCRED-A software package for quantitative risk analysis, Environmental Modelling and Software (proof corrected, in press).

37. Khan, F. I., and Abbasi, S. A., (1999). Modelling and Simulation of Heavy gas dispersion on the basis of modifications in modified plume path theory, *J of Loss Prevention In Process Industries* (in press) .

38. Kletz, T. A., (1986). *What went wrong*, Gulf Publication, London.

39. Kletz, T. A., (1984). The prevention of major leaks - better inspection after construction? *Plant/Operations Prog.*, 3, 19.

40. Langlo Koing, G., and Schatzmawn, (1991). Wind tunnel modelling for heavy gas dispersion", *Atmospheric Environment*, 25 A(7), 1188.

41. Lees, F. P., (1996). *Loss prevention in the process industries*, 1, Butterworths publication, 15/300.

42. Leung, J. C., (1990). Similarity between flashing and non-flashing two-phase flows. *AIChE J.*, 36, 797.

43. Leung, J. C., and Epstein, M. (1990). The discharge of two-phase flashing flow from an inclined duct., *J.Heat Transfer*, 112, 524.

44. Mahieu, A. P., and Ooms, G., (1974). The plume path of vent gases, *Loss Prevention and Safety promotion in process industries*, Elsevier, New York, USA.

45. NEERI (1992). Risk assessment of Reliance petrochemical complex, *National Environmental Research Institute*, Nagpur, India.

46. Nolan, P. F., Petty, G. N., Hardy, N. R., and Bettis, R. J., (1990). Release conditions following loss of containment. *J. Loss Prev. in Process Ind.*, 3(1), 97.

47. Ooms, G., and Tennekes, H., (eds) (1984). *Atmospheric dispersion of heavy gases and small particles*, Springer, Berlin.

48. Pasquill, F., and Smith, F. B., (1983). *Atmospheric diffusion, (third edition)*, John whiley, New York.

49. Pasman, H. J., Duxbury, H. A. and Bjordal, E. N., (1992). Major hazards in the process industries : achievements and challenges in loss prevention, *Journal of Hazardous Materials*, 30, 1.

50. Perry, R. H., Green, D. W., and Maloney, J. O., (1984). *Perry's chemical engineering handbook*, McGraw Hill publication (6th edition), New York.

51. Pietersen, C. M., (1990). Consequence of accidental release of hazardous materials, *Journal of Loss Prev. in Process Ind.*, 3, 136.

52. Picard, D. J., and Bishnoi, (1988). The importance of real-fluid behaviour and non-isentropic effects in modelling decompression characteristics of pipeline fluids for application in ductile fracture propagation analysis, *Canadian J. of Chemical Engineering*, 66, 3.

53. Pilborough, L., (1989). *Inspection of industrial plant, 2nd ed.*, Hampshire: Gower Technical, Aldershot.

54. Predikanis, G. A., and Mayinego, Franz, (1994). Numerical simulation of heavy gas cloud dispersion within topographical complex terrain, *J. Loss Prevention. in Process Ind.*, 7, 391.

55. Prugh, R. W., (1991). Quantitative evaluation of BLEVE hazards, *Journal of Fire Protection Engineers*, 3(1), 9-24.

56. Reed, R. P., and Clark, A. F., (1983). *Materials at low temperatures*, Amer. Soc. of Materials, Ohio, 395.

57. Roberts, C. S., Timmis, R. J., Hackman, M. P., and Willianus, M. L., (1992). *Proc. of the 19th International technical meeting of NATO-CCMS on Air Pollution Modelling and its application (eds. H. Van Dop and G. Kalleos)*, Plenem Press, New York, USA.

58. Shaw, P. and Brisco, F., (1976). *Evaporation from spills of hazardous liquids, United Kingdom atomic energy report*, Report SRD-100, UK.

59. Sumathaipala, K., Venart, J. E. S., and Steward, F. R., (1990b). Two-phase swelling and entrainment during pressure relief valve discharges. *J. Hazardous Materials*, 25(1/2), 219.

60. Sutton, O. G., (1948). The theoretical distribution of air borne pollution from factory chimneys, *J.Royl. Met.Soc.*, 73,426.

61. Technica, (1992). *WHAZAN : A software for hazard assessment*, Technica Ltd., London.

62. TNO, (1991). *EFFECTS : A software for hazard assessment*, TNO Prins Mauritis Research Laboratory, The Netherlands.

63. TNO, (1992). *SAVE : A software for hazard assessment*, TNO Prins Mauritis Research Laboratory, The Netherlands.

64. TPL, (1993). Risk analysis of storage unit of chemical process industry, *Tamilnadu Petrochemical Ltd.*, Manali, Madras, India.

65. Turner, D. B., (1970). Workbook of atmospheric dispersion estimates reports AP-26, *U. S. Env. Prot. Agency report no -NC 277111*, Qeseach Triangle park.

66. Turner, D. B., (1985). Proposed pragmatic methods for estimating plume rise and plume penetration through atmospheric layers, *Atmospheric Environment*, 19, 1215-1218, .

67. Van Ulden, A. P., and Hostlag, A. A., (1985). Estimation of atmospheric boundary layer parameters for diffusion application, *Journal of Climate and Applied Meteorology*, 24, 1196 .

68. Van Ulden, A. P., (1988). The spreading and mixing of a dense cloud in still air. *In Puttock, J.S. 1988b*, 141.

69. Van Ulden, A. P., (1992). A surface-layer similarity model for the dispersion of a skewed passive puff near the ground, *Atmospheric Environment*, 26A, 681.

70. Van Ulden, A. P., (1974). The spreading of a heavy gas released near the ground, *Loss Prevention*, 1, 221. .

71. Vernart, J. E. S., Rutledge, G. A., Sumathipala,K. and Sollows, K., (1993). To BLEVE or not to BLEVE ; Anatomy of Boiling Liquid expanding vapour Explosion, *Process Safety Progress*, 2, 112.

72. VTT, (1993). RISKIT : a software for risk assessment, *The Technical Research Centre of Finland*, Tempere, The Finland.

73. Woodward, J. L., (1990). An integrated model for discharge rate, pool spread, and dispersion from punctured vessels. *J.Loss Prev. in Process Ind.*, 3, 33.

74. Woodward, J. L., and Mudan, K. S., (1991). Liquid and gas discharge rates through holes in process vessels, *J. Loss Prev. in Process Ind.*,4, 161.

Chapter 7

ATTENUATION OF GASEOUS POLLUTANTS BY GREENBELTS

The paper draws attention towards the importance of greenbelts in attenuating gaseous pollutants and presents the methodological, physico-chemical, biological, and horticultural dimensions associated with effective greenbelt design.

We have presented the gist of a system of methodologies developed by us for greenbelt design. A case study demonstrating the applicability of the system has been presented.

INTRODUCTION

In the context of environmental pollution abatement, a greenbelt may be defined as *a strip of trees of such species, and such a geometry, that when planted around a source, would significantly attenuate the air pollution by intercepting and assimilating the pollutants in a sustainable manner* (Ruth and Williams, 1990).

The concept of greenbelt as a source of pollution abatement was recognised initially by three nations: The USA, Britain and Kenya (Ruth and William, 1994; Gareth *et.al.* 1992; Andy, 1991 and Parsons, 1990). Ebenezer Howard, a British social reformer, advanced the concept of greenbelt concept in 1898, in connection with the planning of "new towns" located outside the periphery of London, which was then sprawling far into the countryside. Howard proposed "garden cities" which would that only be free of pollution but would also be antidote of polluted cities, each of which would be surrounded by an agriculture "country belt". It was British architect and planner Raymond Unwin, a town designer and contemporary of Howard's, who actually coined the term *greenbelt* (Ruth and William, 1994; Gareth *et.al.* 1992).

In Britain, Howard's concept took two forms: the greenbelts surrounding the new towns in rural Britain (the first of these was Letchworth, built in 1903); and beginning in the 1930s, the application of the idea to London itself. The *London Green Belt Act* was passed by the British Parliament in 1938. A more elaborate plan was created in 1944 by Patrick Abercrombie, who proposed a belt, five or more miles wide, consisting of both public open spaces and private holdings, that would be regulated so as to preclude runaway suburban development.

The London Green Belt Act was passed by the British Parliament in 1938. A more elaborate plan was created in 1944 by Patrick Abercrombie, who proposed a belt, five or more miles wide, consisting of both public open spaces and private holdings that would be regulated to preclude suburban development.

In the United States, the administration of the then President, Franklin Roosevelt, tried to adapt Howard's new town concept as part of its resettlement program. Three such towns were built with Greenbelt, Maryland, being the best known.

The American concept of greenbelt signified a relatively wide band of rural land or open space surrounding a town or city. There the term has come to mean, generally, any swath of open space separating or interrupting urban development.

The greenbelt town program was one of Frankiln Roosevelt's most innovative and radical intervention in American city building. Its central goal was not to create better urban communities, but rather to generate jobs in a declining national economy. Fortunately, the Resettlement Administration, headed by Rexford G. Tugwell, called on architect/planner Clarence S. Stein to prepare town design guidelines and to serve as planning consultant. Stein settled into an advisory role through which he greatly influenced the character and quality of these communities, especially the best known: Greenbelt Maryland. After World War II, Stein was instrumental in preserving the towns as examples of socially and environmentally responsible community designs and prototypes for a national new town policy. (Parsons, 1990).

In Kenya greenbelt movement has been started through an agro-forestry project founded and run by women association. The movement, started and organised by Professor Wangari Maathi, aims to plant trees for a variety of reasons : to stabilise soil, for use as fuel, for landscape improvement and as a source of income for the women taking part in the scheme. Tree nurseries have been established which issue seedlings that are used to produce greenbelts in both rural and urban areas of the country. These seedlings are provided to women's groups which then oversee the planting and raising of the trees. For every tree that survives more than three months outside the nursery, the woman who plants and cares for it is paid 25 cents (US). Paying female tree-tenders a premium for every seedling that survives not only provides an income for women who would otherwise have none, but has also allowed the Green Belt project to attain a transplant survival rate of around 80 percent. By the end of the 1980s the movement had seen the creation of well over 600 tree nurseries and grown 10 million trees, thanks to the involvement of 50,000 women and children in 3,000 schools (Andy, 1991).

AIR POLLUTION

The history of air pollution is as old as man's invention of fire. But till a few decades ago the carrying capacity of the global atmosphere was sufficient to assimilate the air pollution that was being generated by anthropogenic activities. This situation has deteriorated very fast during the recent years.

Increasing demographic pressure allied with increased developmental activities is generating more copious emission than the air can assimilate (Abbasi, 1998).

Table 7.1. Extent of air pollution in some metropolitan cities 1992* (Abbasi, 1998)

City	SO_2	NO_2	CO	O_3	Pb	SPM
Bangkok	C	C	C	C	B	A
Beijing	A	C	D	B	C	A
Bombay	C	C	C	D	C	A
Buenos Aires	D	D	A	D	C	B
Cairo	D	D	B	D	A	A
Calcutta	C	C	D	D	C	A
Delhi	C	C	C	D	C	A
Jakarta	C	C	B	B	B	A
Karachi	C	D	D	D	A	A
London	C	C	B	C	C	C
Los Angeles	C	B	B	A	C	B
Manila	C	D	D	D	B	A
Mexico City	A	B	A	A	B	A
Moscow	D	B	B	D	C	B
New York	C	C	B	B	C	C
Rio de Janeiro	B	D	C	D	C	B
Sao Paulo	C	B	B	A	C	B
Seol	A	C	C	C	C	A
Shanghai	B	D	D	D	D	A
Tokyo	C	C	C	A	D	C

* *Collected from reports of WHO and UNEP*
A- *Serious problem; WHO guidelines exceed by more than a factor of two*
B- *Moderate to heavy pollution, WHO guidelines exceed by up to a factor of two*
C- *Low pollution, WHO guidelines normally met with*
D- *No data available or insufficient data for assessment*

The situation in the urban areas is particularly grim and is worsening day by day. It has been that levels of one of the common air pollutants–particulate matter–in seven out of eight major cities in India are higher than the permissible limits (for residential areas) of 200 mmg/m³ set by Central Pollution Control Board of India. Recent surveys indicate that other air pollutants–SO_x and NO_x–also exceed permissible limits very often. In a report R.J. Sinha (Sinha, 1993) states that 94% of the traffic constables in Jaipur suffer from one or other disease

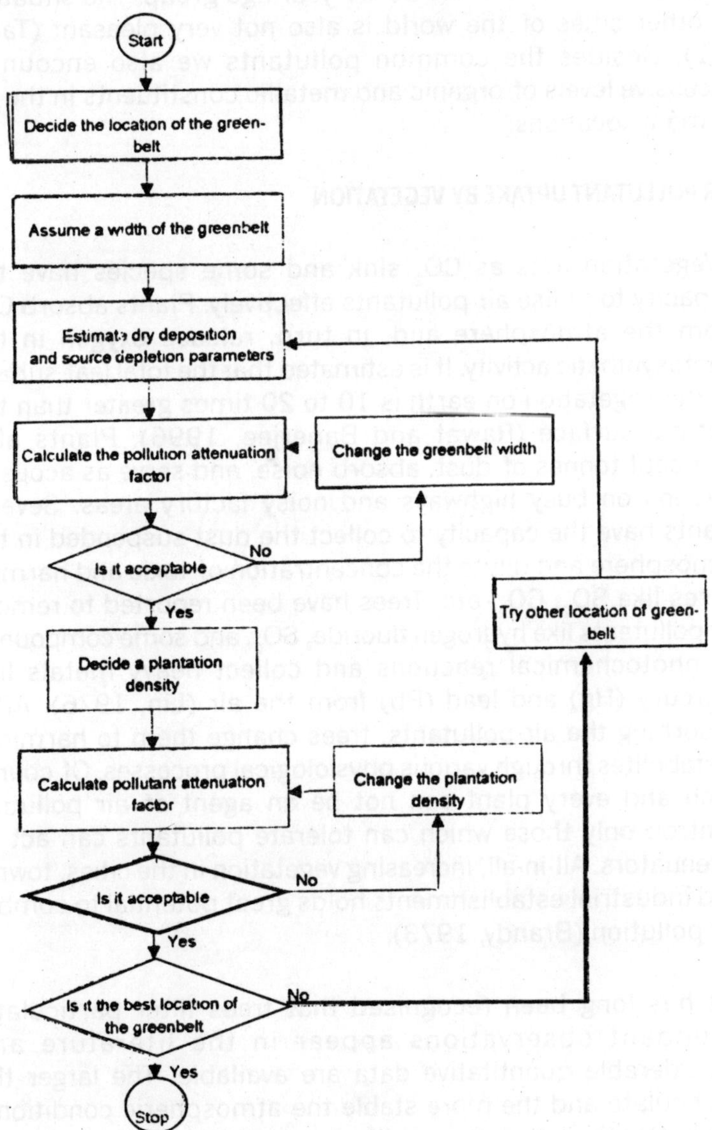

Fig. 7.1: Algorithm for the design of greenbelt

due to air pollution, especially from vehicular emissions. An astonishingly high incidence (55%) of tuberculosis was found among constables in the 20-30 year age group. The situation in other cities of the world is also not very pleasant (Table 7.1). Besides the common pollutants we also encounter excessive levels of organic and metallic constituents in the air in many locations.

AIR POLLUTANT UPTAKE BY VEGETATION

Vegetation acts as CO_2 sink and some species have the capacity to utilise air pollutants effectively. Plants absorb CO_2 from the atmosphere and, in turn, release oxygen in the photosynthetic activity. It is estimated that the total leaf surface of the vegetation on earth is 10 to 20 times greater than the earth's surface (Rawat and Banerjee, 1996). Plants also intercept tonnes of dust, absorb noise, and serve as acoustic screens on busy highways and noisy factory areas. Several plants have the capacity to collect the dust suspended in the atmosphere and dilute the concentration of toxic and harmful gases like SO_2, CO_2, etc. Trees have been reported to remove air pollutants like hydrogen fluoride, SO_2, and some compounds of photochemical reactions and collect heavy metals like mercury (Hg) and lead (Pb) from the air (Lin, 1976). After absorbing the air pollutants, trees change them to harmless metabolites through various physiological processes. Of course each and every plant can not be an agent of air pollution control; only those which can tolerate pollutants can act as attenuators. All-in-all, increasing vegetation in the cities, towns, and industrial establishments holds great potential to combat air pollution (Brandy, 1973).

It has long been recognised that trees filter particulate. Abundant observations appear in the literature and considerable quantitative data are available. The larger the particulate and the more stable the atmospheric conditions, the better the trees screen will operate. With small particulate and increased turbulence, the effectiveness of the screen will decrease.

Table 7.2. Particulate removal rates and land area required for pollutant removal in the vicinity of a 135 megawatt coal burning power station releasing 186 tons of particulate yr^{-1} (Jashnani, 1988).

	deciduous	*conifer*
Particulate removal rate	0.38 tons yr^{-1}	0.14 tons yr^{-1}
Land area required	486 ha	1336 ha

Hill (1971) reported experiential investigations on uptake of ordinary pollutants: SO_2 and NO_2 by alfalfa canopies. Based on pollutant uptake experiments on alfalfa canopies it has been estimated that a continuous cover of assimilating alfalfa canopy under conditions equivalent to those of the studies could remove more than 1/4 ton of NO_2 or SO_2 per square mile per day from air containing an average NO_2 concentration of 6 ppm and SO_2 are of 1-3 ppm (Thakre, 1992).

Present data indicate that vegetation could be an important sink for at least the following air pollutants of major importance: HF, SO_2, NO_2, O_3, Cl_2 and to a lesser extent PAN (Thakre, 1992). Undoubtedly, many others such as HCl could be added to the list. Two important pollutants which are known to not be taken up effectively by plants, are CO and NO. However, oil micro-organisms appear to be a major sink for CO. Nitric oxide, though absorbed slowly by plants, is converted in the atmosphere to other forms which may then be taken up more rapidly. Pollutant uptake by plants is controlled by the interaction of a number of physical, chemical and biological factors that regulate gas exchange processes and pollutant reaction at the sorbing sites. Green growing vegetation in removing unwanted gases and aerosols from the air not only cleanses the general atmosphere of these pollutants but influences most markedly their immediate microenvironment.

Plants remove pollutants from the air in three ways, viz. absorption by the leaves, deposition of particulate and aerosols on leaf surface, and fallout of particulate on the leeward (down wind) side of the vegetation because of the slowing of the air movement (Tewari, 1994). Each one is taken up in detail in subsequent section.

Table 7.3. Solubility in water and uptake rate of pollutants by alfalfa

Pollutant	Uptake rate by alfalfa at 1 ppm (liters min^{-1} m^{-2})	Equivalent deposition velocity (cm sec^{-1})	Solubility at 20°C (cm^3 gas cm^{-3} H_2O)
CO	0.0	0.00	0.02
NO	0.6	0.10	0.05
CO_2	2.0	0.33	0.88
PAN	3.8	0.63	
O_3	10.0	1.67	0.26
NO_2	11.4	1.90	Decomposes
Cl_2	12.4	2.07	2.30
SO_2	17.0	2.83	39.40
HF	22.6	3.77	446

Removal of Particulate Pollutants

Suspended particles in atmosphere are deposited on plant surfaces by three processes: sedimentation under the influence of gravity, impaction under the influence of eddy currents, and deposition under the influence of precipitation. Sedimentation usually results in the deposition of particles on the upper surfaces of plant parts and is most important with large particles. Sedimentation velocity varies with particle density, shape, and other factors. Impaction occurs when air flows fast on obstacle and the air stream divides, but particles in the air tend to continue in a straight path due to their momentum and to strike the obstacle. The efficiency of collection via impaction is the principal means of deposition if (a) particle size is of the order of tens of micrometers or greater, (b) obstacle size is of the order of centimetres or less, (c) approach velocity is of the order of meters per second or more, and (d) the collecting surface is wet, sticky, hairy or otherwise retentive. Ingold (1971) presented data indicating that leaf petioles are considerably more efficient particulate impactors than either twigs (stems) or leaf lamina. For particles of dimension 1·5 μm, impaction is not efficient and interception by fine hairs on vegetation is possibly the most efficient retentive mechanism (Tewari, 1994).

The transfer of particles from the atmosphere to natural surfaces is commonly expressed via deposition velocity. For small particles, for example, condensation aerosols less than 1 µm, deposition velocities are much less than for large particles: for example, spores and pollen 20-40 µm in diameter.

Trace metals, especially heavy metals, are most commonly associated with fine particles in contaminated atmospheres. Trace element investigations conducted in roadside, industrial, and urban environments have dramatically demonstrated the impressive burdens of particulate heavy metals that can accumulate on vegetative surfaces.

Based on a literature survey, particulate removal efficiencies for trees were estimated. Assumptions included: particulate average deposition velocity of 1 cm sec^{-1} for trees and 0.8 cm sec^{-1} for grass and weeds, leaf area index of 5.1 for deciduous trees and 2.3 for conifers, and approximately 2 ha of deciduous tree surface and 1 ha of coniferous tree surface ha^{-1} of land area. Given these assumption particulate removal rates and land area required for removal were estimated (Table 7.2).

Removal of gaseous pollutant

Substantial evidence is available to support the potential that plants is general and trees in particular have function as sinks for gaseous pollutants. The latter are transferred from the atmosphere to vegetation by the combined forces of diffusion and flowing air movement. Once in contact with plants gases may be bound or dissolved on exterior surfaces or be taken up by the plants via stomata. If the surface of the plant is wet and if the gas is water soluble, the former process can be very important. When the plant is dry or in the case of gases with relatively low water solubility, the mechanism is assumed to be the most important (Bennett et al, 1973; Kabel *et. al.*, 1976; William, 1990).

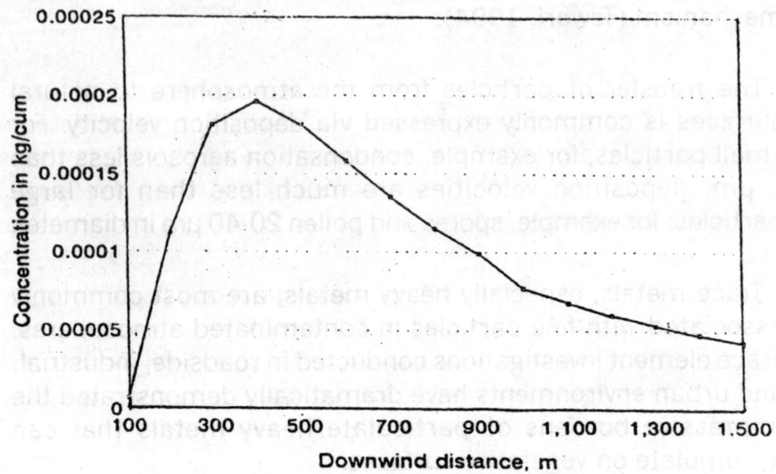

Fig. 7.2: Plot showing variation of concentration along the downwind distance for slightly unstable atmospheric condition during summer season

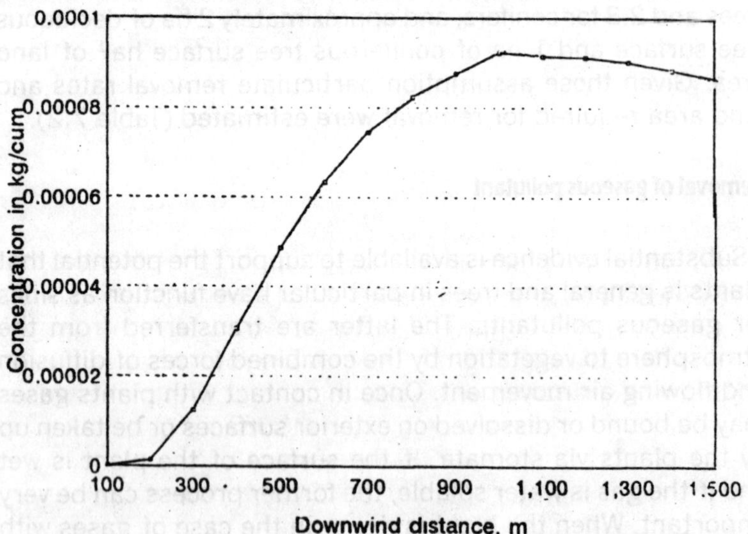

Fig. 7.3: Plot showing variation of concentration along the downwine distance for slightly unstable atmospheric condition during winter season

Plant uptake rates increases as the solubility of the pollutant in water increases. Hydrogen fluoride, sulphur dioxide, nitrogen dioxide, and ozone which are soluble and reactive are readily sorbed pollutants. Nitric oxide and carbon monoxide, which are very insoluble, are absorbed relatively slowly or not at all by vegetation. During daylight periods when plant leaves are releasing water vapour and taking up carbon dioxide, other gases, including trace pollutant gases, in the vicinity of the leaf will also be taken up through the stomata (William, 1990).

The rate of pollutant gas transfer from the atmosphere to interior leaf cells is regulated by a series of resistance conveniently thought of as atmospheric, stomatal, and mesophyllic. Factors controlling atmospheric resistance include wind speed, leaf size and geometry, gas viscosity and diffusivity. Stomatal resistance is regulated by stomatal aperture, which is influenced by water deficit, carbon dioxide concentration, and light intensity. Mesophyllic resistance is regulated by gas solubility in water, gas liquid diffusion, and leaf metabolism (Kabel *et al.*, 1976). Because the rate of pollutant uptake is regulated by numerous forces and conditions, the rate of removal under field conditions is highly variable. If leaf characteristics, wind speed, atmospheric moisture, temperature, and light intensity are quantified, the pollutant uptake rate can be estimated.

The above pollutants uptake is described by general plant uptake process. The fundamental investigations of A. Clyde Hill and Jesse H. Bennett of the University of Utah have made several general conclusions, concerning gaseous pollutant uptake (Hill, 1971; Hill and Chamberlain, 1974; Bennet and Hill, 1975). Their studies have concentrated on alfalfa, oats, barley, and grass. Standard alfalfa canopies removed gaseous pollutants from the atmosphere in rates of the following order: hydrogen fluoride > sulphur dioxide > chlorine > nitrogen dioxide > ozone > peroxyacetylnitrate > nitric oxide > carbon monoxide. In general plant uptake rates increased as the solubility of the pollutant in water increased. Hydrogen fluoride, sulphur dioxide, nitrogen dioxide, and ozone, which are soluble and reactive, were readily absorbed. Nitric oxide and carbon monoxide, which are very insoluble, were absorbed relatively slowly or not at all (Table 7.3). The rate of pollutant removal found to increase linearly as the concentration of the pollutant increased over the ranges of concentration that are encountered in ambient air and that were low enough not to cause stomatal closure.

Table 7.4. Parameters considered for computation of optimum width of Greenbelts

Symbols	*Parameters*
Separation distance between source and Greenbelt (m)	x_1
Width of Greenbelt (m)	X_2
Height of Greenbelt (m)	h
Pollution attenuation coefficient (Af)	
Dry deposition velocity (ms^{-1}) in absence of Greenbelt	Vdp
Deposition velocity onto plant canopy	Vdg

Under growth chamber conditions, wind velocity, canopy height, and light intensity were shown to affect the rate of pollutant removal by vegetation. As previously stressed, light plays a critical role in determining physiological activities of the leaf and stomatal opening, and as such exerts a great influence on foliar removal of pollutants. Under conditions of adequate soil moisture, however, pollutant uptake by vegetation was judged almost constant throughout the day, as the stomata were fully open. Pollutants were absorbed most efficiently by plant foliage near the canopy surface where light-mediated metabolic and pollutant diffusivity rates were greatest. Sulphur and nitrogen dioxides were taken up by respiring leaves in the dark, but uptake rates were greatly reduced relative to rates in the light. The Removal of sulphur dioxide is also described by tree uptake process. Because of its high solubility in water, large amounts of sulphur dioxide are absorbed to external tree surfaces when they are wet. In the dry condition, sulphur dioxide is readily absorbed by tree leaves and is rapidly oxidised to sulphate in mesophyll cells. At low uptake rates sulphur dioxide is presumed to be oxidised about as rapidly as it is absorbed (Bennett and Hill, 1973; William, 1990).

Roberts (1974) measured sulphur dioxide sorption by single leaves or shoots of several 1 year-old seedlings of numerous woody species. All species examined were capable of reducing high ambient levels within test chambers. Relatively little work has been reported on trees as sinks for hydrogen fluoride is very water soluble and very reactive, and can be adsorbed onto plant surfaces and absorbed through stomata. Leaves exposed to hydrogen fluoride may accumulate fluoride to one million times the ambient concentration. Nitrogen dioxide dissolved in water yields nitrate and nitrate ions in solution. The latter can be reduced to ammonia in leaf cells (Bennett and Hill, 1973).

A FEW IMPORTANT FACTS RELATED TO POLLUTION ATTENUATION THROUGH VEGETATION

1. The interception and retention of atmospheric particles by plants is highly variable and is primarily dependent on:
   - Size, shape, wetness, and surface texture of the particles.
   - Size, shape, wetness, and surface texture of the intercepting plant part.
   - Micro-and ultramicro-climatic conditions surrounding the plant.
2. More is known concerning the physical-mechanical aspects of particle deposition under controlled conditions than is known about the relative capture and retention efficiencies of various plants and different species under natural conditions.
3. Generally, greater leaf surface roughness increases particle capture efficiency for particles approximately 5 μm (and less) in diameter. Smooth leaved species (for example, horse chestnut and yellow poplar) are less efficient than rough leaved species (for example, elm and hazel).
4. Surface roughness acts to decrease the stability of the boundary layer (region of retarded air flow) surrounding the leaf and thus acts to increase particle impaction. Leaf hairs and leaf veins are principal contributors to surface roughness.

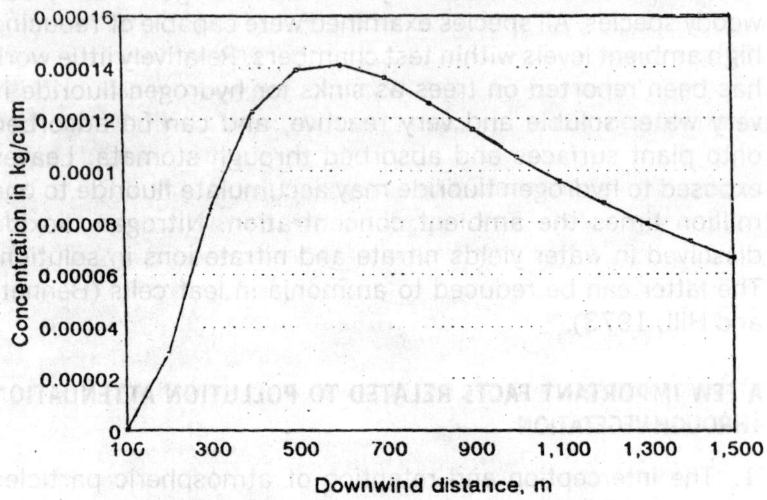

Fig. 7.4: Plot showing variation of concentration along the downwide distance for slightly unstable atmospheric condition during rainy season

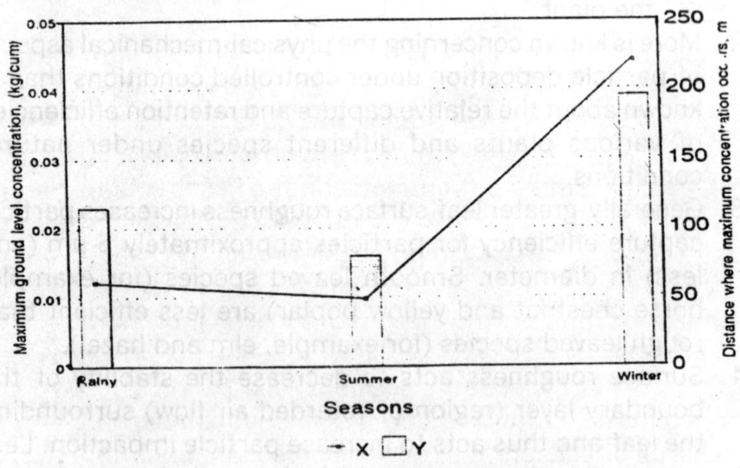

Fig. 7.5: Variai on of maximum ground level concentration (X) and distance where this concentration (Y) occurs with seasons

5. Smaller leaves are generally more efficient particle collectors than larger leaves.
6. Particle deposition (but probably not retention) is heaviest at the leaf tip and along leaf margins where a turbulent boundary layer is present. Leaves with complex shapes and large circumference/area ratios collect particles most efficiently.
7. Increased wind speeds and increased particle size typically increase particulate deposition velocities.
8. Deposition velocities to petioles and stems are generally many times greater than deposition velocities to leaf laminas. Collection of atmospheric particles by leafless deciduous species in the winter may remain quite high duc to twig and shoot impaction.
9. Conifers are generally effective particulate sinks than deciduous species.
10. Mechanisms by which particles are re-suspended or otherwise removed from tree surfaces must be investigated more thoroughly.
11. Leaf with abundant trichomes (leaf hairs) are found to be particulate accumulator.

GREENBELT (GB) AND ITS DESIGN

A large number of gaseous and particulate air pollutants are emitted in the air environment. The physical and chemical properties and effects of these pollutants vary a great deal individually and synergistically. The nature and quantum of pollutant depends on the type of industry and the kind of raw material and energy used in its operation.

Eventhough urban forestry and greenbelt have a common objective of reducing atmospheric air pollutant load, their definitions and mode of achieving this objectives are widely different. For example, urban forestry is centred around mass human population e.g. residential areas, whereas greenbelt is centred around the source of pollution such as industries, power plants, transport system, etc. However the purpose of vegetation in both the cases are same (Rawat and Banerjee, 1996).

Table 7.5. Trees suitable for planting in the Greenbelt along with salient features

Name of the plant species	Height Architecture	Canopy Collecting efficiency	Dust Tolerance Index (APTI)	Air Pollution
Albizzia lebbek	Tall	Round	Moderate	***
Azadirachta indica	Tall	Semi-erect	Fair	**
Pithecolobium dulce	Tall	Round	Moderate	***
Ficus glomerata	Tall	Round	Moderate	***
Ficus infectoria	Tall	Round	Moderate	***
Polyalthia longitolia	Tall	Erect	Moderate	**
Tectona grandis	Tall	Erect	Moderate	**
Terminalia arjuna	Tall	Erect	Moderate	**
Bauhinia purpurea	Medium	Semi-erect	Good	**
Butea Monosperma	Medium	Semi-erect	Good	*
Cassia fistula	Medium	Round	Fair	***
Lagerstroemia flosreginae	Medium	Semi-erect	Moderate	**
Saraca indica	Medium	Round	Fair	*
Thespesia populnea	Medium	Round	Moderate	**
Acacia arabica	Dwarf	Round	Good	**
Diospyros embryopteris	Dwarf	Round	Moderate	***
Thevetia nerifolia	Dwarf	Round	Fair	*
Parkinsonia aculeta	Dwarf	Semi-erect	Good	*

*** High ** Medium * Low

The development of GB, by using pollution tolerant plants can significantly contribute towards air quality improvement. This involves selecting suitable plant species, determining climatic parameters, studying wind and temperature profiles, nature of pollutants to be ameliorated, and general landscape of the locality. The design of the GB and its composition may vary from place to place and industry to industry. A general social forestry or plantation type approach will not be of much help in industrial plantations (Rao, 1980,83).

The planning of GB, shelter-belt, or pollution-sinks also involves facets of bioaesthetics. Accordingly the selection of plant species may involve numerous plant characteristics, like tolerance, canopy structure, foliage form, height of plant and its overall flowering and production potential. This involves careful scrutiny of plants in nature as well as in horticultural conditions, in order to assess their suitability and performance in a stressed ecological situation of polluted environment.

The physical state of pollutants may be particulate or gaseous. The particulate ones may be either settable or suspended (SPM). In either case, they may eventually fallout on surfaces of materials, plants and animals. The gaseous pollutants also get absorbed on surfaces. The effect of the pollutant on the impinging surface, is a function of the degree of toxicity of the pollutant.

The pollutants thus falling out may remain suspended for some time in the air-shed but these eventually get deposited either as wet deposition or dry deposition on surfaces of vegetation, soil, water, buildings and other properties. These may also be deposited on outer surfaces of animal bodies or inhaled into their lungs.

The effect of these pollutants, either adsorbed on the surface or absorbed inside the system of plants will depend on the surface characteristics of the impinging surface and chemistry of the pollutant. In case of plants all those external and internal factors which affect the stomatal aperture will also affect the level of pollution interacting on plants. Several methods have been developed to evaluate the suitability of plants for using them for purposes mentioned above. Biomonitoring of air pollutants through the use of plants, microbes and animals has now become a standard procedure in the study of air pollution ecology (Sivasamy and Srinivasan, 1996).

Objectives of GB design

GB development envisages a multiplicity of objectives encompassing the micro-level air pollution abatement to enhancement of socio-economic values of the region.

- The prime objectives of GB is attenuation of air and noise pollution. It comes to the immediate rescue during accidental release/explosion minimising the risk to a considerable level.

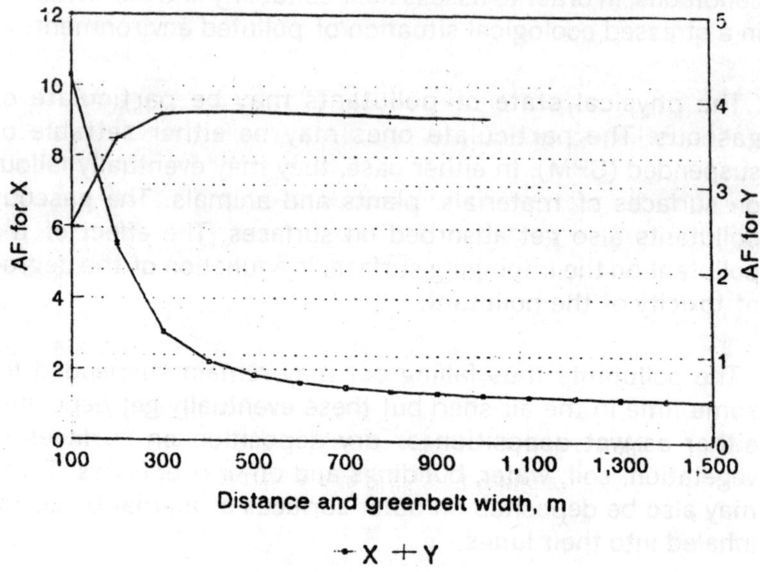

Fig. 7.6: The Variation of AF as a function of distance between green belt and pollutant source (X), and greenbelt width (Y) for rainy season.

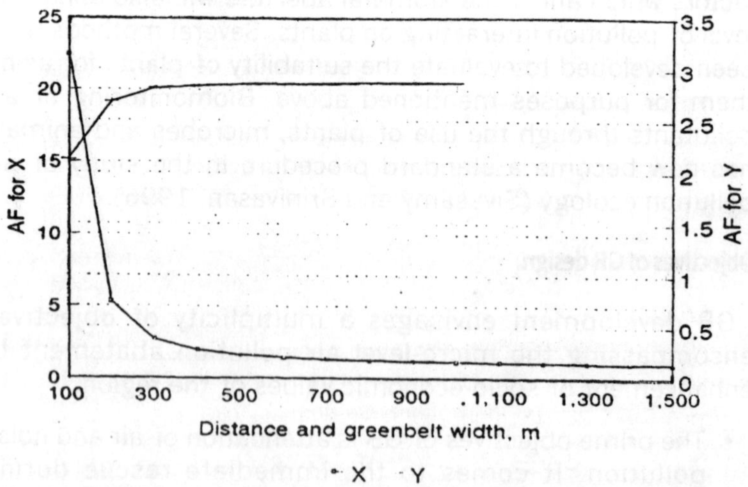

Fig. 7.7: The Variation of AF as a function of distance between green belt and pollutant source (X), and greenbelt width (Y) for Summer season.

- GB can serve as a measure for soil protection for erosion losses, enhance the aesthetic value and beautify the landscapes.
- Development of GB can help generate employment avenues and thus, involve the mass participation in environmental protection activity. This will increase the socio-economic standards of habitants and also help maintain the harmonious structuring of the human living areas and industries and other anthropogenic activities.

GB development plan mainly depends upon :

i) Nature and extent of pollution load
ii) Assimilative capacity of the ecosystem
iii) Climatic factors
iv) Soil and water quality.

Government of India has made it mandatory to have GB areas around the new as well as existing industries. However, no specific norms regarding the width of GB and pollution potential activity have been promulgated so far. Thus, width of the GB varies from industry to industry. In Germany and Netherlands, there are fixed criteria for width of GB to be developed around the identified activity zone as and depending on the source. Thus, in Germany the width of GB varies from 100 meters around centres of urban areas that is commercial centres to 2000 meters around heavy industries, specially situated in isolation because of their heavy pollution potential. While in Netherlands, it varies from more than 500 m for heavy industry to 50 meters for light and non-polluting industries. In India, many agencies, government and NGOs are recommending the GB development around the industrial complexes.

Designing of green belt is a very specialised task. This needs careful consideration of the local agro-climatic conditions, source and type of pollutants and finally selection of the right types of tree species. Planning is to be done in such a way that greenbelt is developed within a short period and remains effective over the years. The complete process of optimal design of greenbelt is presented in Figure 7.1.

Table 7.6. Values of Greenbelt design parameters for Kirumambakkam industrial estate

Parameters	Values
Distance from pollutant source to greenbelt	100 m
Greenbelt width	200 m
Value of K(se/st)	0.9
Pitch	Triangular
Inter tree spacing for tall tree	25-30 ft.
Inter tree spacing for middle height tree	30-40 ft.
Inter tree spacing for shrub	·25 ft.
Pollution attenuation factor	3.0
Tree species	as given in Table 7

The effectiveness of a GB for interception and retention of atmospheric particles depends on several factors viz. shape, size, wetness, surface texture and nature (Solubility and insolubility) of the particles/pollutants as well as intercepting plant parts. Damp surface of the plants enhances pollutant removal rate by 10% because under such conditions stem, branches, twigs and leaves are engaged in absorption process. Light has also got a pronounced effect in foliar removal of pollutants by influencing physiological activities and stomatal opening (Smith, 1981). Under urban environment, moisture restricts absorption of gaseous pollutants by limiting stomatal opening (Ahmad *et. al.*, 1989,1991).

Important parameters to be considered during design of GB

i) Height and canopy area of trees
ii) Mean wind velocity and direction
iii) Distance from source/Location of sources of pollutants
iv) Pollutant concentration
v) Nature of pollutants
vi) Dry deposition velocity of plants (specific to pollutants and plants)
vii) Topography and size of the land available.

Meteorological Considerations

Meteorological conditions are very important in regulating the transport, dispersion and fate of air pollutants in the atmosphere. Stable atmospheric layers with frequent inversions are the unfavourable meteorological conditions, as they help in accumulation of pollutants in a localised area. On the other hand, atmosphere instability and turbulence promote greater dispersal of pollutants, thus helping in reducing the adverse impacts of toxic emissions.

The pollutants which are added into the air environment are greatly affected both in quantity and quality by the meteorological parameters prevailing in the area. The important meteorological parameters are solar radiation, rainfall, temperature, humidity, wind direction and speed and conditions of temperature inversion. Besides the above, topography of the area plays an important role in dispersion, diffusion, dilution and general transport and fallout of the pollutants. A given concentration of a pollutant may attain different ambient concentrations after a lapse of time under different meteorological conditions. Therefore, emission rate being the same, ambient concentration of pollutants may differ from area to area because of varying climatic conditions (Rao, 1980).

Land availability becomes a major constraint in GB development around a source of pollution. The sources can be point, line or area and in every case the criteria for GB development will vary depending on the source strength. Quantitative assessment for the land requirement for GB development has to be on systematic and scientific basis so that the authorities responsible can be convinced and at the same time optimum returns are obtained through the developmental activities.

GREENBELT (GB) DESIGN BASED ON POLLUTION ATTENUATION COEFFICIENT

We have developed an elaborate system of methodologies for greenbelt design (Khan and Abbasi, 1999). This includes a) assessment of the pattern of dispersion of air pollutants exiting from a stack (or a number of stacks) to determine the location of greenbelt for optimal pollutant attenuation,

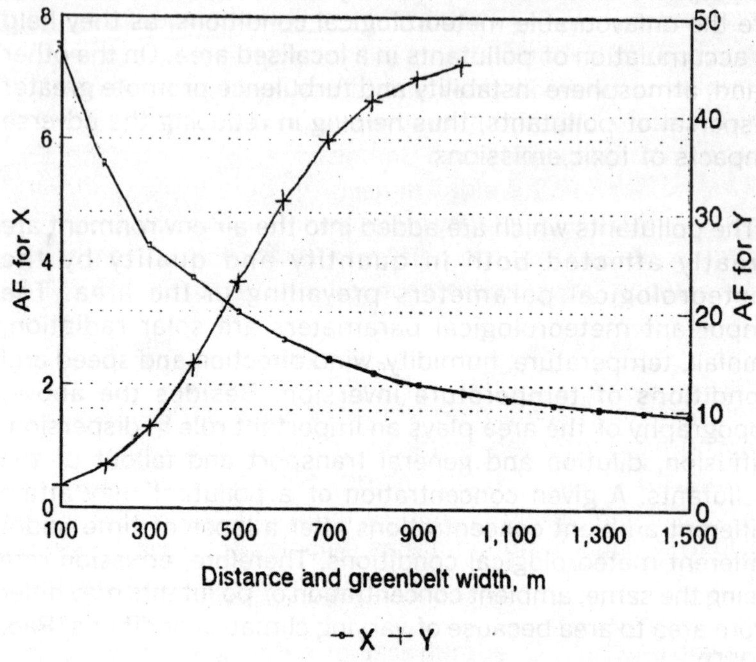

Fig. 7.8: The Variation of AF as a function of distance between green belt and pollutant source (X), and greenbelt width (Y) for Winter season.

b) assessment of the appropriate width of greenbelt, density of plantations in it, tree height, and the overall shape of the greenbelt, and c) selection of tree species suitable for different agro-climatic conditions and for different pollutants.

For this system, we have built upon the pioneering work of Sharma and coworkers (1992, 1994) who had proposed a model for greenbelt design. We have also drawn upon the experimental work done over several years by Ahmad and co-workers (1989,1991), Singh and Rao (1982,83), and William and co-workers (1990,1994). The gist of our approach is presented below:

When a parcel of pollutant travels through a GB it obeys the exponential law for dry removals.

$$Q_x = Q_c * \exp^{(-l'x)}$$

where,

Q_x = Pollution travelling through greenbelt at distance, x
Q_c = Mass flux entering the greenbelt
= Pollution attenuation coefficient
$= K V_d / U_c$
V_d = Dry deposition velocity of pollutant for vegetative canoipy,ms^{-1}
P_c = Foliage surface area of a single tree
P_t = Foliage surface area density of tree

Attenuation factor Af is a measure of effectiveness of GB and can be defined as the ratio of mass flux travelling the same distance in absence of GB (Q_{WB}) and through GB (Q_B)

$$A_f = Q_{WB}/Q_B$$

$$A_f = \frac{F_D (X_1 + X_2)}{F_D X_1 [\mathrm{erf}\{h_e/ÖÖ26z(X_1)\} e^{-l'} X_2] + \mathrm{erfc}\{h_e/ÖÖ26z(X_1)\} F'_D . X_2}$$

where,

X_2 = Width of GB (m)
h_e = Effective height of GB (m)
A_f = Pollution attenuation coefficient (m^{-1})
X_1 = Distance between pollution source and GB (m)

$F_D (X_1 + X_2)$ and $F'_D (X_1)$ are the plume depletion factors due to dry deposition of pollutant on natural surface.

$F'_D X_2$ = Plume depletion factor for the distance above GB

Parameters considered for computation of optimum width of GB along with their symbols in equation are given in Table 7.4.

SELECTION OF TREES FOR GB

As elaborated in preceding chapters, the effectiveness of the GB depends on the selection of the right type of the tree species tolerant to the particular pollutants of that area. An ideal tree for planting in the GB should have following characters (Sharma *et. al.,* 1991,1994) :

- Fast growth rate for quick development of canopy,
- Strong branches for durable canopy to withstand storm,
- Large leaf size for more retention of pollutants,
- Dense foliage for better trapping of pollutants,
- Long life span for extended life of the GB.

It is necessary to know the pollution tolerance level of the trees before selecting them for planting in green belt. Singh and Rao (1983) have worked out a formula of Air Pollution Tolerance Index (APTI) on the basis of leaf parameters to evaluate the tolerance level of the trees. It is suggested that trees having high APTI value are to be planted in the green belt for minimising gaseous pollutants (a few is presented in Table 7.5). On the other hand, for minimising dust pollution trees having high dust trapping ability are to be selected (a few been named in Table 7.5).

Further, depending upon the topo-climatological conditions and regional ecological status selection of appropriate plant species for this purpose should be based upon the following criteria : the plants should be—

i. fast growing,
ii. with thick canopy cover,
iii. preferably perennial and evergreen,
iv. with large leaf area index,
v. indigenous,
vi. resistant to specific air pollutants,
vii. able to maintain the ecological and hydrological balance of the region.

Depending upon the topo-climatological conditions, type of available land and cure the scheme for tree plantation should to be vigorously implemented.

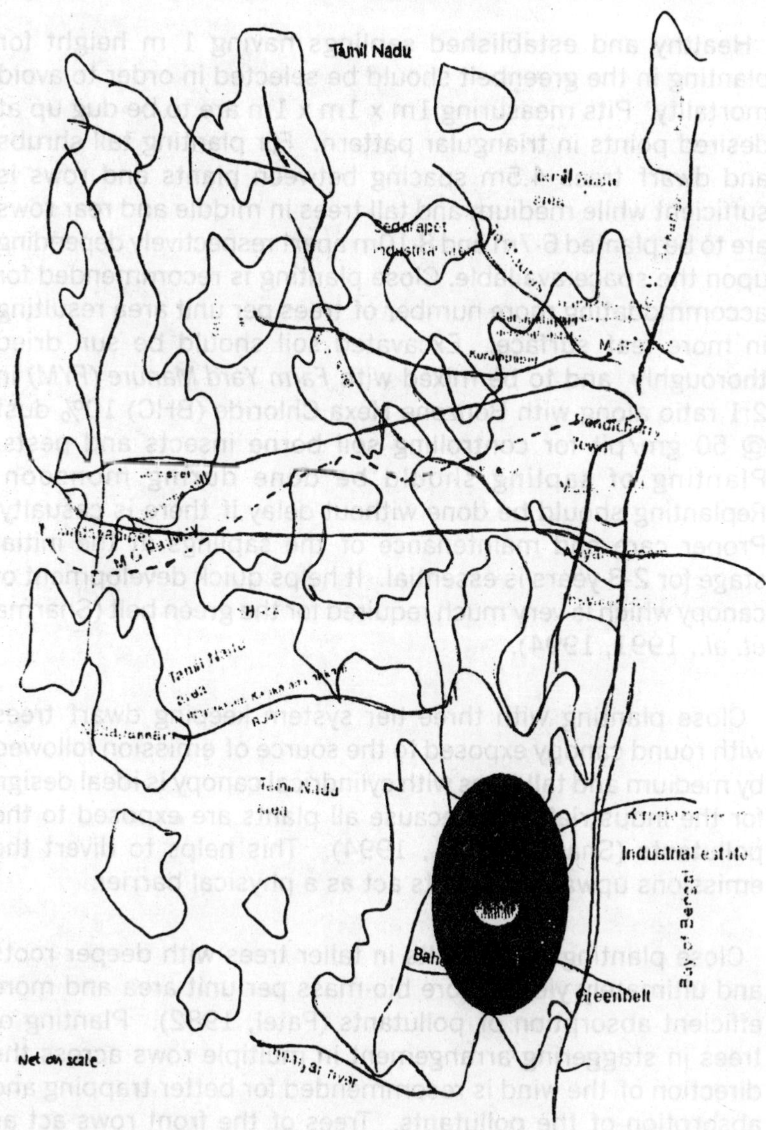

Fig. 7.9: Proposed greenbelt around the Kirmabukkam industrial estate

PLANTING AND MANAGEMENT OF THE GB

Healthy and established saplings having 1 m height for planting in the greenbelt should be selected in order to avoid mortality. Pits measuring 1m x 1m x 1m are to be dug up at desired points in triangular pattern. For planting tall shrubs and dwarf trees 4.5m spacing between plants and rows is sufficient while medium and tall trees in middle and rear rows are to be planted 6-7m and 8-10m apart respectively depending upon the space available. Close planting is recommended for accommodating more number of trees per unit area resulting in more leaf surface. Excavated soil should be sun dried thoroughly and to be mixed with *Farm Yard Manure (FYM)* in 2:1 ratio along with Benzene Hexa Chloride (BHC) 10% dust @ 50 gm/pit for controlling soil borne insects and pests. Planting of sapling should be done during monsoon. Replanting should be done without delay if there is casualty. Proper care and maintenance of the saplings at the initial stage for 2-3 years is essential. It helps quick development of canopy which is very much required for the green belt (Sharma *et. al.*. 1991, 1994).

Close planting with three tier system keeping dwarf trees with round canopy exposed to the source of emission followed by medium and tall trees with cylindrical canopy is ideal design for the industrial area because all plants are exposed to the pollutants (Sharma *et. al.,* 1994). This helps to divert the emissions upward as plants act as a physical barrier.

Close planting also results in taller trees with deeper roots and ultimately yields more bio-mass per unit area and more efficient absorption of pollutants (Patel, 1982). Planting of trees in staggering arrangement in multiple rows across the direction of the wind is recommended for better trapping and absorption of the pollutants. Trees of the front rows act as absorptive layer while the core area (rear rows) cleans the air. The width of the outer area of the plantation should be 3-4 times wider than the core area depending upon the availability of the space (Nayar, 1985). For designing green belt as city lungs in urban areas, the pattern of planting should be a little different from industrial areas. Dwarf trees and shrubs in multiple rows should be planted all along the periphery, flanked by medium and tall trees gradually towards centre so that all the plants can intercept from different directions.

Finally the prime consideration for recommending GB plantation scheme are :

i. Nature of the pollutant
ii. Emission level-height and source strength
iii. Maximum impact zone

CASE STUDY

Kirumambakkam Industrial Estate

Kirumambakkam is one of the main industrial estate of Pondicherry state and is located in remote area (far from the town population). It has good open space for further expansion and greenbelt plantation. In Kirumambakkam industrial estate the main pollutants emitted from the various sources are SO_2, NO_x, SPM, H_2S, Cl_2 and H_2SO_4 mist. It is has been observed from the air quality study of the area that the area is highly polluted necessitating control measures including greenbelt. Design green belt is one such effort. The present study (design of green belt has been conducted taking SO_2 as reference pollutant with a source strength of 45 kg/s.

Using dispersion models earlier proposed by these authors (modified plume path theory, detailed at Khan and Abbasi, 1997,1998)) dispersion of pollutant (SO_2) has been assessed for various seasons as a function of downwind distance. The resulting concentration profiles are presented in Figures 7.2 to 7.4. It is evident that during summer the dispersion is fastest (Figure 7.2) while it is slowest during winter (Figure 7.3). The values of maximum ground level are seen highest during winter that too at a larger distances from the release source. It is expected be so because if the rate of polltant dispersion is low, the maximum ground level concentration shall occur at larger distance from the source than when the dispersion is swift (Figure 7.5).

Table 7.7: Tree species recommended for Greenbelt plantation for the present study area

TREES :

Tall Trees

1. Azadirachta indica (Neem)
2. Tamarindus indica(Tamarind)
3. Ficus religiosa (Peepal)
4. Mangifera indica (Mango)
5. Tectona grandis (Teak)

Medium Dwarf Trees

1. Butea monosperma
2. Poinciana regia (Gulmohar)
3. Parkinsonia aculeta
4. Thevetia nerifolia
5. Acaccia arabica(KateriaBabu)

SHRUBS :

1. Bougainvillea (Baganvillas)
2. Calotropis Procera (Madar)
3. Ipomoea fistula (Behaya)
4. Nerium odorum (Lal kaner)
5. Thevetia nerifolia (Peela kaner)

HERBS :

1. Vinca rosea
2. Cynodon dactylon
3. Ipomaea cornea
4. Achyranthes aspera (Latjira)
5. *Solanum xanthocarpum (Bhatkatauja)*

Design of GB around Kirumambakkam industrial area

Various greenbelt design parameters have been estimated using the methodology and model proposed by these authors (Khan and Abbasi, 1999). Design calculations have been carried out for the three different seasons and finally the design parameters have been decided based on these results.

Pollution attenuation factor, is the most important parameter in deciding the location of the greenbelt, the greenbelt width, the plantation density, and the height of trees. Figures 7.6 to

7.8 depict profiles of AF as a function of distance between greenbelt and source, and plume width for the three different seasons. Figure 7.6 indicates that for the rainy season, a greenbelt of 200m would be optimal (AF value of 7.25). For the south-west wind direction a width of 150 m would be required. Thus, a greenbelt of 200 m width would be required to opimally attenuate the air pollutants in the rainy season.

Figures 7.7 and 7.8 reveal that greenbelt widths of 200m and 500m would be optimal for summer and winter seasons. The location of greenbelt is decided based on AF and the occurrence of maximum concentration (Figure 7.5 to 7.6). Analysis predicts a distance of 100m would give optimal results considering the occurrence of maximum concentration during summer season (~85m).

An analysis to decide the plantation density reveals that K (ratio of foliage area) value of 0.9 would give most favourable results. The final values of these design parameters are presented in Table 7.6.

Canopy and selection of tree species

A study has been conducted to decide the canopy of the tree plantation and the spacing between the trees. It is suggested that a triangular pitch canopy should be developed and the spacing of 25-30 ft. should be maintained between trees of middle height and the tall ones. Location of Kirumambakkam industrial area and according proposed greenbelt is depicted in Figure 7.9.

We recommend that a combination of tall trees, trees of middle height, and shrubs be planted to capture maximum pollutants. Four rows of tall tree, six rows of middle height trees and eight rows of shrubs are recommended. Five different trees species in each category feasible to plant in the present geological and climatic conditions are presented in Table 7.7.

REFERENCES

1. Abbasi, S.A. (1998). Environmental pollution and its control, *Cogent International, Pondicherry,* ix+445 pages.

2. Ahmad, K.J., Mohd. Yunus, S.N.Singh, Kanti Srivastava, Nandita Singh, Vivek Pandey and Jyoti Mishra, (1989). Study of plants in relation to air pollution. Environmental Botany Laboratory, *National Botanical Research Institute,* Lucknow.

3. Ahmad, K.J., Yunues Mohd., Sing, S.N., Srivastava, K., Singh, N., Panday Vivek and Mishra, J (1991). Air Pollution and plants. *CSIR News (8), 170-174.*

4. Andy, C. (1991). Dictionary of Environment and development. *Earth scan publications Ltd.*, London.

5. Bennett, J.H. and A.C. Hill. (1973). Absorption of gaseous air pollutants by a standardised plant canopy. *J.Air Pollution Control Assoc., 23, 203–06.*

6. Bennett, J.H., A.C. Hill and D.M. Gates. (1973). A model for gaseous pollutant sorption by leaves. *J. Air Pollution Control Assoc., 23, 957–62.*

7. Brandy, V. (1973). Air pollution, *Harcourt Brace Jovanowitch Inc., New York,* 300.

8. Gareth, J., Alan, R., Jean, F. and Graham, H. (1992). Environmental Science *Pub: Harper Perennial (A Division of Haper Collins Publishers).* 564 pages.

9. Hill, A.C. (1971). Vegetation: A sink for atmospheric pollutants. *J. Air Pollution Control Assoc.,* 21, *341–46.*

10. Hill, A.C. and E.M. Chamberlain Jr. (1974). The removal of water soluble gases from the atmosphere by vegetation. *Atmospheric- Surface Exchange of Particulate and Gaseous Pollutants Symp. Richland, WA, Sept. 4–6, 12 pp.*

11. Ingold, C.T. (1971). Fungal Spores. *Clarendon Press, Oxford, 302 pages.*

12. Jashnani, I. (1988). A Study of the Feasibility of Using Trees to Reduce Pollutants Resulting from the Proposed Coal Conversion of Unit No. 2 H.A. Wagner Power Plant, *Engineering and Computer Services*, Columbia, 63–69.

13. Kabel, R.L., O'Dell, R.A., Taheri, M. , and Davis, D.D. (1976). A preliminary model of gaseous pollutant uptake by vegetation. *Centre for Air Environment Studies, Publ. No. 455-76, Pennsylvania State Univ., University Park, PA.*

14. Khan, F.I., and Abbasi, S.A. (1997). Modelling of Heavy gases dispersion using modified plume path model, J of Loss Prevention in Process Industries, 11 (2).

15. Khan, F.I. and Abbasi, S.A. (1998). Risk assessment in chemical process industries : advanced techniques, *Discovery Publishing House*, New Delhi, XI+423.

16. Khan, F.I. and Abbasi, S.A. (1999). Greenbelts for pollution abatement : concepts, design applications, *Discovery Publishing House*, (in press).

17. Lin, D.A. (1976). Air pollution-Threat and Responses, *Addison Wesley Publishing Company,* London.

18. Nayar, M.P. (1985). Tree conopies. *Air pollution and plants- A state of the art report.* Ministry of Environ. and Forests, *New Delhi*.

19. Parsons, K.C., (1990). Clarence Stein and the Greenbelt Towns Settling for Less, *American Planning Association (APA) Journal,* Spring, 121–34.

20. Patel, V.J. (1982). Utilisation of industrial waste through high density plantation. Proceedings of National Conference on energy plantation, Bangalore, 1–12.

21. Rao, D.N. (1980). Ecological implication of Urban Industrial pollution, *In rural habitat transformation in World frontiers* (Eds. R.L. Singh), Varanasi, 84-95.

22. Rao, D.N. (1983) Sulphur dioxide pollution versus plant injury with special reference to fumigation and precipitation. *Proc.Syomp. on "Air pollution Control" Vol. 1,* Indian Association for Air Pollution Control, New Delhi, 91–96.

23. Rawat, J.S. and Banerjee, S.P. (1996). Urban forestry for improvement of Environment, *Journal of Energy Environment Monitor* 12(2), 109–16.

24. Roberts, B.R. (1974). Folliar sorption of atmospheric sulfur dioxide by woody plants. *Environ. Pollu., 7, 133–40.*

25. Ruth, A.E, and William, R.E. (1994). The Encyclopedia of the Environment. The Rene Dubor Centre for Human Environments, *Houghton Miffin Company, USA.*

26. Sharma, S.C., Sharga, A.N. and Roy, R.K. (1991). landscaping of indutrial regions. *Horticulture-New Technology and Applications.* Kluwer Academic Pub., *The Netherlands.*

27. Sharma, S.C., Sharga, A.N. and Roy, R.K. (1994). Abatement of industrial pollution by landscaping. *Indian J. of Environ. Protection. 14(2), 95-97.*

28. Singh, S.K. and Rao, D.N. (1983). Evaluation of plants for their tolerance to air pollution. *Proceedings of symposium on Air Pollution Control,* 1 , 218–24.

29. Singh, S.K. and Rac, D.N.(1982). The influence of plants for their tolerance to air pollution. *In proceedings of symposium on Air Pollution Control. Vol. I,* Indian Association for Air Pollution Control, New Delhi, 218–24.

30. Sinha, R.J., (1993). *The Environmentalist,* 13, 111–15.

31. Sivasamy, N. and Srinivasan, V.(1996). Environmental pollution and its control by trees. I.A.R.I, Pusa, New Delhi.

32. Smith, W.H. (1981). Air-pollution and forests: Interaction between air contaminants and forests ecosystems. *Springer-Verlag,* New York.

33. Tewari, D.N. (1994). Urban Forestry. *Indian Forester,* 120(8), 647–57.

34. Thakre, R. (1992). Environmental criteria for siting industry and greenbelt, *In An introduction to air pollution by Trivedi, R.K and Goel, P.K., National Environmental Engineering Research Institute,* Nagpur.

35. William, H.S. (1990). Interaction Between Air contaminants and Forest Ecology. *In Air Pollution and Forest. (second edition), Springer-verlag, New York.*

Chapter 8

TORAP - A NEW TOOL FOR CONDUCTING RAPID RISK ASSESSMENT IN PETROLEUM REFINERIES AND PETROCHEMICAL INDUSTRIES

The use of a new computer-automated tool TORAP (Tool for Rapid risk Assessment in Petroleum refinery and Petrochemical industries) is demonstrated through a rapid and quantitative risk assessment of a typical petroleum refinery. The package has been applied for an appraisal of the risks of accidents (fires, explosions, toxic release) posed by different units of the refinery, and to identify steps to prevent/manage accidents.

The studies reveal that TORAP enables a user to quickly focus on the accidents likely to occur, and enables forecasting the nature and impacts of such accidents. This information is directly utilizable in identifying 'soft' spots and in taking appropriate remedial measures to prevent or control accidents.

The special attributes of TORAP are: a) wide range of applications - achieved by incorporating models capable of handling all types of industrial fires and explosions, b) sophistication - brought about by including state-of-the-art models developed by these authors and others, c) user-friendliness - achieved by incorporating on-line help, graphics, carefully formatted output, and, above all, an automatic module with which even a lay user can conduct risk assessment. The entire package, especially its automatic module, is supported by an extensive knowledge-base built into the software.

INTRODUCTION

The science of risk analysis which has emerged to forecast the likelihood of accidents, assess the consequences of the likely accidents, work out the strategies to prevent the accidents and also to cushion the adverse impacts if an accident does occur.

A total risk assessment exercise covering all steps (Greenberg and Crammer, 1992; Khan and Abbasi, 1995a, 97a) exhaustively from beginning to end is expensive in terms of time as well as monetary and personnel inputs. It often becomes necessary to conduct *rapid risk assessment* (RRA) to draw the same conclusions that a full fledged risk assessment would lead to, albeit with lesser (yet practicable) accuracy and precision (Khan and Abbasi, 1997b,1997c,1997d,1997e).

In this chapter we describe a software package, and the system of methodologies on which the package is based, for conducting RRA in petroleum refinery and petrochemical industries. The package is named TORAP and is code in Visual C++ on Windows 95 platform.

TORAP improves upon the existing packages in the following areas :

a) ***wider applicability***; TORAP incorporates a larger number of models to handle a larger variety of situations specific to petroleum refineries and petrochemical industries;
b) ***greater sophistication***; more precise, accurate, and recent models have been incorporated in TORAP than handled by existing packages. This is specifically done to incorporate the complexity involved in the refineries and petrochemical industries.
c) ***greater user-friendliness***;
d) ***scope for assessing second and higher order accidents***; whereas the existing RRA packages are capable of handling only the primary accidents, TORAP has provision for assessing the likelihood of secondary and higher order accidents triggered by the primary event.

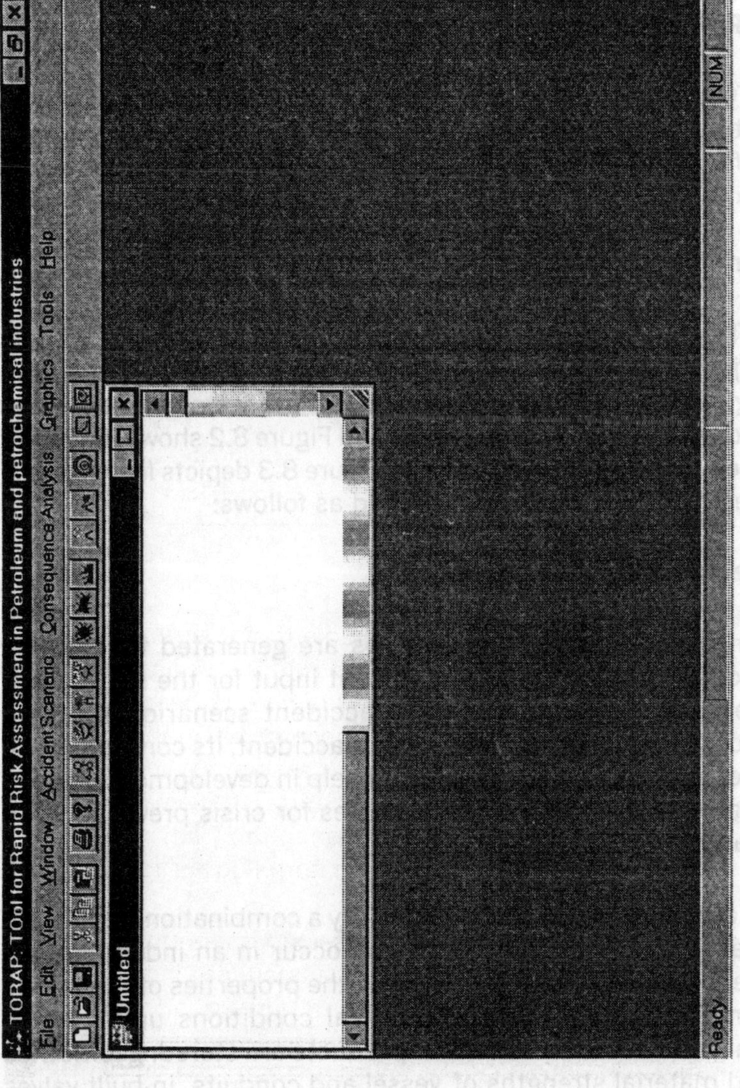

Fig. 8.1: Main menu and availabe options in TORAP

The chapter also illustrates the applicability of the new package in a real-life situation.

TORAP METHODOLOGY

TORAP enables accident simulation and damage potential estimation of petroleum and petrochemical industries. The software has been developed in object-oriented architecture using Visual C++ as a coding tool. The software is compatible on WINDOWS operating environments. It is operable on computers with a minimum of 8MB RAM and 32 MB ROM.

The sequence of actions or main steps involved in TORAP, its object architecture, and information pathways are depicted in Figures 8.1–8.3. Figure 8.1 represents the sketch of main menu and available options of TORAP and Figure 8.2 shows the object oriented architecture of TORAP. Figure 8.3 depicts five essential steps of TORAP, briefly described as follows:

The accident scenario generation step

In this step accident scenarios are generated for the unit under study. It is a very important input for the subsequent steps. The more realistic the accident scenario, the more accurate is forecasting the type of accident, its consequences, and associated risks. This would help in development of more appropriate and effective strategies for crisis prevention and management.

Each accident scenario is basically a combination of different likely accidental events that may occur in an industry. Such scenarios are generated based on the properties of chemicals handled by the industry, physical conditions under which reactions occur or reactants/products are stored, geometries and material strengths of vessel and conduits, in-built valves and safety arrangements etc. External factors such as site characteristics (topography, presence of trees, ponds, rivers in the vicinity, proximity to other industries or neighbourhoods etc.) and meteorological conditions are also considered.

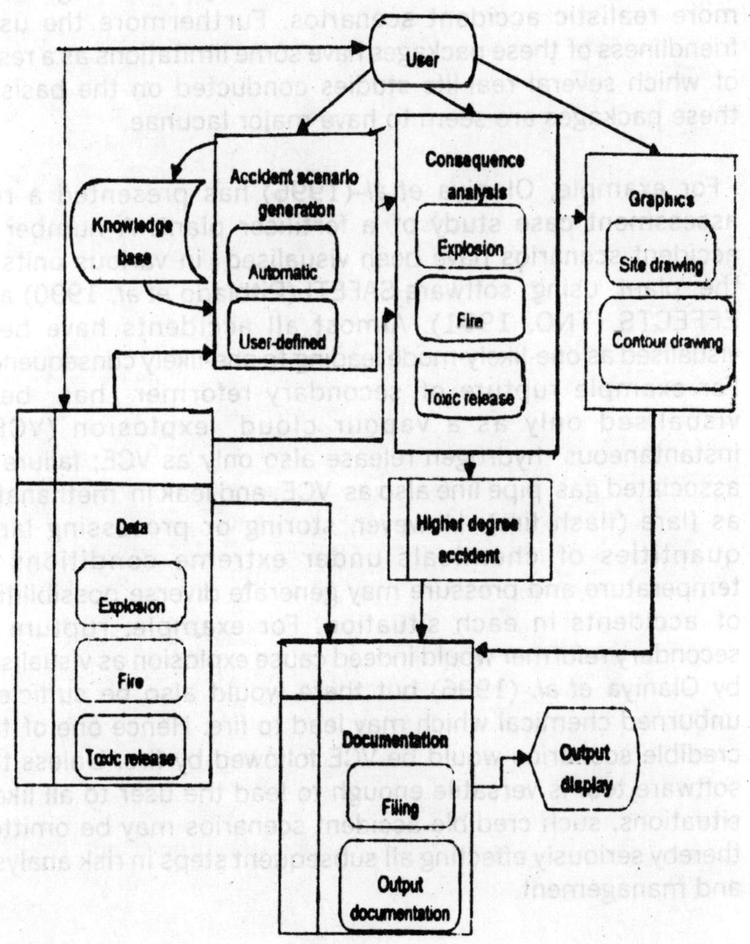

Fig. 8.2: Object-oriented architecture of TORAP

In the available software packages such as WHAZAN (Technica, 1992), EFFECTS (TNO,1991), RISKIT (VTT,1993), and SAVE (TNO,1992), this concept of risk assessment has been used to some extent. But the level of sophistication needs to be enhanced subsequently by using advanced models of thermodynamics, heat transfer, and fluid dynamics to generate more realistic accident scenarios. Furthermore the user-friendliness of these packages have some limitations as a result of which several real-life studies conducted on the basis of these packages are seem to have major lacunae.

For example, Olaniya *et al*–(1996) has presented a risk assessment case study of a fertiliser plant. A number of accident scenarios have been visualised in various units of the plant using software SAFETI (Pitblado *et al*, 1990) and EFFECTS (TNO, 1991). Almost all accidents have been visualised as one-likely-mode leading to one-likely consequence. For example rupture of secondary reformer has been visualised only as a vapour cloud explosion (VCE), instantaneous hydrogen release also only as VCE; failure of associated gas pipe line also as VCE, and leak in methanator as flare (flash fire). However, storing or processing large quantities of chemicals under extreme conditions of temperature and pressure may generate diverse possibilities of accidents in each situation. For example, rupture of secondary reformer would indeed cause explosion as visualised by Olaniya *et al.* (1996) but there would also be sufficient unburned chemical which may lead to fire. Hence one of the credible scenarios would be VCE followed by fire. Unless the software tool is versatile enough to lead the user to all likely situations, such credible accident scenarios may be omitted thereby seriously effecting all subsequent steps in risk analysis and management.

We have also come across several other reports (NEERI,1992; CISRA,1993; TPL,1993) in which one of the existing risk assessment packages have been used. In all these reports several credible accident scenarios have been left unconsidered indicating a lack of rigor and/or user-friendliness of the packages. For example in NEERI (1992), which has considered

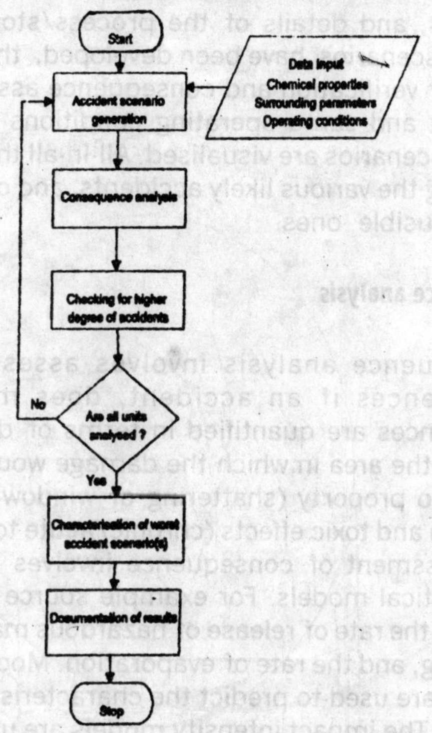

Fig. 8.3: Procedure of TORAP

jet fire in fuel storage vessel, and fire ball formation due to rupture of propylene storage vessel, BLEVE in fuel storage vessel, and CVCE/BLEVE followed by fire ball in propylene storage vessel have been overlooked. In CISRA (1993), which has dealt with instantaneous release of chlorine, and fire ball due to release of hydrogen from the storage vessel, the scenarios pertaining to BLEVE followed by fire ball due to release of hydrogen and continuous release of chlorine from vent valve have been omitted; TPL (1993) reports release and fire of ethylene and allyl chloride in the form of flash fire, but has missed out the credible scenario of explosion in ethylene vessel as BLEVE.

In TORAP all credible accident scenarios are carefully generated based on the properties of chemicals, operating conditions, and details of the process/storage units. Once accident scenarios have been developed, they are processed for further verification and consequence assessment. For the same unit and same operating conditions several plausible accident scenarios are visualised. All-in-all this option helps in simulating the various likely accidents, and characterising the worst plausible ones.

Consequence analysis

Consequence analysis involves assessment of likely consequences if an accident, does materialise. The consequences are quantified in terms of damage radii (the radius of the area in which the damage would readily occur), damage to property (shattering of window panes, caving of buildings) and toxic effects (chronic/acute toxicity, mortality). The assessment of consequence involves a wide variety of mathematical models. For example source models are used to predict the rate of release of hazardous material, the degree of flashing, and the rate of evaporation. Models for explosions and fires are used to predict the characteristics of explosions and fires. The impact intensity models are used to predict the damage zones due to fires, explosion and toxic load. Lastly toxic gas models are used to predict human response to different levels of exposures to toxic chemicals. A list of models included in TORAP for consequence estimation are given in Table 8.1. Several different types of explosion and fire models such as confined vapour cloud explosion (CVCE), partially-confined vapour cloud explosion (PVCE), boiling liquid vapour cloud explosion (BLEVE), vented explosion, pool fire, flash fire, jet fire and fire ball are included. Likewise, models for liquid release and two phase release have been incorporated. A special feature of TORAP is that it is able to handle dispersion of heavy (heavier-than-air) gases, as-light-as-air and lighter-than-air gases. A brief description of different types of accident events is presented in a subsequent sections.

Checking for higher degree of accidents

An accident in a unit caused by another accident in another unit is termed as a 'second order accident' or 'secondary accident'. If the secondary accident causes another accident in a third unit such an accident is termed 'third order' or 'tertiary' accident. This phenomenon is more probable to occur in petroleum refineries and petrochemical industries as these industries handle large variety of hazardous chemicals that also in large quantity. The most recent disaster at Visag in November 14, 1997 further proves this concept. It is a speciality of TORAP that the package is capable of simulating second and higher order accidents. To do this, it uses models developed by Pieterson (1985, 1990), Clancey (1977), Eissenberg *et al.* (1975), Fowcett and Wood (1993) and Khan and Abbasi (1997a). If the probability of occurrence of secondary accident is higher than a minimum value, the package will estimate the damage potential of the secondary accident and its likelihood of causing a third degree accident, and so on. To estimate the probability of occurrence and damage characteristics, the package uses information related to operating condition of secondary unit, chemical properties and topographical/ meteorologival parameters (wind velocity, roughness, obstruction etc.).

Characterisation of worst accident scenario

Arriving at worst accident scenario is the last step in the TORAP algorithm. The step determines the worst accident scenario based on the results of consequence analysis. This step holds the key to the final objective of the risk analysis excessive -devising strategies to avert a crisis or to minimise its adverse impact if the crisis does take place. It is possible that more than one 'worst accident scenario' emerges from the TORAP analysis because more than one sequence of events can lead to identical magnitudes of 'worst' damage. In such situations the control strategies would be developed by keeping all the `worst' scenarios in the view.

Table 8.1. List of different models used in TORAP

Event	*Model incorporated*	*Reference*
Explosion		
BLEVE	thermodynamic and heat transfer model	Baker *et al.*,1983; Martinsen *et al.*,1986; Prugh,1991; Vernart *et al.*,1993
VCE/PVCE	condensed vapor cloud explosion	Kletz,1977; Lees,1996; Prugh,1987 Van den Beg and coworkers, 1989,1993
CVCE	vapor cloud explosion in confinement (vessel or building)	Lees,1996; Baker *et al.*,1983 Davenport,1986; CCPS, 1994; Greenbook,1992
Vented explosion	release with momentum, combustion flame propagation, and shock wave/ overpressure impulse models	Harrison and Cyyre, 1987; Cates and Samules, 1991 Cates and Bimson, 1992;
Fire		
Flash fire	flare model, fire torch and spontaneous combustion	Kayes,1986; Greenberg and Crammer,1992 Fowcett and Wood,1993; CCPS, 1994
Pool fire	combustion of liquid pool	Kayes,1986; Greenbook,1992; Davies,1993
Fire ball	spontaneous combustion of vapor cloud	Kayes,1986; Roberts,1983; Pieterson, 1990 Baker *et al.*,1983; Davies,1993.
Jet fire	Continuous release, combustion, flame propagation and heat tansfer models	Crocker and Napier,1986; Cowley and– Pritchard 1991; Johnson *et al.*, 1994

{Cont.}......

Toxic release		
Light gas dispersion	Gaussian dispersion model with Erbink modification	Pasquill and Smith,1983; Gifford,1961 Turner,1970,1985; Erbink,1993
Heavy gas dispersion	Box model, modified plume path theory for heavy gas	Van Ulden,1974,1985; Deaves,1992 Erbink,1995; Khan and Abbasi,1997b
Domino effect (higher order accident)		Greenbook,1992; Lees,1996; Khan and Abbasi,1997a

DESIGN AND APPLICATION OF TORAP

In the following section TORAP is described in detail *vis a vis* design and application.

Data module

The main purpose of the module is to collect all relevant information needed for the execution of other modules. This module consists of three main objects derived from the main DATA object: toxic release, fire, and explosions. The explosion object is further divided into sub objects such as BLEVE, CVCE, VCE and vented explosion. The fire object branches into flash fire, pool fire and fire ball. The object oriented architecture of DATA module is shown in Figure 8.4.

Accident scenario module

This module, dealing with the generation of accident scenarios, is based on the advanced concepts of hazard assessment proposed by Arendt (1990), Papazoglou *et al.* (1992), Vernart *et al.* (1993) and Khan and Abbasi (1995b,1995c). The accident scenarios are generated based on chemical properties, operating conditions, and details of the process/storage units. The logistics involve in the development of accident scenarios are shown in Figure 8.5. For the same unit and same operating conditions various plausible accident scenarios can be visualised. Thus, this option helps in simulating the various likely accidents, and characterising the worst plausible ones. This module consists of two submodules (objects)–*automatic* and *user defined*.

The automatic submodule

This is a derived object to the main accident scenario object. It deals with the knowledge base which decides the accident scenario for a set of information provided by the user. The knowledge base is a compendium of conditions and facts stored in if-and-else reasoning sequences. The information

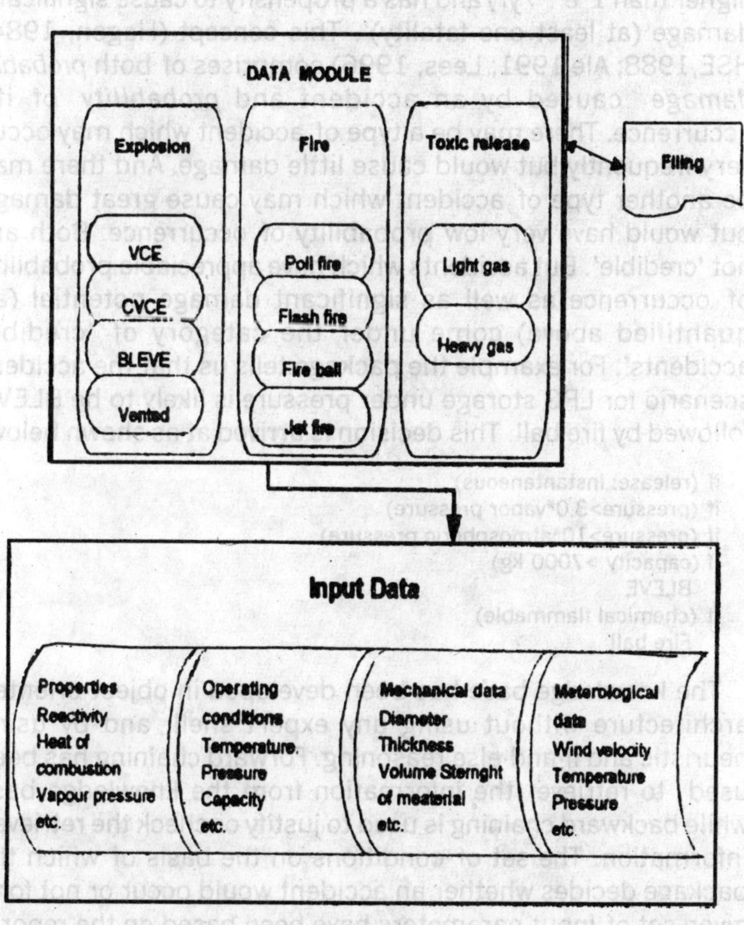

Fig. 8.4: Object-oriented architecture of Data Module

provided by the user is passed on to the knowledge base which examines whether the information satisfies the conditions necessary for a 'credible accident'; the latter is defined as 'the accident which is within the realm of possibility (*i.e,* probability higher than $1*e^{-06}$ /yr) and has a propensity to cause significant damage (at least one fatality)'. This concept (Hagon, 1984; HSE,1988; Ale 1991; Lees, 1996) comprises of both *probable damage* caused by an accident and *probability* of its occurrence. There may be a type of accident which may occur very frequently but would cause little damage. And there may be another type of accident which may cause great damage but would have very low probability of occurrence. Both are not 'credible'. But accidents which have appreciable probability of occurrence as well as significant damage potential (as quantified above) come under the category of 'credible accidents'. For example the package tells us that the accident scenario for LPG storage under pressure is likely to be BLEVE followed by fire ball. This decision is arrived at as shown below:

```
if (release: instantaneous)
if (pressure>3.0*vapor pressure)
if (pressure>10*atmospheric pressure)
if (capacity >7000 kg)
   BLEVE
if (chemical flammable)
   Fire ball
```

The knowledge base has been developed in object-oriented architecture without using any expert shell, and by using heuristic and if-and-else reasoning. Forward chaining has been used to retrieve the information from the knowledge base while backward chaining is used to justify or check the retrieved information. The set of conditions on the basis of which the package decides whether an accident would occur or not for a given set of input parameters have been based on the reports of past accidents (Lees,1996; Pietersen,1985) and data generated by controlled experiments simulating accidents.

The user defined submodule

In this option, the user defines the accident scenario on the basis of his/her knowledge and experience. For example, failure of a liquefied petroleum gas storage vessel can be visualised through various accident scenarios such as: leakage followed by PVCE/VCE, fire accompanied by CVCE, jet fire, etc.

The user considers these possibilities and chooses one or more of the likely modes of accidents. These decisions become inputs to the subsequent analysis by TORAP. For example; accidental release of propylene under pressure may be visualised by following accident scenarios;

i) CVCE followed by flash fire or fire ball.
ii) VCE followed by pool fire or flash fire.
iii) BLEVE followed by fire ball.
iv) Fire ball.

Of these accident scenarios, the first one has the maximum damage potential. The probability of consequent damage can be made low by: a) providing high pressure safety valve, b) making provision for the availability of sufficient quantity of inert gas in times of emergency to dilute the concentration of propylene in atmosphere, and c) providing means for rapid cooling of the vessel when such need arises. The other three scenarios can likewise be developed and studied for evolving appropriate damage prevention/control strategies.

Consequence analysis module

This module consists of state-of-the-art mathematical models for simulating the accidents chosen as credible in the previous step (Figure 8.6). This module works out the scale and the characteristics (type of accident, damage potential, percentage of lethality, and damage radii) of the accidents, the types of damaging impacts (shock-waves, heat loads, missiles, toxic dispersion etc.) they may cause, and their area of impact. The output of this module quantifies impacts such as peak overpressure, shock-wave velocity, shock wave duration, heat load, missile velocity, toxic load, damage radii of different impacts, and probabilities of causing lethality. The output of the consequence analysis has been so formatted that it can be directly used in reports without editing. Moreover, using these results makes it easy to draw damage/risk contours.

The mathematical models used in this module are listed in Table 8.1. Brief explanatory notes on the phenomena simulated by these models are presented below:

Toxic release submodule

This submodule assesses the consequences of release of toxic gas/vapour. It simulates different types of release scenarios such as; continuous release, two-phase release, and instantaneous release. In conducting dispersion studies it takes into consideration, the densities of the gases or gas-air mixtures (because of the pronounced influence density exerts on the shape of the plume). The models can thus simulate dispersions of heavier-than-air, as light as air, and lighter-than-air gases/gas-air plumes. This module first estimates the concentration profile of the toxic gas that would develop consequent dispersion under the given meteorological conditions. It then works out the areas of toxic impact and the extents of toxicity that would be caused on the basis of exposure-based-toxicity data.

This submodule can handle the following options: heavy gas dispersion, light gas dispersion, lethality estimation and, damage estimation. Brief descriptions of these options are presented below:

Option heavy gas

This option estimates dispersion characteristics (concentration profile, distance travelled by the cloud/plume, and the dimensions of the cloud/plume) of gases having effective density higher-than-air. It uses BOX model for instantaneous release and PLUME (heavy gas) model for continuous release (Bington,1986; Van Ulden,1974,1985; Deaves,1992; Erbink,1995; Khan and Abbasi,1997b) to estimate gas concentrations and other dispersion

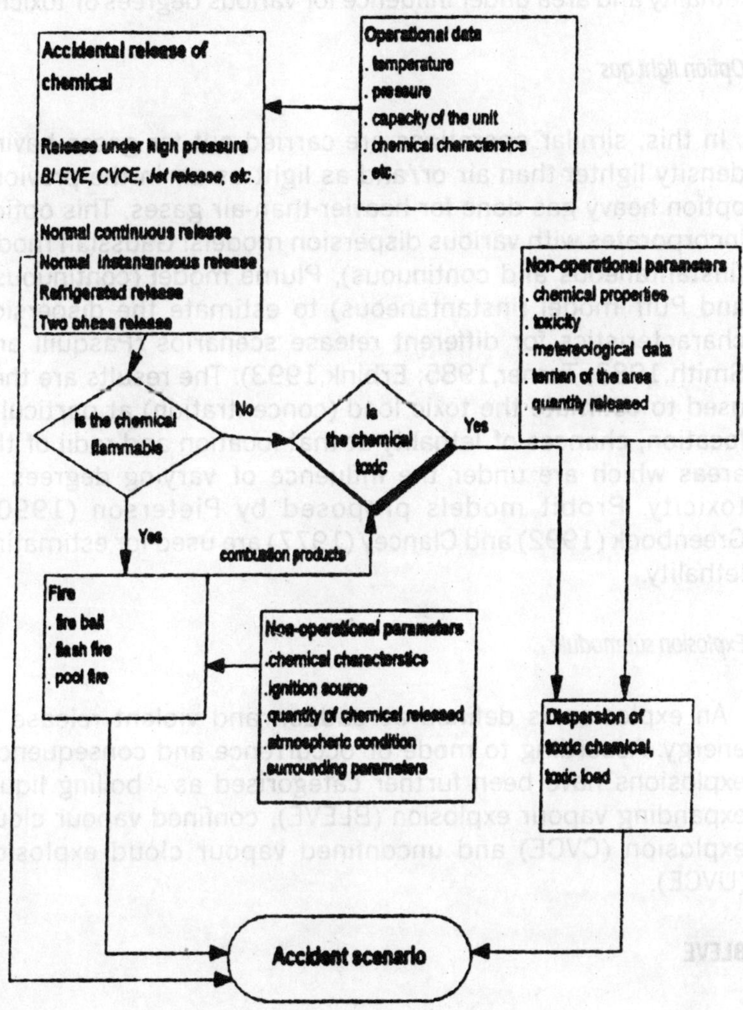

Fig. 8.5: The logistics associated with generation of accident scenarios.

characteristics. The results are then passed on to damage estimation options to calculate the percent likelihood of lethality and area under influence for various degrees of toxicity.

Option light gas

In this, similar operations are carried out for gases having density lighter-than-air or/and as light as air in the previous option heavy gas done for heavier-than-air gases. This option incorporates with various dispersion models: Gaussian model (instantaneous and continuous), Plume model (continuous), and Puff model (instantaneous) to estimate the dispersion characteristics for different release scenarios (Pasquill and Smith,1983; Turner,1985; Erbink,1993). The results are then used to estimate the toxic load (concentration) at particular location, chances of lethality at that location and radii of the areas which are under the influence of varying degrees of toxicity. Probit models proposed by Pieterson (1990), Greenbook (1992) and Clancey (1977) are used for estimating lethality.

Explosion submodule

An explosion is defined as sudden and violent release of energy. According to mode of occurrence and consequence, explosions have been further categorised as · boiling liquid expanding vapour explosion (BLEVE), confined vapour cloud explosion (CVCE) and unconfined vapour cloud explosion (UVCE).

BLEVE

A sudden release of pressurised gas or boiling liquid processed or stored under high pressure leads to BLEVE. As highly energised molecules (due to high pressure) have high. tendency to escape, a sudden release leads to a very fast movement (expansion) of molecules which in turn results in the generation of shock (blast) waves. If the material is flammable then there are chances of fire too. The velocity of the blast wave in BLEVE ranges from 330 to 450 m/s and generates a positive overpressure of 0.5 to 1 atm. The duration of dynamic pressure and shock-wave is of the order of a few seconds. In general, damaging effect of BLEVE is restricted to areas of 200-700 m radii.

VCE

The delayed spontaneous ignition of a vapour cloud of flammable chemical in a unconfined or semi-unconfined (congested boundary) space results in vapour cloud explosion (VCE) or unconfined/partialy-confined vapour cloud explosion (PVCE). Until the early 1980s a vapour cloud explosion (VCE) was generally referred to as an unconfined vapour cloud explosion (UVCE). However, since in combustion of a vapour cloud the overpressure tends to occur due to the presence of structures and obstacles and of partial confinements, the term 'unconfined' is now generally omitted.

A vapour cloud explosion is one of the most serious hazards in the process industries because a vapour cloud may drift some distance from the point where the leak has occurred before exploding; it may thus, threaten areas lying far away from the source of the vapour cloud.

Two approaches are used to assess vapour cloud explosions: one is the TNT equivalent model and the other is multi-energy method (Van den Berg and co-workers, 1984, 1985, 1991, 1993). They are briefed here:

In the TNT equivalent model the explosion is taken to be equivalent to that of a TNT explosion. This model is empirical but, within its limitations, has served the study of explosions rather well.

The TNT equivalent model is based on a single parameter: the mass of TNT. It is made more flexible by taking into consideration a second parameter -the height above ground at which the explosion occurs. Higher the point of explosion

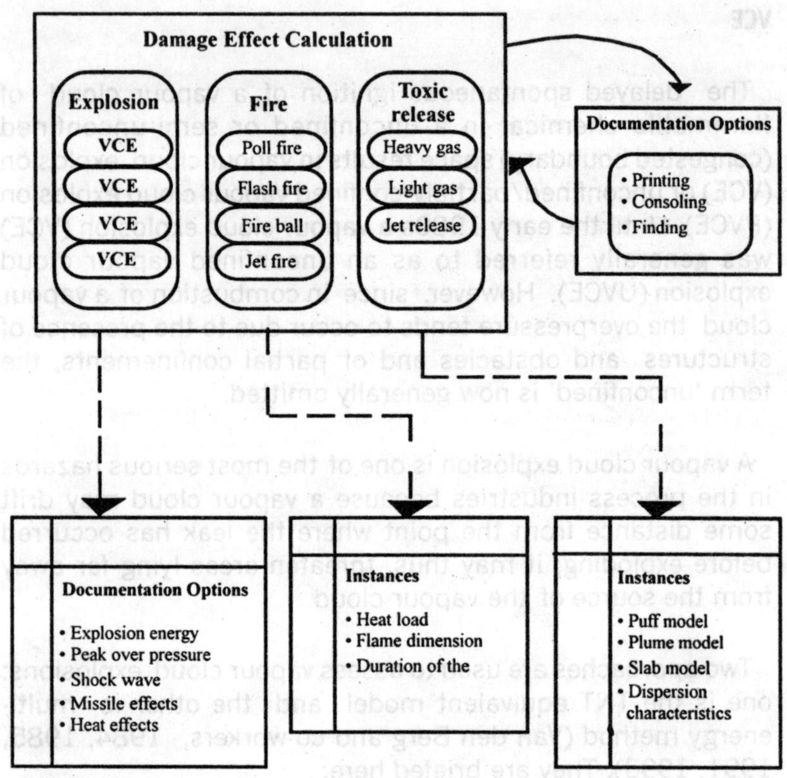

Fig. 8.6: Object-oriented architecture of Consequence Analysis module.

lower is the overpressure near the epicentre. The use of an arbitrary assumed explosion height helps in obtaining better fits to overpressure assessed from damage in actual explosion.

The TNO vapour cloud explosion model is the multi-energy (ME) model described by Van den Berg (1984), Van den Berg, (1985), Van den Berg *et al.*, (1991), Van den Berg and Lennoy (1993). This model allows the peak overpressure, peak dynamic pressure and duration time to be estimated.

The model recognises the role of partial confinement in vapour cloud explosions. It is assumed in the model that the explosion in those parts of the cloud which are confined is of much higher strength than in those where it is unconfined. The method involves estimating the combustion energy available in the various parts of the cloud and assigning to each part an initial strength.

CVCE

An Explosion in a confined space such as a vessel or a pipe triggered by excessive pressure development either due to a runaway reaction process, overfilling, or absorption of heat from external sources - is called as CVCE. Liquids of low boiling points, flammable gases, or highly reactive chemicals processed under extreme conditions, are most likely to generate CVCE. A CVCE occurs when the pressure in a confinement reaches a critical limits beyond the safety level. For example, a vessel will explode when the pressure goes beyond its design or bursting pressure. CVCE differs from BLEVE in three respects. First in CVCE the explosion occurs within a confinement, while in BLEVE the material expands outside the boundary of the confinement (vessel) and can drift away before exploding. Secondly CVCE occurs at very high pressure, quite higher than the pressures adequate for BLEVE. Lastly CVCE could be more disastrous than BLEVE as it generates shock-waves of higher speeds and greater overpressure. The impact of CVCE can be observed over a much larger distance (1 to 3 km). Due to its large area of impact and more severe

Table 8.2. Output of TORAP for an accident scenario (scenario 1)

Parameters			*Values*
Explosion : CVCE			
Energy released during explosion	(kJ)	:	3.21e+09
Peak over pressure	(kPa)	:	600.0
Variation of over pressure in air	(kPa/s)	:	512.0
Shock velocity of air	(m/s)	:	753.0
Duration of shock wave	(ms)	:	88
Missile characteristics			
Initial velocity of fragment	(m/s)	:	408.0
Kinetic energy of fragment	(kJ)	:	4.16e+05
Fragment velocity at study point	(m/s)	:	400.0
Penetration ability at study point (based on empirical model)			
Concrete structure	(m)	:	0.2735
Brick structure	(m)	:	0.3495
Steel structure	(m)	:	0.0408
Damage Radii (DR) for various degree of damage due to over pressure			
DR for 100% complete damage	(m)	:	101

DR for 100% fatality or 50% complete damage	(m)	:	154
DR for 50% fatality or 25% complete damage	(m)	:	228

Damage radii (DR) for the varying degree of damage due to missile

DR for 100% damage or 100% fatality	(m)	:	2770
DR for 50% damage or 100% fatality	(m)	:	2886
DR for 100% fatality or 10% damage	(m)	:	2989

Fire : Fire Ball

Radius of the fire ball	(m)	:	195.0
Duration of the fire ball	(s)	:	80.0
Energy released by fire ball	(kJ)	:	4.63e+09
Radiation heat flux	(kJ/sq.m)	:	1770313.0

Damage Radii (DR) due to thermal load

DR for 100% fatality/damage	(m)	:	405
DR for 50% fatality/damage	(m)	:	507
DR for 100% third degree of burn	(m)	:	585
DR for 50% third degree of burn	(m)	:	751

Higher order accident

The probability of leading secondary accident		:	0.57

shock-waves CVCE has a greater potential of causing secondary, tertiary and higher order accidents.

Vented explosion

Vented explosion is a phenomenon observed because of the formation of fire torch in a vessel due to ignition of a flammable chemical (gas or liquid) at release point. In this situation the flame front may move backwards in the vessel with high speed and ignite the contents of the vessel resulting in excessive pressure development leading to an explosion. In vented explosion flame front speed may reach over 50 m/s. It is differentiated from CVCE by its mode of initiation and its damaging effects. Eventhough vented explosion also generates shock wave, over pressure, heat load, and missiles, its damaging effects are generally confined to smaller areas than the impacts of CVCE. The intensity of the damaging effects are also lower: a vented explosion typically generates shock (blast) wave of 330 to 500 m/s and overpressure of 0.3 to 10 atm.

Fire submodule

Uncontrolled combustion of any chemical in the presence of air is termed as fire. According to the mechanism of formation and the broad shapes it attains, the fire can be classified into three main types: pool fire, flash fire, and fire ball.

Release of low boiling/non boiling liquid from a vessel may give rise to a pool of liquid which on ignition would yield *pool fire*. In certain situations a pool fire may also generate an explosive vapour cloud by supplying the required heat of evaporation to a liquid pool. However, possibility of occurrence of this phenomenon is limited to the boiling liquid processed/ stored under liquefied or refrigerated conditions. There may be different ways of initiation of a pool fire but the ultimate destructive impact of a pool fire is caused by its heat load.

An instantaneous combustion of flammable gas or high boiling liquid (liquid of high vapour pressure) on ignition causes *flash fire*. Flash fire generally occurs when the quantity of chemical is not high enough to form an explosive cloud. The low flammability characteristics and rate of release of the chemical restrict flame speed precluding generation of a blast wave. However, the heat load generated by flash fire is quite high and its damaging effect can be observed over long distances. According to the different modes of release and ignition, flash fires can be characterised as *flare*, *fire torch* etc.

Burning of a flammable gas issuing from a pipe or other orifice at the point of exit leads to jet fire. Jet flames have been involved in a number of accidents. Perhaps the most dramatic were the large jet flames from the gas riser at Piper Alpha (Less, 1996).

Jet flame can occur in chemical process industries, either by design or by accident. They occur intentionally in burners and flares. Ejection of flammable fluid from a vessel, pipe or pipe flange can give rise to a jet flame if the material ignites. An intermediate situation, and one which particularly concerns the designer, is where the jet flame results from ignition of flammable material vented from a pressure relief valve. Scenarios involving jet flames are not easy to handle, since a large jet flame may have a substantial 'reach', sometimes up to 50 m or more.

Compared to pool fire and fire balls, jet flames can occur in more diverse ways. The scenario most often considered is that of a vertical flame or an upward pointing jet, in calm conditions or in wind. But a jet may point upwards not vertically but at an angle, in such a case there may be a variety of wind directions, confluent with, opposed to or across the jet. There can also be a horizontal jet too, on which the wind may act in tandem, opposed to, or across it.

A spontaneous ignition of vapour cloud not having sufficient energy to explode leads to a *fire ball*. This phenomenon is

Table 8.3. Output of TORAP for an accident scenario (scenario 2)

Parameters			*Values*
Explosion : BLEVE			
Total energy released	(kJ)	:	3.6e+07
Peak over pressure	(kPa)	:	483.0
Variation of over pressure in air	(kPa/s)	:	358.0
Shock velocity of air	m/s)	:	691.0
Duration of shock wave	(ms)	:	47
Missile characteristics			
Initial velocity	(m/s)	:	55.0
Kinetic energy of fragment	(kJ)	:	7.55e+03
Fragment velocity at study point	(m/s)	:	54.0
Penetration ability at study point (based on empirical models)			
Concrete structure	(m)	:	0.0135
Brick structure	(m)	:	0.0173
Steel structure	(m)	:	0.0055
Damage Radii (DR) for various degree of damage due to over pressure			
DR for 100% complete damage	(m)	:	34

DR for 100% fatality or 50% complete damage	(m)	:	51
DR for 50% fatality or 25% complete damage	(m)	:	75
Damage radii (DR) for the varying degree of damage due to missile			
DR for 100% damage or 100% fatality	(m)	:	1808
DR for 50% damage or 100% fatality	(m)	:	1924
DR for 100% fatality or 10% damage	(m)	:	2027
Fire : Fire Ball			
Radius of the fire ball	(m)	:	195.0
Duration of the fire ball	(s)	:	80.0
Energy released by fire ball	(kJ)	:	5.84e+09
Radiation heat flux	(kJ/sq.m)	:	2229564.0
Damage Radii (DR) due to thermal load			
DR for 100% fatality/damage	(m)	:	454
DR for 50% fatality/damage	(m)	:	569
DR for 100% third degree of burn	(m)	:	656
DR for 50% third degree of burn	(m)	:	843
Higher order accident			
The probability of leading secondary accident		:	0.31

generally observed for high boiling yet highly flammable liquids stored or processed under extreme conditions of temperature and pressure. In some cases (high capacity, stable conditions) these clouds may also generate blast waves. The fire ball is different from the flash fire in terms of flame speed, minimum capacity of chemical required, mode of release, and ignition. The destructive ability of fire ball is very high as the heat load generated by it is of the order of 1000 kJ/m^2. The fire ball radii generally vary from 100 meter to 300 meter and its impact duration from few seconds to few minutes. The fire ball characteristics depend on the type and the mass of the chemical involved.

Higher order accident submodule

This submodule of the consequences analysis module analyse the damage potential of the primary event at the point of location of the secondary unit and checks for the likelihood of occurrence of the secondary accident. This module has an independent set of information which should be provided by the user. This information, pertains to the operational details of the secondary unit, chemicals used, meteorological conditions, and topological characteristics. This submodule consists of various sets of mathematical models (Pietersen,1985; Clancey,1977; Kletz,1977; Greenbook,1992; Prugh, 1987; Khan and Abbasi,1997a) to estimate the probability of occurrence of secondary accidents due to the impact of primary ones. If the probability of secondary accident is estimated and found credible, the unit is processed for consequences estimation in a manner similar to the study of the primary accident. The same procedure can be repeated for higher order accidents.

Graphics module

This module enables visualisation of risk contours in the context of the site of accidents. The option has two facilities: (i) site drawing, and (ii) contour drawing.

The site drawing option enables the user to draw any industrial site layout using freehand drawing or using any already defined drawing tool. The contour drawing option has the facility for drawing various damage/risk contours over the accident site.

Documentation module

This module deals with handling of data file, scenario file, output file, and flow of information. This module also works as 'information manager': it provides the necessary information to each module and submodule to carry out desired operations, and stores the results in different files. Besides this, it also provides all commonly used file operations such as copying, deleting, consoling and printing.

The applicability of the software is demonstrated with an illustrated example of its use in RRA as follows:

AN ILLUSTRATION OF *TORAP* APPLICATION IN CONDUCTING RRA

Problem Statement

Rapid risk assessment study has been carried out for a petroleum refinery storage unit. The units stores wide variety of chemicals, in large quantities under extreme conditions of temperature and pressure. Here, we are presenting detailed study of five chemicals namely Propylene, Benzene, LPG (liquefied petroleum gas), Kerosene, and ammonia.

Accident scenario generation

The storage units the refinery possess the following three types of hazard: explosion hazard (due to Propylene, Benzene, and LPG), fire hazard (due to Propylene, Benzene, LPG, and Kerosene), and toxic hazard (due to ammonia). The plausible accident scenarios envisaged in these units are as follows:

Table 8.4. Output of TORAP for an accident scenario (*scenario 3*)

Parameters			*Values*
Explosion : BLEVE			
Total energy released	(kJ)	:	1.3e+08
Peak over pressure	(kPa)	:	600.0
Variation of over pressure in air	(kPa/s)	:	472.0
Shock velocity of air	(m/s)	:	753.0
Duration of shock wave	(ms)	:	58
Missile characteristics			
Initial velocity	(m/s)	:	106.0
Kinetic energy of fragment	(kJ)	:	2.76e+04
Fragment velocity at study point	(m/s)	:	103.0
Penetration ability at study point (based on empirical models)			
Concrete structure	(m)	:	0.0358
Brick structure	(m)	:	0.0457
Steel structure	(m)	:	0.0105
Damage Radii (DR) for various degree of damage due to over pressure			
DR for 100% complete damage	(m)	:	51

DR for 100% fatality or 50% complete damage	(m)	:	78
DR for 50% fatality or 25% complete damage	(m)	:	116

Damage radii (DR) for the varying degree of damage due to missile

DR for 100% damage or 100% fatility	(m)	:	2119
DR for 50% damage or 100% fatility	(m)	:	2235
DR for 100% fatality or 10% damage	(m)	:	2339

Fire : Fire Ball

Radius of the fire ball	(m)	:	195.0
Duration of the fire ball	(s)	:	80.0
Energy released by fire ball	(kJ)	:	5.60e+09
Radiation heat flux	(kJ/sq.m)	:	2139252.0

Damage Radii (DR) due to thermal load

DR for 100% fatility/damage	(m)	:	445
DR for 50% fatility/damage	(m)	:	557
DR for 100% third degree of burn	(m)	:	643
DR for 50% third degree of burn	(m)	:	826

Higher oreder accident

The probability of leading secondary accident		:	0.15

Propylene storage : scenario 1

An excessive pressure development in the storage vessel of propylene (under high pressure and temperature) leads to confined vapour cloud explosion (CVCE). The vapour cloud generated by CVCE on ignition turns into a fire ball and consequently damages the other storage vessels and nearby process units.

Benzene storage : scenario 2

An excessive high pressure build-up in the storage vessel of benzene either due to overpressurization or runaway reaction would cause release of benzene as BLEVE (maximum reported incidents). The unburned vapour cloud on ignition leads to fire ball.

LPG storage : scenario 3

An instantaneous release of liquefied petroleum gas (LPG) under high pressure, leads to BLEVE and, as the chemical is highly flammable, the released cloud on meeting an ignition source turns to a fire ball.

Kerosene storage : scenario 4

An instantaneous release of kerosene either through vent valve or through any accidental opening causes the formation of pool of liquid which on meeting ignition source burns as pool fire.

Ammonia storage : scenario 5

Catastrophic rupture of a vessel storing ammonia in pressurised condition.

These scenarios have been processed for damage potential estimation through TORAP.

Hazard quantification

The results of the calculations for different accident scenarios are summarised below:

Table 8.2 presents the summary of calculation (output of TORAP) for *scenario* 1. The missiles generated by CVCE may hit nearby targets and can lead to secondary explosions/fires or toxic releases. The vapour cloud generated by CVCE on ignition may cause a fire ball and hence severe heat radiation effect. The shock wave generated due to CVCE can cause injury as well as second order accidents by seriously damaging other vessels. It has been estimated that shock waves with 50% probability of causing injury would be observed over an area of ~500 m radius. The heat radiation effect with 50% probability of lethality would be observed over an area of ~300 m radius and missile effects with 50% chances of damage (without considering probability of hitting) would be observed across ~750 m radius.

As per *scenario* 2 BLEVE would generate shock waves as well as missiles. In addition, there would be secondary impact of the released material getting ignited and forming fire ball thereby generating additional heat load. The output of TORAP for this scenario is presented in Table 8.3. The lethal impact of shock-waves, heat load, and missiles would go upto and beyond a radius of ~350m.

Table 8.4 presents the summary of calculation (results of TORAP-) for scenario 3 (accident in LPG storage vessel). The vapour cloud generated by BLEVE on ignition would cause a fire ball and hence result in severe heat radiation effect. It has been estimated that shock waves with 50% probability of causing injury would be propagated over an area of 20 m radius which is smaller compared to the impacts areas of other accidents considered in this study. However, the heat radiation effect with 50% probability of lethality would reach over an area of ~384 m radius.

The instantaneous release from kerosene tank as per scenario 4 would generate liquid kerosene pool (Table 8.5). The released chemical on meeting ignition source would generate heat load (pool fire). There is sufficient probability that part of the liquid kerosene evaporate either due to atmospheric evaporation or due to heat supplied by burning of liquid kerosene. The vapour cloud form due to this evaporation may cause flash fire. We have studied this possibility using TORAP. It suggest that quantity of vapour generated is not enough to cause flammable cloud. Thus, no flash fire. However, heat load generated due to pool fire having propensity to cause damage would be observed over an area of ~250 m radius.

The accident scenario for ammonia release has been identified as instantaneous release (boiling liquid expanding vapour explosion-BLEVE) of ammonia under slightly stable conditions (most frequently observed stability class). Eventhough ammonia vapour has density lighter-than-air, light gas dispersion model can not be used because in the present case study, ammonia is stored in refrigerated condition and an instantaneous release (as BLEVE) would lead to the formation of vapour cloud consisting of liquid ammonia droplets. Thus, the effective density of the cloud will be higher-than-air. TORAP also confirms this phenomenon and suggests two phase release with shock wave development and non-buoyant dispersion.

Table 8.6 shows the output of TORAP; it specifies accidental release scenario as-*instantaneous two phase release of gas* (flashing followed by evaporation) under sub-normal condition. It also presents the output of dispersion of heavy-gas module using BOX model. It suggests that at a distance of 500 meter from the accident site the concentration of gas would be 2.64E-06 kg/m^3, which is a lethal dose considering that STEL (short term exposure limit) lethality concentration 0.027*E-09 kg/m^3 (Fowcett and Wood, 1993). The concentration profile for ammonia with distance suggests that up to a distance of 1500 meters the concentration of ammonia would be above permissible limit vis a vis STEL (15 minutes exposure) for instantaneous exposure.

Table 8.5. Output of TORAP for an accident scenario (scenario 4) in Benzene storage vessel

Parameters			*Values*
Fire : Pool fire			
Instantaneous Model			
Radius of the pool fire	(m)	:	8.0
Burning area	(sq.m)	:	202.0
Burning rate	(kg/s)	:	34.0
Heat flux	(kJ/sq.m)	:	238399.0
Damage Radii (DR) due to thermal load			
DR for 100% fatality/damage	(m)	:	149
DR for 50% fatality/damage	(m)	:	186
DR for 100% third degree of burn	(m)	:	215
DR for 50% third degree of burn	(m)	:	276
Higher order accident			
The probability of leading secondary accident		:	0.01

Risk estimation

Using the results obtained by TORAP and the probability of occurrence of the accident scenarios, individual fatal risk factors have been estimated. The probability of occurrence of an individual event has been adopted from literature and the industry concern. The risk factor is a direct representation of the threat (taking into consideration both damage potential and probability of occurrence) to an individual in an area.

The study reveals that *scenario 5* represents the worst likely disaster within the realm of credibility. It has the largest area-of-lethal-impact (shock wave over an area of radius ~100 m and lethal concentration across an area of radius ~1500 m). Further, the most thickly populated areas (including residential areas) lie within its range. If one considers the cumulative effects, the *scenarios* 1 would come out as the worst, as more intense impacts (in terms of heat radiation, shock waves, and missiles) are observed per unit area in these scenarios.

Scenarios 2, and 3 also have the potential to lead secondary accidents as severe heat load generated in these would encompass other storage vessels. All-in-all, *scenario* 5 is the worst as far as primary effects are concerned whereas *scenario* 1 is the worst in terms of its potentiality of causing cascading (domino) effects.

Table 8.6. Output of TORAP for an accident scenario (*scenario 5*) in Ammonia storage vessel

Parameters			*Values*
Toxic Release and Dispersion			
Heavy Gas Dispersion characteristics			
Box Instantaneous: Model			
Concentration at distance 200	(kg/cu.m)	:	1.76e-06
Concentration at cloud axis	(kg/cu.m)	:	1.62e-01
Value of source height	(m)	:	7.00
Damage radii (DR) for various degree of damage :			
DR for 100% lethality	(m)	:	6713
DR for 50% lethality	(m)	:	8630
DR for 10% lethality	(m)	:	15461
Puff characteristics:			
Puff concentration at centre of cloud	(kg/cu.m)	:	7.77e-03
Concentration at cloud edges	(kg/cu.m)	:	7.74e-03
Distance along down wind	(m)	:	200.0000
Dosage at study point	(kg/cu.m)	:	0.4327
Higher order accident			
The probability of leading secondary accident		:	0.0

SUMMARY AND CONCLUSION

This chapter presents a new Tool–TORAP–for rapid risk assessment of petroleum refinery and petrochemical industries. The special attributes of TORAP are: a) wide range of applications–achieved by incorporating models capable of handling all types of industrial fires and explosions, b) sophistication–brought about by including state-of-the-art models developed by these authors and others, c) user-friendliness–achieved by incorporating on-line help, graphics, carefully formatted output, and, above all, an automatic module with which even a lay user can conduct risk assessment. The entire package, especially its automatic module, is supported by an extensive knowledge-base built into the software.

The use of a new computer-automated tool TORAP (TOol for Rapid risk Assessment in Petroleum refinery and Petrochemical industries) is demonstrated through a rapid and quantitative risk assessment of a storage unit of typical petroleum refinery. The studies reveal that TORAP enables a user to quickly focus on the accidents likely to occur, and enables forecasting the nature and impacts of such accidents. This information is directly utilizable in identifying 'soft' spots and in taking appropriate remedial measures to prevent or control accidents.

REFERENCES

1. Ale, B. J. M., (1991). Risk analysis and risk policy in the Netherlands and EEC, *J Loss Prevention in Process Industries*, 4(1), 58.

2. Arendt, J. S., (1990). Using quantitative risk assessment in the chemical process industries, *Reliability Engineering and System Safety*, 29, 133–49.

3. Baker, W. E., Cox, P. A., Westin, P. S., Kulesz, J. J., and Strelow, R. A., (1983). Explosion hazard and evaluation, *Elesvier*, Amsterdam.

4. Bington, P. W. M., (1986). Heavy gas dispersion from source inside building or in wake, *Proceeding I Chem E Symposium on refinement of estimate of the consequence of heavy toxic release*, Manchester.

5. Cates, A. T., and Samuels, B., (1991). A simple assessment methodology for vented explosions. *J. Loss Prevention in Process Industries*, 4, 287.

6. Cates, A. T., and Bimson, S. J., (1991). Fuel and obstacle dependency in pre-mixed transient deflagration, *13th ICODERS*, Nagoya, Japan.

7. CCPS, (1994). Fire and explosion guidelines, Centre for Chemical Process Safety, *AICh E*, New York.

8. CISRA, (1993). Risk assessment of Chloralkali industry, *Cell for Industrial Safety and Risk Analysis*, CLRI ·Madras, India.

9. Clancey, V. J., (1977). Dangerous clouds, their growth and explosive properties, *Chemical Process Hazards*, 6, 111–23.

10. Cowley, L. T., and Pritchard, M. J., (1991). Large-scale natural gas and LPG jet fires and thermal impact on structures, *Gastech 90*, 36.

11. Crocker, W. P., and Napier, D. H., (1986). Thermal radiation hazards of liquid pool fires. In *Hazards*, IX, 159.

12. Davenport, J. A., (1986). Hazards and protection of pressure storage of liquefied petroleum gases. *Loss Prevention and Safety Promotion*, 5, 22.1.

13. Davies, P. A., (1993). A guide to the evaluation of condensed phase explosion, *Journal of Hazardous Materials*, 33, 1-33.

14. Deaves, D. M., (1992). Dense gas dispersion modelling (HEAVG-PLUME), *Journal of Loss Prevention Process Industries*, 5(4), 219–22.

15. Eissenberg, N. A., Lynch, C. J., and Breeding, R. J., (1975). Vulnerability Model: A Simulation System for Assessing Damage Resulting from Marine Spills, *Report CG-D-136-75 (Enviro control)*, Rockville M. D., USA.

16. Erbink, J. J., (1993). The advanced Gaussian model STACKS, *Procd. of ERCOFTAC work shop on Inter comparison of Advanced Practical Short range Atmospheric Dispersion Modelling*, Manno, Switzerland.

17. Erbink, J. J., (1995). Turbulent diffusion model from tall stacks, *Ph.D. thesis submitted to VRIJ University*, The Netherlands.

18. Fowcett, H. H., and Wood, W S, (1993). Safety and accident prevention in chemical operation, (third ed.), *John willey*, New York.

19. Gifford, F. A., (1961). Use of routine meteorological observation for estimate atmospheric diffusion, *Nuclear Safety*, 24, 67.

20. Greenberg, H. R., and Crammer, J. J., (1992). Risk Assessment and Risk Management in Chemical Process Industries, *Van Norstand Reinhold*, New York.

21. Greenbook, (1992). Methods for determining of possible damage to people and objects resulting from release of hazardous materials, *Report CPR-16E*, Voorburg, Warrington, UK.

22. Hagon, D. O., (1984). Use of frequency-consequence curves to examine the conclusion of published risk analysis and to define broad criteria for major hazard installations, *Chem Engng Res Dev*, 62, 381.

23. Harrison, A. J., and Keyre, J A, (1987). 'External explosions' as a result of combustion venting. *Combustion Science Technology*., 52, 91.

24. HSE, (1988). The tolerability of risks formation from nuclear power stations, Health and Safety Executive , *HM Stationary Office*, London.

25. Johnson, A. D., Brightwell, H. M., and Carsley A. J., (1994). A model for predicting the thermal radiation hazards from large-scale horizontally released natural gas jet fires. *Hazards*, XII, 123.

26. Kayes, P. J., (1986). Manual of industrial hazard assessment technique, *Technica Ltd*., London.

27. Khan, F. I., and Abbasi, S. A., (1995a). Risk Analysis : An optimum approach, *research report to CPCE / RA 011/ 95*, Pondicherry University.

28. Khan, F. I., and Abbasi, S. A., (1995b). Anatomy of industrial accidents, *Chemical Engineering World*, September, Bombay.

29. Khan, F. I., and Abbasi, S. A., (1995c). Accident simulation in CPI using TORAP, *Indian Journal of Chemical Technology*, 3, 338–44.

30. Khan, F. I., and Abbasi, S. A., (1997a). Models for domino effect analysis in chemical process industries, *Process Safety Progress (accepted; in press)*.

31. Khan, F. I., and Abbasi, S. A., (1997b). Modelling and simulation of heavy gas dispersion on the basis of modification in plume path theory, *Atmospheric Environment*, (in press).

32. Khan , F. I., and Abbasi, S. A., (1997c). MAXCRED : a tool for rapid risk assessment in chemical process industries, *Enviornmental Modelling and Software*, (in press).

33. Khan, F. I., and Abbasi, S. A., (1997d). Risk analysis of epichlohydrin manufacturing industry using new computer automated tool MAXCRED, *J Loss Prevention in Process Industries*, 10(2), 91.

34. Khan, F. I., and Abbasi, S. A., (1997e). Risk analysis of a cloralkali industry situated in densely populated area, *Process Safety Progress*, 16(3), 172.

35. Kletz, T. A., (1977). Unconfined vapour cloud explosions, *AIChE Loss Prevention*, 11, 50.

36. Lees, F. P., (1996). Loss prevention in CPI, *Butterworths*, London.

37. Martinsen, W. E., Johnson, D. W., and Terell, W. F., (1986). BLEVE : Their causes effects and prevention, *Hydrocarbon Processing*, 65(11), 141.

38. NEERI (1992). Risk Assessment of Reliance Petrochemical Complex, *National Environmental Research Institute*, Nagpur, India.

39. Olaniya, R. S., Mathurkar, H. N., and Deshpande, A. W., (1996). Probabilistic risk assessment of fertilizer plants, 3 87-94.

40. Papazoglou, I. A., Nivoliantou, Z., Aneziris, O., and Christou, M., (1992). Probabilistic safety analysis in chemical process industries, *Journal of Loss Prevention Process Industries*, 5, 3-15.

41. Pasquill, F., and Smith, F. B., (1983). Atmospheric diffusion, *John willey*, New York.

42. Pitblado, R. M., Shaw, S. J., and Stevens, G. C., (1990). The SAFETI risk assessment package and case study application, *Safety and loss prevention in the chemical and oil industries*, 51.

43. Pietersen, C. M., (1985). Analysis of the LPG-incident in San Juan Ixhuasapec- Mexico city, *TNO report*, The Netherlands.

44. Pietersen, C. M., (1990). Consequence of accidental release of hazardous materials, *Journal of Loss Prevention Process Industries*, 3, 136–41.

45. Prugh, R. W., (1987). Evaluation of unconfined vapour cloud explosions, *Proceedings of International Conference of Vapour Cloud Modelling*, Boston.

46. Prugh, R. W., (1991). Quantitative evaluation of BLEVE hazards, *Journal of Fire Protection Engineers*, 3(1), 9-24.

47. Roberts, A. P., (1983). Thermal radiation hazard from release of LPG from pressurized storage, *Fire and Safety Journal*, 4, 197-212.

48. Technica, (1992). *WHAZAN* : A software for hazard assessment, *Technica Ltd.*, London.

49. TNO, (1991). *EFFECTS* : A software for hazard assessment, *TNO Prins Mauritis Research Laboratory*, The Netherlands.

50. TNO, (1992). SAVE : A software for hazard assessment, *TNO Prins Mauritis Research Laboratory*, The Netherlands.

51. TPL, (1993). Risk analysis of storage unit of chemical process industry, *Tamilnadu Petrochemical Ltd.*, Manali, Madras, India.

52. Turner, D. B., (1970). Workbook on atmospheric dispersion estimates report Ap-26, *US EPA*, Qeseach Triangle park, NC.

53. Turner, D. B., (1985). Pragmatic method for the estimating boundary layer parameter for diffusion calculation, *Boundary Layer Metrology*, 8, 93-100.

54. Van den Berg, A. C., (1984). Blast effects from vapour cloud explosion, *Ninth Int. Sym. on The Prevention of Occupational Accidents and Diseases in the Chemical Industry*, Lucerne.

55. Van den Berg, A. C., (1985). The multi-energy method. A framework for vapour cloud explosion blast prediction. *J. Hazardous Materials.*, 12(1), 1.

56. Van den Berg, A. C., (1989). REAGAS - A Code for Numerical Simulation of 2–D Reactive Gas Dynamics. *Rep. PML-1989-in-48*, TNO Prins Maurits Lab. Rijswijk, The Netherlands.

57. Van den Berg, A. C., and Lennoy, A., (1993). Methods for vapour cloud explosion blast modelling, *J Hazardous Materials*, 34, 171.

58. Van den Berg, A. C., Van Wingerden, C. J. M., and The, H. G., (1991). Vapour cloud explosion : experimental investigation of key parameters and blast modelling, *Trans I Chem Engg*, 69B.

59. Van Ulden, A. P., (1974). On spreading of a heavy gas cloud released near ground, *Loss Prevention and Safety Promotion*, 1, 221.

60. Van Ulden, A. P., (1985). Estimate of atmospheric boundary layer parameters for diffusion application, *Journal of Climatology and Applied Meteorology*, 25, 1609.

61. Vernart, J. E. S., Rutledge, G. A., Sumathipala, K., and Sollows, K., (1993). To BLEVE or Not to BLEVE ; Anatomy of boiling liquid expanding vapour explosion, *Process Safety Progress*, 2, 112–23.

62. VTT, (1993). *RISKIT* : A software for risk assessment, *The Technical Research Centre of Finland*, Tempere, The Finland.

Chapter 9

CASE STUDY: APPLICATION OF A NEW COMPUTER AUTOMATED TOOL *TORAP* IN CONDUCTING RISK ASSESSMENT OF A PETROCHEMICAL INDUSTRY

The chapter describes the application of a new computer automated tool, developed by us, in the risk analysis of a typical chemical industry engaged in the manufacture of linear alkyl benzene. Using the tool —a comprehensive software package TORAP (TOol for rapid Risk Assessment of petroleum refineries and Petrochemical industries)—nine different scenarios, one for each storage unit, have been studied. It is observed that the accident scenario for chlorine (instantaneous release followed by dispersion) leads to the largest area-under-lethal-impact while the accident scenario for propylene (CVCE followed by fire ball) forecasts the most intense damage per unit area. The accidents involving propylene, benzene, and fuel oil have high possibility of causing *domino/secondary* accidents as their destructive impacts (shock waves, heat load) would envelop other storage and process units.

Besides demonstrating the utilizability of TORAP, this study also focuses attention on the need to bestow greater effort towards risk assessment/crisis management. The authors hope that the study would highlight the severity of the risk posed by the industry and would thus generate safety consciousness among plant managers. The study may also help in developing accident-prevention strategies and installation of damage control devices.

INTRODUCTION

In a recent communication[1,4] (Khan and Abbasi[9a,b], 1997[3,4]) we have discussed how serious accidents can take place in chemical process industries—in the form of explosions, fires and release of toxic chemicals. More often than not, such accidents take heavy toll of property and human lives. The accidents—if they involve release of large quantities of toxic chemicals—also contaminate large areas and render them useless for several years[1,2].

A brief review of some of the recent accidents which have occurred in different industries is presented below to illustrate the myriad ways in which human or equipment failures can cause an accident leading to loss of lives and property[2,3].

Illustrative case studies

On 21 September, 1921 evening two explosions occurred at the Oppau works of Badische Aniline and Sodafabrik (BASF) in the span of 3 seconds. The explosions created a mammoth crater of 80 m diameter, destroyed the plant, and 700 of the 1000 houses nearby. The explosion was caused by the detonation of some 4500 t of a 50:50 mixture of ammonium sulfate and ammonium nitrate. It was set off by blasting powder, which was being used to break up storage piles of material which had become caked. Exactly the same procedure had been carried out without any mishap some 16000 times previously!

Even houses in the adjacent city of Ludwigshafen and in the Mannheim area were damaged. Walls were dislocated and windows broken. At these places and at Heidelberg, which is about 14 miles from Oppau, the effect of the explosion was first felt by two very heavy earthquake-like shocks. In Mannheim some seconds later and in Heidelberg 82 seconds after the shocks there came an enormous rush of air which broke windows and doors and caused damage to gas holders, oil tanks, and many river barges. The sound of the explosion and the earth shocks reached as far as Bayreuth, at a distance of 145 miles, and the air pressure wave caused considerable damage in Frankfurt, which is about 53 miles from the scene of the explosion. The explosion killed 430 people, including 50 people in the village[2].

Table 9.1. The main units of linear alkyl benzene (LAB) plant

Units	Reference used *in Figure 2*	Operation
Prefractionation unit	A	Fractionation
Rerun column	B	Fractionation
Stripping unit	C	Seperation
Hydrotreater unit	D	Reaction
Molex unit	E	Extrative seperation
PACOL unit	F	Reaction
Alkylation unit	G	Reaction
Propylene chlorination	H	Reaction
AllylChloride purification	I	Fractionation
Chlorine absorption in water	J	Absorption
Chlorohydrination	K	Reaction
Propylene bullets	L	Storage
Chlorohydrin	M	Storage
Fuel oil tanks	N	Storage
Chlorinated organic tanks	O	Storage
Chlorine storage	P	Storage
Benzene storage	Q	Storage
N-Paraffine storage	R	Storage
Kerosine storage	S	Storage

On 4 January 1966 a spillage occurred in a refinery in France when an operator was draining water from a 1200 m pressurized propane sphere. The propane vapor spread over a radius of 150 m and was ignited by a car on the road. The pool of propane below the sphere engulfed the vessel in flames. The resultant boiling-liquid-expanding-vapor explosion (BLEVE) killed the fireman and 17 others. The conflagration took 48 hours to control and caused extensive damage to the refinery.

In June 1974 a 16" elbow of a pipe carrying potassium carbonate solution in a fertilizer plant at Tamilnadu, India, ruptured suddenly, splashing the hot solution into the nearby

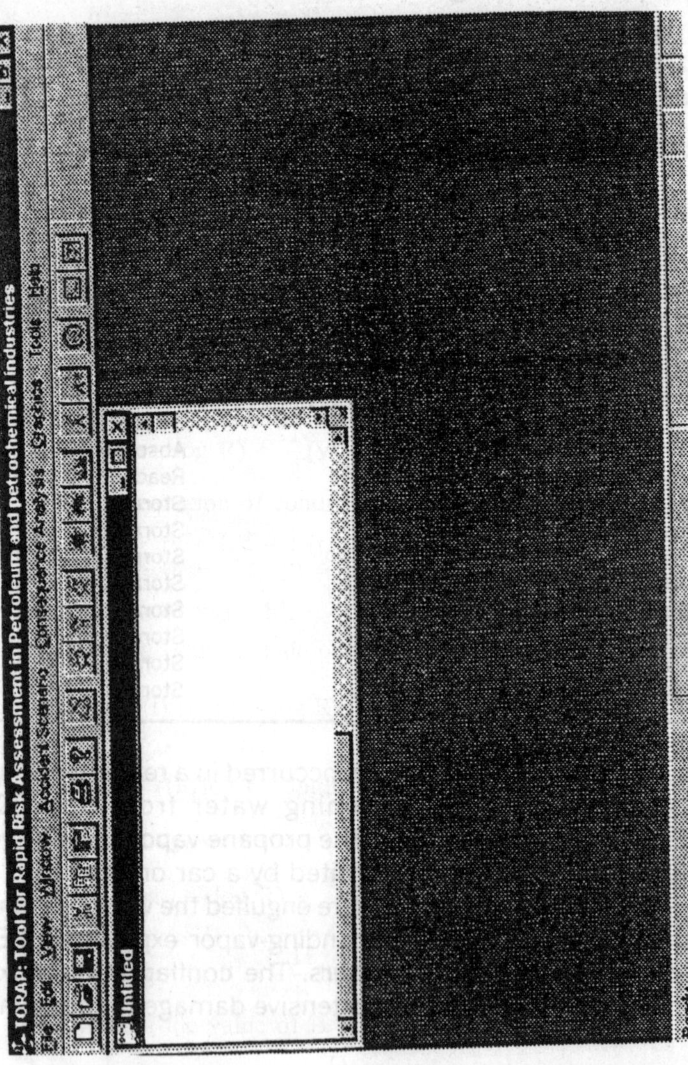

Fig. 9.1: Main menu and available options in TORAP

control room. The toughened glass panes shattered; 8 persons died in the control room instantaneously, one died in the hospital and others sustained grievous injuries[2].

In March, 1984, an explosion in a refinery at Kerala destroyed fire tender along with the shed wherein it was housed, besides a chemical warehouse, cooling tower and other facilities. Later investigations revealed many shortcomings in the plant layout.

On 12 December, 1987, a crude oil storage tank in a refinery at Maharashtra, India, started boiling over, spilling the contents on the dike around it. Emergency services and fire brigades, who were alerted, tried to evacuate the contents. After 4 hours of pumping out, the tank caught fire and exploded, spilling the contents. Eight hours of vigorous fire fighting had to be carried out before the fire could be controlled. There was extensive damage to the property. A liberal sizing of the dike and providing a separate dike for a large tank like this would have helped to prevent the spread of fire to other tanks.

An accident took place on 18 April, 1989, in a 14" natural gas pipeline in a gas company in India. The pipeline was carrying compressed natural gas at a pressure of about 295 - 298 psig from the compressor station to various consumers. The spot of the accident was about 730 ft. from the compressor station. A security personnel heard a loud sound at about 0950 hrs and saw a huge cloud of black smoke emanating from the ruptured pipeline which caught fire immediately. The flame rose as high as 150 ft during the initial stage.

The fire damaged buildings consisting of the general stores and the office of the materials department. Two employees died and six others received burn injuries. Investigations revealed that the portion of the pipeline which had blown off was extensively corroded as compared to other portions of the pipeline. The underground pipeline was close to the materials department where old lead cells were stored. The corrosion could be due to the leakage of spent weak acid which seeped through the ground and corroded the buried pipeline[1].

Sao Paulo accident : On February 25, 1984, at least 508 people, most of them young children, were killed in Sao Paulo (Brazil) when a 2-ft diameter gasoline pipe ruptured and 700 tons of gasoline spread across a strip of swamp. The cause of the pipe rupture was not reported, though, it was said to have been brought up to pressure above safe limits. It was also stated that there was no way of monitoring the pressure in the pipe line[1].

Bhopal disaster : In Bhopal, India, the worst-ever disaster in chemical process industries took place on december 3, 1984. A leak of methyl isocyanate (MIC) from the Union Carbide Factory, where it was used as an intermediate in the manufacture of a pesticide, spread beyond the plant boundary and caused death by poisoning of over 2,500 people -injuring about 10 times as many .

According to press reports, the contents of the MIC storage tank got overheated and boiled, causing the relief valves to lift. The discharge of vapor—about 25 tons—was too great for the capacity of the scrubbing system. The escaping vapor spread beyond the plant boundary where a shanty town had sprung up. The cause of the overheating was contamination of the methyl isocyanate, by water or other materials, and several possible mechanisms were suggested. According to some reports cyanide was produced. Had Union Carbide conducted risk analysis, (specifically maximum credible accident analysis) during the design of the MIC system or even later, it would have learnt that in the event of a MIC leak the scrubbing system would be inadequate. This would have enabled the industry to install better emergency handling systems, thereby saving thousands of lives [3].

Pepcon explosion : On May 4, 1988, a massive explosion destroyed Pacific Engineering and Production Company,

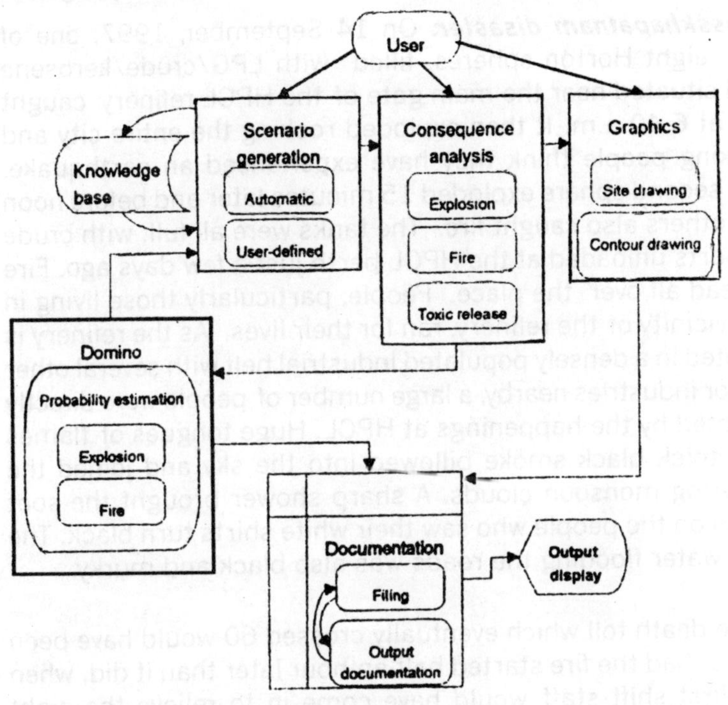

Fig. 9.2: Architecture of TORAP

(PEPCON) plant near Henderson, about 12 miles south of Las Vegas, USA.

Maharashtra accident : On 5 November, 1990, an explosion at the offside battery of compressors at a gas cracking plant in Maharashtra, India, killed 35 persons, besides causing heavy damages to property and business interruption losses. Among the deficiencies in the layout, identified after the disaster, was the location of a contractor's shed dangerously close to the gas compressors. Less publicized, but perhaps of greater consequence, was the lack of facility to shut down the flow of hydrocarbon at the site itself. The plant personnel had to run to the control room faster than the vapor that followed them to close the feed valve[3].

***Visskhapatnam disaster*:** On 14 September, 1997, one of the eight Horton spheres, filled with LPG/crude/kerosene and situated near the main gate of the HPCL refinery, caught fire at 6.40 a.m. It then exploded rocking the entire city and making people think they have experienced an earthquake. The second sphere exploded 15 minutes later and before noon the others also caught fire. The tanks were all full, with crude imports unloaded at the HPCL berth just a few days ago. Fire spread all over the place. People, particularly those living in the vicinity of the refinery, ran for their lives. As the refinery is located in a densely populated industrial belt with several other major industries nearby, a large number of people were directly effected by the happenings at HPCL. Huge tongues of flames and thick black smoke billowed into the sky and joined the hovering monsoon clouds. A sharp shower brought the soot down on the people who saw their white shirts turn black. The rain water flooding the roads was also black and murky.

The death toll which eventually crossed 60 would have been higher had the fire started half-an-hour later than it did, when the first shift staff would have come in to relieve the night shift. And Sunday being a holiday, the administrative personnel, who number over 200, were saved as they were not on duty[5].

Prevention/Control of accidents

In view of the serious and ever - increasing risk posed by the chemical process industries as elucidated above, it is necessary to conduct risk assessment[5-9] and develop strategies to prevent the accidents or, if preventive measures fail, to cushion their adverse impacts.

Risk assessment involves the following essential steps[7,8,9].

a) to identify vulnerable spots or 'high risk' points in an industry;
b) to simulate accidents and assess the damage they may cause;

c) to use the results of the previous step in identifying the priority areas where preventive measures need to be introduced;
d) to develop disaster management plans based on 'b' and 'c' above.

Table 9.2. Operating conditions of different storage units

Hazardous chemicals	No. of Tanks	Capacity of each tank *(Ton/m³)*	Operating pressure *(atm)*	Operating temperature *(°C)*
Benzene	2	400 m³	4.55	40
Normal paraffine	2	200 T	2.15	25
Propylene	2	60 T	15.20	40
Chlorine	2	100 m³	4.25	20
Allylchloride	2	50 m³	1.00	35
Kerosene	2	450 m³	1.15	30
Fuel oil	1	400 m³	1.00	40
Chlorohydrine	1	100 m³	1.22	40
Chlorinated organics	2	200 m³	1.22	40

It is evident that the step 'b' is very important in quantifying the risk posed. We have, therefore, developed techniques and tools for conducting risk assessment in chemical process industries. These include rapid risk assessment using a computer automated tool MAXCRED (Khan and Abbasi[10-13], 1996; 1997), and its advanced version MAXCRED-II (Khan and Abbasi[14,15], 1997), optimizing HAZOP (hazard and operability study) procedure (Khan and Abbasi[16-18], 1997), and safety-based design (Khan and Abbasi[19], 1997).

We have recently enhanced the capabilities of MAXCRED-II in several ways which include a) handling of cascading/domino effect, b) faster processing, c) more efficient error handling d) more visual appealing. This has been done by employing advanced concepts of software engineering (event-based design, object oriented database, etc.). The resulting computer automated tool TORAP enables quantitative simulation of accidents in chemical process industries. It enables forecasting of the type of accidents and the types and extents of damage such accidents would cause. Once this information is available by conducting simulations with the aid of TORAP, it becomes easy to devise necessary accident prevention and damage control strategies on the basis of the characteristics of the industrial site.

THE SOFTWARE PACKAGE TORAP

TORAP is a software package developed at risk assessment division of centre for pollution control and energy technology. The package enables simulation of accidents and estimation of their damage potential. TORAP has been developed with an intention to provide a more versatile and accurate tool for rapid risk assessment than is possible with existing packages. It may be seen that earlier version of TORAP has significantly greater capabilities than other commercial packages whereas TORAP further improves the its sophistication by incorporating domino/cascading effect, and implementation of advance concept of software engineering[11-14]. A brief description of the contents and capabilities of TORAP is given below:

Software and hardware features

Coding medium	:	Visual C++
Working environment	:	WINDOWS
Main menus and system architecture	:	as in Figure 9.1 and Figure 9.2 respectively
Basic algorithm	:	as in Figure 9.3
System	:	PC/AT-486 or higher
Minimum RAM needed	:	4 MB
Operating time	:	The entire accident simulation exercise beginning from keying in the input data to obtaining easy-to-use tabulated/graphic print-outs is approximately 20 minutes.

TORAP has five main modules (options) : scenario generation, consequence analysis, domino, documentation, and graphics. In the scenario generation module accident scenarios are generated for the unit under study. It is a very important input for the subsequent steps. More realistic the accident scenario, more accurate is the forecast of the type of accident, its

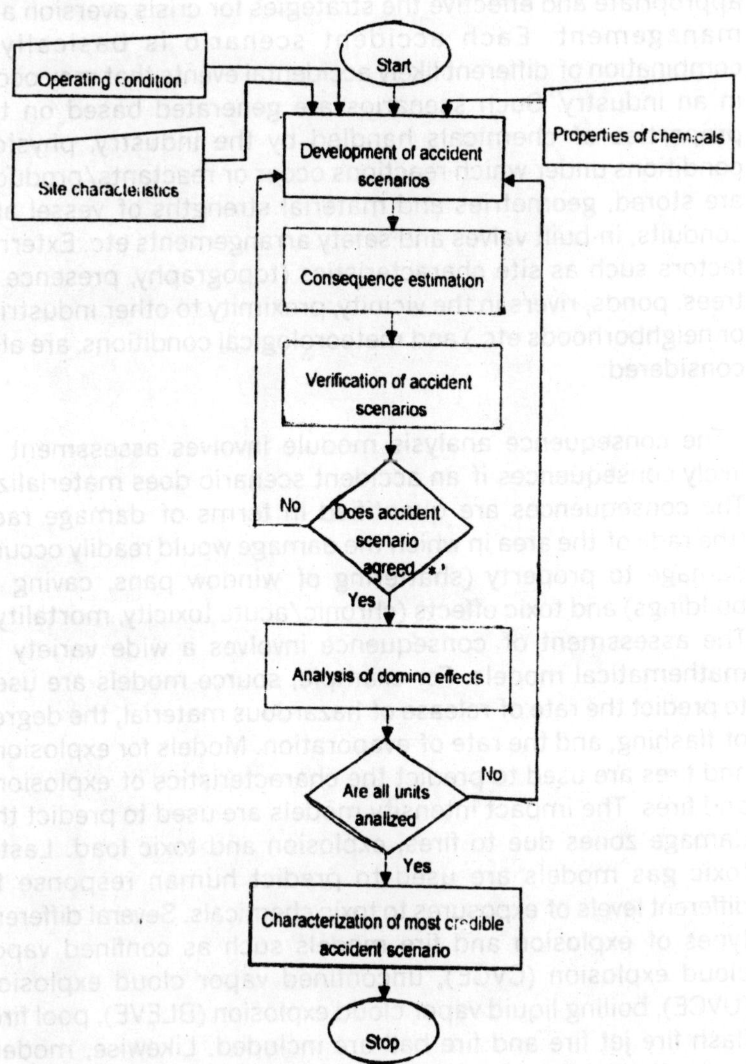

Fig. 9.3: Sequence of steps involve in TORAP

consequences, and associated risks; consequently more appropriate and effective the strategies for crisis aversion and management. Each accident scenario is basically a combination of different likely accidental events that may occur in an industry. Such scenarios are generated based on the properties of chemicals handled by the industry, physical conditions under which reactions occur or reactants/products are stored, geometries and material strengths of vessel and conduits, in-built valves and safety arrangements etc. External factors such as site characteristics (topography, presence of trees, ponds, rivers in the vicinity, proximity to other industries or neighborhoods etc.) and meteorological conditions are also considered.

The consequence analysis module involves assessment of likely consequences if an accident scenario does materialize. The consequences are quantified in terms of damage radii (the radii of the area in which the damage would readily occur), damage to property (shattering of window pans, caving of buildings) and toxic effects (chronic/acute toxicity, mortality). The assessment of consequence involves a wide variety of mathematical models. For example, source models are used to predict the rate of release of hazardous material, the degree of flashing, and the rate of evaporation. Models for explosions and fires are used to predict the characteristics of explosions and fires. The impact intensity models are used to predict the damage zones due to fires, explosion and toxic load. Lastly toxic gas models are used to predict human response to different levels of exposures to toxic chemicals. Several different types of explosion and fire models such as confined vapor cloud explosion (CVCE), unconfined vapor cloud explosion (UVCE), boiling liquid vapor cloud explosion (BLEVE), pool fire, flash fire jet fire and fire ball are included. Likewise, models for liquid release and two phase release have been incorporated. A special feature of TORAP is that it is able to handle dispersion of heavy (heavier-than-air) gases as well as lighter-as-air and lighter-than-air gases. A brief description of different types of accident events is presented in a subsequent section.

Domino module analyzes the damage potential of the primary event at the point of location of the secondary unit and checks for the likelihood of the occurrence of the secondary accident. If the probability of the secondary accident is sufficiently high than the appropriate accident scenarios are developed and analyzed for consequences[20].

Table 9.3. The output of TORAP for *scenario 1*

Parameters			*Values*
Distance of study from accident epicenter	(m)	:	200
Explosion : CVCE			
Energy released during explosion	(kJ)	:	1.82E+8
Peak over pressure	(kPa)	:	534.1
Variation of over pressure in air	(kPa/s)	:	444.4
Shock velocity of air	(m/s)	:	999.3
Duration of shock wave	(ms)	:	145.6
Missile characteristics			
Initial velocity of fragment	(m/s)	:	1195.6
Kinetic energy of fragment	(kJ)	:	8.75E+05
Fragment velocity at study point	(m/s)	:	787.7
Penetration ability at study point (based on empirical model)			
Concrete structure	(m)	:	0.775
Brick structure	(m)	:	0.792
Steel structure	(m)	:	0.094
Fire : Fire ball			
Radius of the fire ball	(m)	:	176.4
Duration of the fire ball	(s)	:	93.4
Energy released by fire ball	(kJ)	:	2.46e+07
Heat flux	(kJ/sq.m)	:	857.5

The graphics module enables visualization of risk contours in the context of the site of accidents. The option has two facilities: (i) site drawing, and (ii) contour drawing. The site drawing option enables the user to draw any industrial site layout using freehand drawing or using any already defined drawing tool. The contour drawing option has the facility for drawing various damage/risk contours over the accident site. The contours can be drawn in different shapes and sizes as per the requirement of the user.

The documentation module of TORAP mainly deals with handling of different files such as: data file, scenario file, output file and flow of information. This object works as 'information manager': it provides the necessary information to each module and sub-module to carry out desired operations, and stores the results in different files.

All-in-all TORAP is a versatile tool for risk assessment and is envisaged to be self contained in the sense that it does not need other packages for data analysis or graphics support.

TYPES OF ACCIDENTS

Four types of accidents[5,8,9] are possible - (i) fire, (ii) explosion, (iii) toxic release and dispersion, and (iv) a combination of i, ii, and iii.

Boiling liquid expanding vapor cloud explosion (BLEVE)

BLEVE[9] is a phenomenon which results from sudden release of gas or liquid stored at temperatures above their boiling points. At the vent or release point, a sudden decrease in pressure results in explosive vaporization of the stored material leading to a blast effect. The magnitude of BLEVE mainly depends on the material capacity and its rate of release.

Unconfined vapor cloud explosion (UVCE)

UVCE[9,20] generally occur when sufficient amount of flammable material (gas or liquid having high vapor pressure) gets released and mixes with air to form a flammable cloud such that the average concentration of the material in the cloud is higher than the lower limit of explosion. The resulting explosion has high potential of damage as it occurs in an open space covering large areas. The intensity of explosion mainly depends on the quantity of material released and the strength of the ignition source.

Table 9.4. The output of TORAP for *scenario 2*

Parameters			*Values*
Distance of study from accident epicenter	(m)	:	200
Explosion : BLEVE			
Total energy released	(kJ)	:	5.67E+05
Peak over pressure	(kPa)	:	65.4
Variation of over pressure in air	(kPa/s)	:	37.4
Shock velocity of air	(m/s)	:	105.6
Duration of shock wave	(s)	:	55.1
No missile effect			
Toxic release and dispersion			
Heavy gas dispersion characteristics			
BOX instantaneous: model			
Concentration at distance 200 m	(kg/cu.m)	:	1.76E-04
Concentration at cloud axis	(kg/cu.m)	:	1.45E-02
Value of source height	(m)	:	5.0
Puff characteristics:			
Puff concentration at centre of cloud	(kg/cu.m)	:	8.15E-04
Concentration at cloud edges	(kg/cu.m)	:	8.15E-05
Distance along downwind	(m)	:	200.0
Dosage at study point	(kg/cu.m)	:	0.074

The explosive power of a UVCE can be expressed in terms of blast wave characteristics (overpressure, overpressure-impulse, reflected pressure, duration of shock wave etc.). The peak overpressure is a very important parameter; its magnitude depends on the speed of flame propagation. Any obstruction in flame propagation enhances the blast effect.

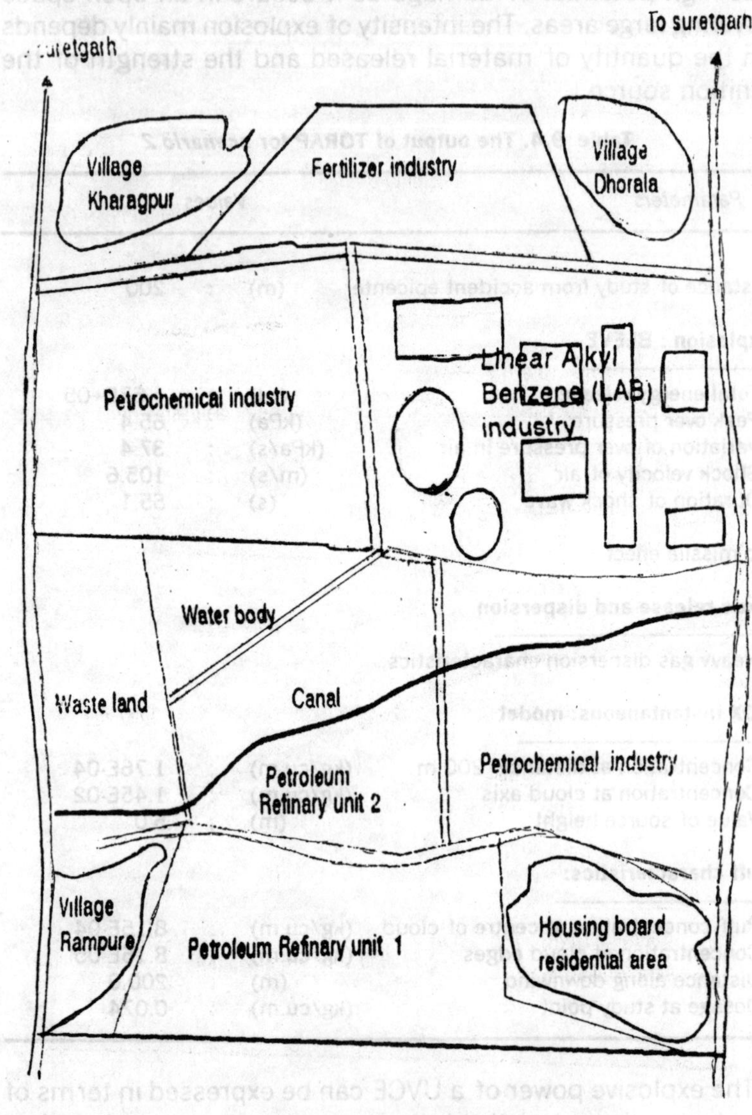

Fig. 9.4. Layout of the study area showing location of industry and their surroundings

Confined vapour cloud explosion (CVCE)

CVCE[9,10,20], as the name suggests, is a vapour explosion occurring in one or other type of confinement. Explosions in vessels and pipes are examples of CVCE. Excessive generation of high pressure in the confinement leads to this type of explosion. It also has high potential of causing damage as it may generate fragments (missiles) propelled at high velocities which can cause newer accidents. The energy delivered to the fragments by blast wave causes the fragments to become air-borne and to act as missile. The missiles are characterized by velocity, weight and penetration strength. However, the cumulative effect of CVCE depends upon the mass of material involved in the explosion and the explosion pressure.

Pool fire

Continuous release of flammable liquid results in pool fire[9,10,22]. The characteristics of such fire mainly depend on the duration of release, saturation pressure, and flammable properties of materials.

Flash fire

Flash fire[9,22] occurs mainly due to the instantaneous release of material having boiling point lower than atmospheric temperature. It does not explode when the material release rate and flame speed are not high enough. However, it spreads quickly throughout the flammable zone of the vapor cloud.

Jet fires

Burning of a flammable gas issuing from a pipe or other orifice at the point of exit leads to jet fire.

Jet flames have been involved in a number of accidents. Perhaps the most dramatic were the large jet flames from the gas riser at Piper Alpha[9,20,21].

Jet flame can occur in chemical process industries, either by design or by accident. They occur intentionally in burners and flares. Ejection of flammable fluid from a vessel, pipe or pipe flange can give rise to a jet flame if the material ignites. An intermediate situation, and one which particularly concerns the designer, is where the jet flame results from ignition of flammable material vented from a pressure relief valve. Scenarios involving jet flames are not easy to handle, since a large jet flame may have a substantial 'reach', sometimes up to 50m or more.

Compared to pool fire and fire balls, jet flames can occur in more divers ways. The scenario most often considered is that of a vertical flame on an upward pointing jet, in calm conditions or in wind. But a jet may point upwards not vertically but at an angle, in such a case there may be a variety of wind directions, confluent with, opposed to or across the jet. There can also be a horizontal jet, on which also the wind may act in tandem, opposed to it, or across.

Fire ball

An instantaneous ignition of flammable vapour cloud would lead to the formation of a fire ball[10,11,20]. The radius of fire ball, its radiation heat intensity, and temperature in the fire ball depend upon the dimension of the flammable cloud as well as the mass of the vapor released in the cloud. A very high temperature of 500 to 1500 K is developed in the confines of fire ball. It is potentially the most disastrous of industrial fires that may be caused by highly flammable gases stored or processed under pressure.

The damage associated with such fires may be assessed on the basis of the dose of heat radiation received from them in a given time interval.

Table 5. The output of TORAP for scenario 3

Parameters			*Values*
Distance of study from accident epicenter	(m)	:	200
Explosion : UVCE			
Total energy released by explosion	(kJ)	:	6.52E+06
Peak over pressure	(kPa)	:	135.7
Variation of over pressure in air	(kPa/s)	:	63.4
Shock velocity of air	(m/s)	:	178.4
Duration of shock wave	(ms)	:	79.4
Missile characteristics:			
Initial velocity of missile	(m/s)	:	344.3
Kinetic energy associated with missile	(kJ)	:	1988775.3
Fragment velocity at study point	(m/s)	:	143.5
Penetration ability at study point (based on empirical model)			
Concerete structure	(m)	:	0.035
Brick structure	(m)	:	0.037
Steel structure	(m)	:	0.002
Fire : Pool fire			
Instantaneous model			
Radius of the pool fire	(m)	:	5.0
Burning area	(sq.m)	:	76.4
Burning rate	(kg/s)	:	37.1
Heat flux	(kJ/sq.m)	:	1211.1

ILLUSTRATIVE EXAMPLE : APPLICATION OF TORAP

A risk assessment study has been carried out for a chemical industry situated in a congested industrial complex (Figure 9.4) engaged in manufacturing of linear alkyl benzene (LAB) and some by products such as, chlorinated propylene, and chloroydrin. During the manufacture of LAB industry deals with various hazardous chemicals such as benzene, propylene, n-paraffins, chlorine, and kerosene at elevated temperatures and pressures. The industry and nearby areas are threatened because of different type of hazards (explosion, fire, toxic

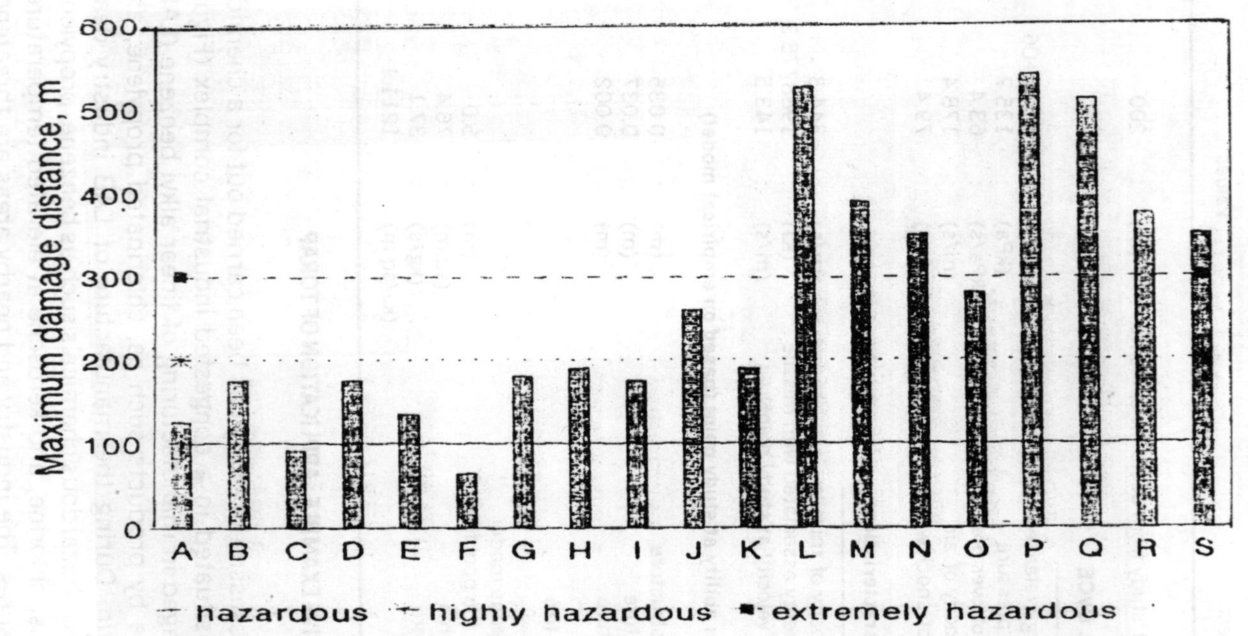

Fig. 9.5: Damage potential of different units in linear alkyl benzene plant; the elaborations of AB, ... S are given in Table 1.

gaseous release, and corrosive liquid release) present in the industry.

Brief process description

The raw materials (kerosene, chlorine, n-paraffins, etc.) are stored in tanks in central tank farm. Kerosene, the main raw material is pumped to prefractionation unit, where it is preheated and is then fed to the stripper unit. Here, components C9 and less are stripped at the top and C10 and above are obtained at the bottom. The bottom stream from the stripper is sent to the rerun column, where the C10 to C13 are separated out. These are mixed with hydrogen and sent to catalytic bed reactor at an elevated temperature. Here, the sulfur, nitrogen, and olefins, which are poisonous to the catalyst are converted into H_2S, NH_3 and paraffins. The treated feed is then sent to the Molex unit for the separation of paraffins. The normal and non-normal paraffins thus obtained are purified by extraction process. The normal paraffins are then sent to Pacol unit, where they are dehydrogenated at elevated temperatures (480 °C) over a catalyst bed of Platinum (pt) on alumina. The cracked products are separated in a stripper column and passed on to alkylation unit. Benzene is added to the mono-olefins in the presence of hydrofluoric acid which catalyses the reaction to form linear alkyl benzene and other products. The main units of linear alkyl benzene plants are listed in Table 9.1.

We have conducted risk analysis of the entire plant. In order to optimize the time and effort, a hazard identification and ranking exercise was first carried out. The exercise is meant to identify the type of hazards present and a rough estimation of their hazard potential. The units are accordingly characterized and ranked in priority for the comprehensive study. Dow's Fire and Explosion Index[22] has been used for this step and a summary of results thus obtained are presented in Figure 9.5. It is clear from the figure that storage of propylene, benzene, and chlorine are the most hazardous, followed by the storage of kerosene, fuel oil, n-paraffins, and allylchloride. Other process units: Pacol, Alkylation, Hydrotreater, Molex, and Chlorohydrination are comparatively less hazardous. All these units have been further subjected to TORAP for hazard assessment and damage potential estimation.

Table 9.6. The output of MAXCRED for *scenario 4*

Parameters			*Values*
Distance of study from accident epicenter	(m)	:	200
Fire : Pool fire			
Continuous model			
Burning area	(sq.m)	:	77.0
Burning rate	(kg/h)	:	15314.9
Heat flux	(kJ/sq.m)	:	994.7

For the sake of brevity, the present paper gives details of the study of only the most hazardous units in the plant (storage unit).

The storage unit comprises of the vessels storing various chemicals at elevated temperature and pressure (Table 2). Each storage vessel (for each hazardous chemical) has been subjected to TORAP for hazard assessment and damage potential quantification. To felicitate understanding of the use of TORAP, the results of the study are presented in steps that follow the algorithm of TORAP.

Accident scenarios generation

Based on the history of major accidents in process industries and the authors' experience, following scenarios have been visualized for accidents in different storage units[9-13,20-21,23].

Table-9.7. The output of MAXCRED for *scenario 5*

Parameters			*Values*
Distance of study from accident epicenter	(m)	:	200
Fire : Pool fire			
Instantaneous model			
Radius of the pool fire	(m)	:	5.0
Burning area	(sq.m)	:	71.1
Burning rate	(kg/s)	:	36.2
Heat flux	(kJ/sq.m)	:	1210.0

Propylene storage : scenario 1

An excessive pressure development in the storage vessel of propylene (under high pressure and temperature) leads to confined vapour cloud explosion (CVCE). The vapour cloud generated by CVCE on ignition turns into a fire ball and consequently damages the other storage vessels (chlorine, allylchloride, benzene, paraffins, fuel oil).

Chlorine storage : scenario 2

A sudden release of pressurized chlorine triggers a boiling liquid expanding vapour explosion (BLEVE) and dispersion of toxic vapor cloud; in other words explosive release followed by toxic dispersion[11,12].

n-paraffin storage : scenario 3

Unconfined vapour cloud explosion (UVCE) occurs in the n-paraffin unit accompanied by a pool fire. This shall happen if a small leak in the storage vessel releases material at moderate flow rate and forms a vapor cloud in the atmosphere which on ignition leads to UVCE. Due to the heat generated by UVCE or by external source of ignition the remaining material in the vessel or dike catches fire causing a pool fire. A UVCE followed by pool fire can damage neighbouring vessels, due to excess shock waves and heat load and trigger secondary and higher order accidents.

Allylchloride storage : scenario 4

Instantaneous release followed by pool fire in the allylchloride storage unit.

Fuel oil storage : scenario 5

There is a pool fire in the fuel oil storage unit.

Chloroydrin and Chlorinated organic storage : scenarios 6 and 7

Release of chloroydrin and chlorinated organic is followed by evaporation and pool fire.

Benzene storage : scenario 8

An excessive high pressure development in the storage vessel of benzene leads to BLEVE (maximum reported incidents). The unburned vapor cloud released due to explosion on ignition leads to fire ball.

Kerosene storage : scenario 9

An instantaneous release of kerosene (due to high pressure development) either through vent valve or through any accidental opening causes the vessel to explode in the manner of a BLEVE. As the released chemical is flammable it turns to flash fire on meeting an ignition source.

These scenarios have been processed for damage estimation through TORAP.

Hazard quantification

The results of the calculations for different accident scenarios are summarized below:

Table 9.8. The output of TORAP for *scenario 6*

Parameters			*Values*
Distance of study from accident epicenter	(m)	:	200
Fire : Pool fire			
Continuous model			
Burning area	(sq.m)	:	77.0
Burning rate	(kg/h)	:	12253.7
Heat flux	(kJ/sq.m)	:	385.5

As per *scenario* 1 (Table 9.3) the missiles generated by CVCE may hit nearby targets and can lead to secondary explosions or toxic releases. The CVCE vapour cloud on ignition may cause a fire ball and hence severe heat radiation load. The shock wave from the CVCE can cause injury as well as second order accidents by seriously damaging other vessels. It has been estimated that shock waves with 50% probability of causing injury would be observed over an area of ~500 m radius. The heat radiation effect with 50% probability of lethality would be observed over an area of ~300 m radius and missile effects with 50% chances of damage (without considering probability of hitting) would be observed across ~750 m radius.

As per *scenario* 2, there would be sudden release of chlorine. The consequent BLEVE would lead to shock waves, and toxic releases (Table 9.4). In this scenario the damage potential of the shock waves is estimated to be comparatively lesser than *scenario* 1. 50% damage —causing shock wave would be operative over an area of ~250 m radius. As chlorine is non-combustible, no heat radiation effects would be observed, but a build-up of lethal concentration would take place over an area of ~1500 m radius.

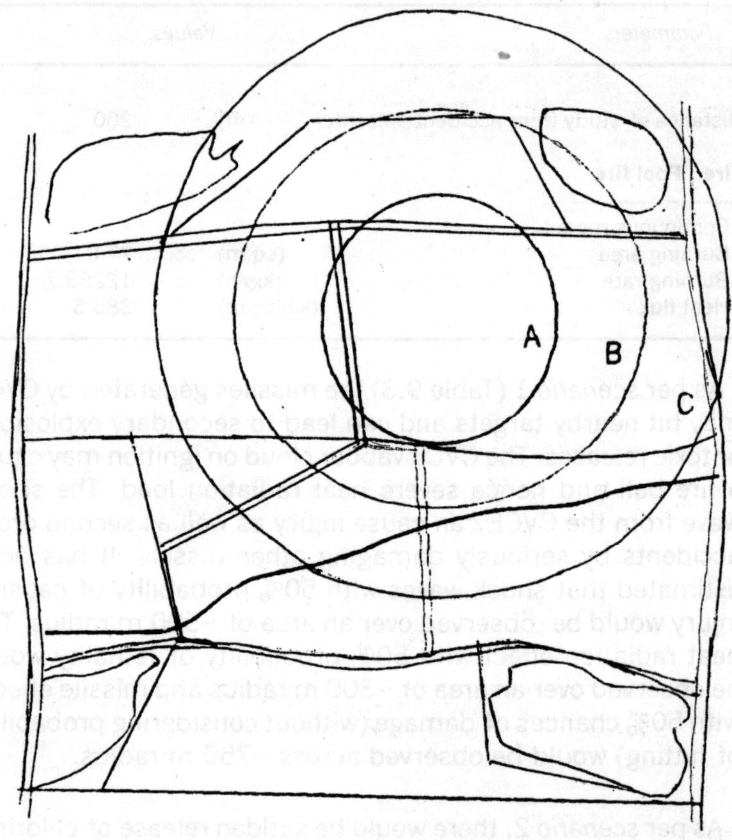

Fig. 9.6: Risk contours indicating the impact area for an accident occurring in propylene storage vessel; due to severe risk (A), high risk (B), and moderate risk (C).

An UVCE as per *scenario* 3 would give rise to heat radiation effects, shock waves, and missile effects. In addition, there would be secondary impact of burning of released material in the dike/vessel which would again lead to additional heat load. In *scenario* 3, missile effect is not significant (Table–5). However, shock waves of intensity high enough to damage all objects coming in their way would be persistent in an area of ~375 m radius. The combined impact of heat radiation, UVCE, and pool fire would be lethal over an area of ~500 m radius. Due to intense heat load some of the chemical would evaporate and disperse causing its build up to a lethal concentration over an area of ~200 m radius.

The model results for *scenario* 4 are presented in Table, 9.6. Lethal heat load would be observed over the study area; the maximum damage distance would envelope an area of ~400m radius.

The output of TORAP for *scenario* 5 (fuel oil) is tabulated in Table 9.7. Lethal heat load would be observed over an area of ~525 m radius.

The results of *scenarios* 6 and 7 have been presented in Tables 9.8 and 9.9 respectively. In both cases the lethal heat load (50% chances of fatality) is confined to an area of ~200 m radius.

As per *scenario* 8 BLEVE would generate shock waves as well as missiles. In addition, there would be secondary impact of the released material getting ignited and forming fire ball thereby generating additional heat load. The output of TORAP for this scenario is presented in Table 9.10. The lethal impact of shock-waves, heat load, and missiles would go upto and beyond a radius of ~350m.

The instantaneous failure of (boiling liquid expanding vapour explosion) kerosene tank as per *scenario* 9 would generate shock-waves and missiles (Table 9.11). The released chemical on meeting ignition source would generate heat load (flash fire). The heat load having propensity to cause damage would be observed over an area of ~250 m radius.

Risk estimation

The individual risk factors for fatality have been estimated using the results obtained by TORAP and the probability of occurrence of the accident scenarios, individual fatal risk factors have been estimated[1,24-26]. The risk factor is a direct representation of the threat (taking into consideration both damage potential and probability of occurrence) to an individual in an area.

Table 9.9. The output of TORAP for *scenario 7*

Parameters			*Values*
Distance of study from accident epicenter	(m)	:	200
Fire : Pool fire			
Continuous model			
Burning area	(sq.m)	:	77.0
Burning rate	(kg/h)	:	13144.3
Heat flux	(kJ/sq.m)	:	477.6

In order to enable visualization of accident scenarios, their risk contours have been drawn over the study area (Figures 9.6–9.14). It may be seen that the risk contours for *scenarios* 1-3, (Figures 9.6–9.8) are extending beyond the boundaries of the industry and enveloping other industries and nearby populated areas. The risk contours for *scenarios* 4 to 7 are confined to the campus of the plant (Figures 9.9–9.12). The risk contours for *scenarios* 8 and 9 extend beyond the boundary of factory (Figures 9.13 and 9.14) and encompasses other industries (storage vessels of other industries). This may cause secondary accidents, the impacts of which may go beyond the industrial complex premise.

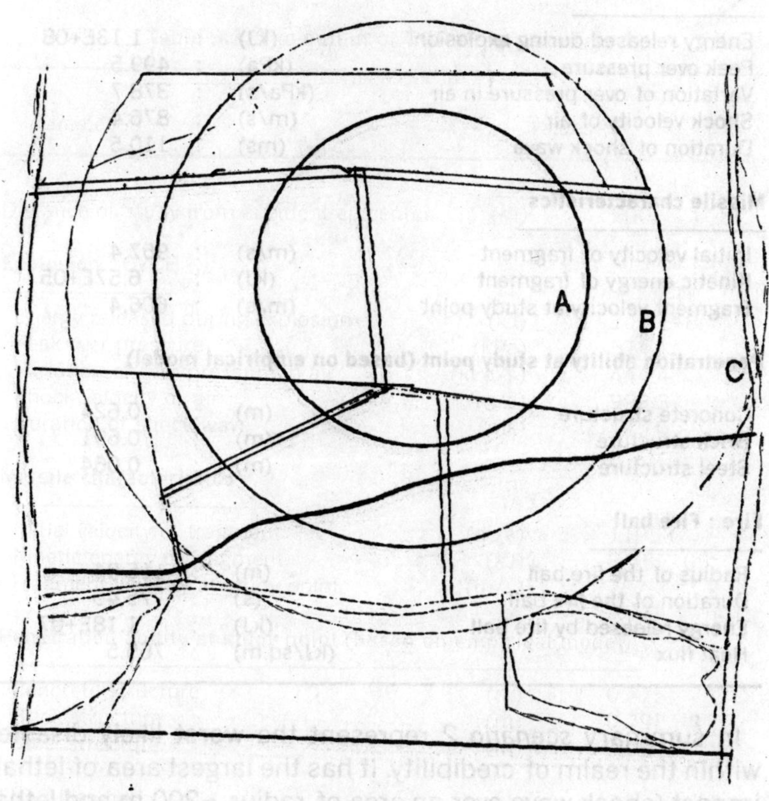

Fig. 9.7: Risk contours indicating the impact area for an accident occurring in chlorine storage vessel; due to severe risk (A), high risk (B), and moderate risk (C).

Table 9.10. The output of TORAP for *scenario 8*

Parameters			*Values*
Distance of study from accident epicenter	(m)	:	200
Explosion : BLEVE			
Energy released during explosion	(kJ)	:	1.13E+08
Peak over pressure	(kPa)	:	499.5
Variation of over pressure in air	(kPa/s)	:	378.7
Shock velocity of air	(m/s)	:	876.4
Duration of shock wave	(ms)	:	110.5
Missile characteristics			
Initial velocity of fragment	(m/s)	:	957.4
Kinetic energy of fragment	(kJ)	:	6.57E+05
Fragment velocity at study point	(m/s)	:	606.4
Penetration ability at study point (based on empirical model)			
Concrete structure	(m)	:	0.624
Brick structure	(m)	:	0.691
Steel structure	(m)	:	0.064
Fire : Fire ball			
Radius of the fire ball	(m)	:	115.04
Duration of the fire ball	(s)	:	75.45
Energy released by fire ball	(kJ)	:	1.18E+07
Heat flux	(kJ/sq.m)	:	765.5

In summary *scenario* 2 represent the worst likely disaster within the realm of credibility. It has the largest area-of-lethal-impact (shock wave over an area of radius ~200 m and lethal concentration across an area of radius ~1500 m). Further, the most thickly populated areas (including residential areas of Kharagpur and Dhorala) lie within its range. If one considers the cumulative effects, the *scenarios* 1 and 8 would come out as the worst, as more intense impacts (in terms of heat radiation, shock waves, and missiles) are observed per unit area in these scenarios. Of the nine most credible scenarios, *scenarios* 1 and 8 are the most likely to cause cascading effects as missiles, shock waves, and radiation effects would be generated simultaneously and other industries or units dealing with flammable and toxic materials are situated within the striking distance of the primary accident. *Scenarios* 3, 4, 5 and 9 also have the potential to lead secondary accidents as severe heat load generated in these would encompass other storage vessels. All-in-all, *scenario* 2 is the worst as far as primary effects are concerned whereas *scenarios* 1 and 8 are the worst in terms of its potentiality of causing cascading (domino) effects.

Table 9.11. The output of TORAP for *scenario 9*

Parameters		*Values*	
Distance of study from accident epicenter	(m)	:	200
Explosion : BLEVE			
Total energy released	(kJ)	:	5.45E+05
Peak over pressure	(kPa)	:	62.6
Variation of over pressure in air	(kPa/s)	:	35.1
Shock velocity of air	(m/s)	:	97.4
Duration of shock wave	(ms)	:	47.1
No missile effect			
Fire : Flash fire			
Volume of vapor cloud	(cub.m)	:	11629.2
Duration o fire	(sec)	:	3876.3
Heat flux	(kJ/sq.m)	:	441.9

SUMMARY AND CONCLUSION

A new software package TORAP has been developed for performing quantitative risk analysis. The package generates different credible accident scenarios, and quantifies the damage they can cause. This information can then be used in developing strategies for preventing accidents and to dampen their adverse impacts if the accidents do take place.

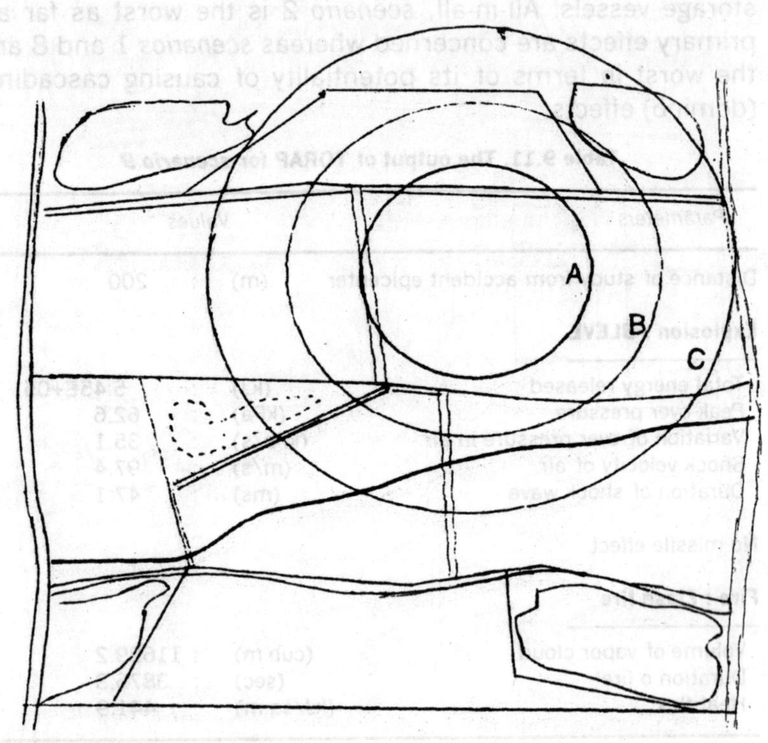

Fig. 9.8: Risk contours indicating the impact area for an accident occurring in paraffine storage vessel; due to severe risk (A), high risk (B), and moderate risk (C).

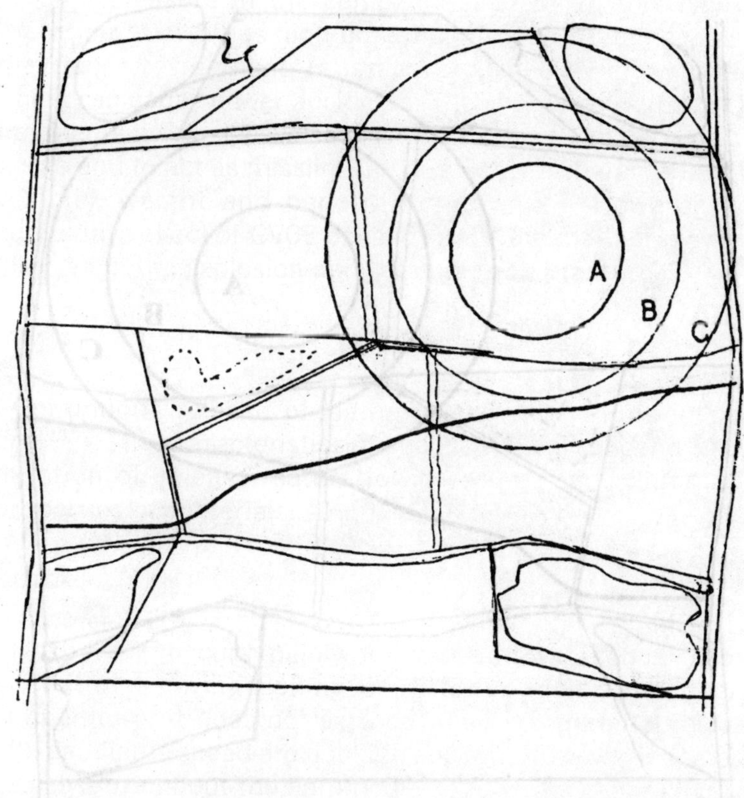

Fig. 9.9: Risk contours indicating the impact area for an accident occurring in allyl chlorine storage vessel; due to severe risk (A), high risk (B), and moderate risk (C).

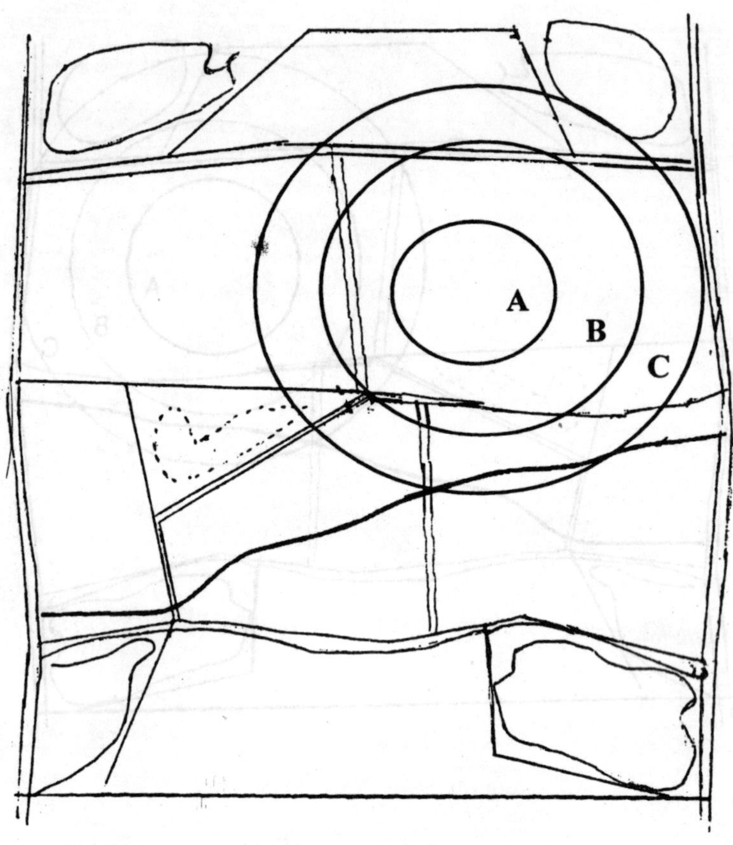

Fig. 9.10: Risk contours indicating the impact area for an accident occurring in fuel oil vessel; due to severe risk (A), high risk (B), and moderate risk (C).

Fig. 9.11: Risk contours indicating the impact area for an accident occurring in chlorineduring storage vessel; due to severe risk (A), high risk (B), and moderate risk (C).

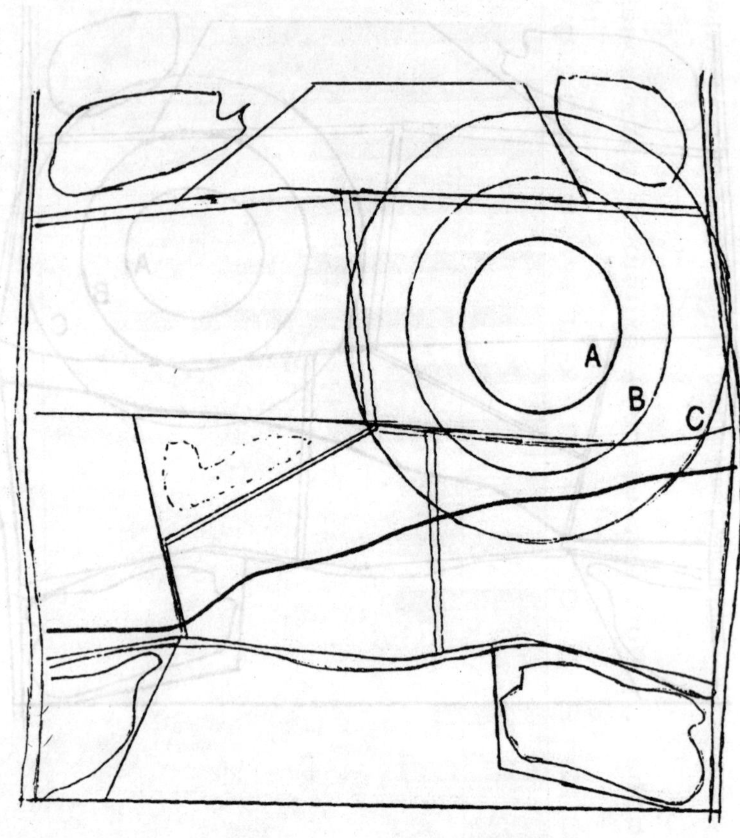

Fig. 9.12: Risk contours indicating the impact area for an accident occurring in chlorinated hydrocarbon storage vessel; due to severe risk (A), high risk (B), and moderate risk (C).

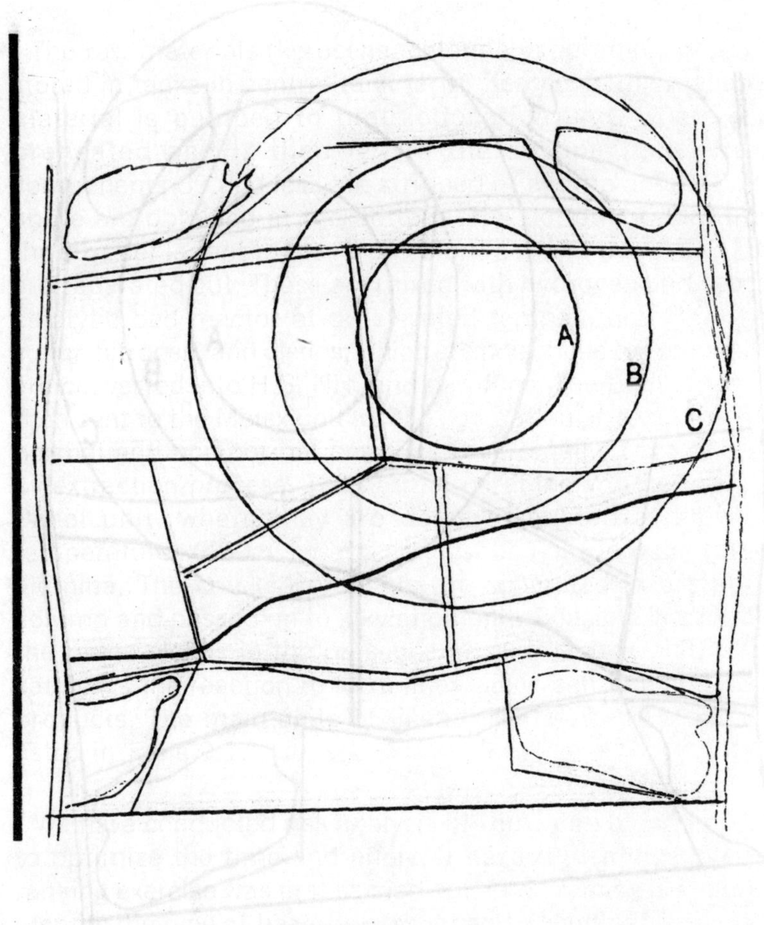

Fig. 9.13: Risk contours indicating the impact area for an accident occurring in benzene storage vessel; due to severe risk (A), high risk (B), and moderate risk (C).

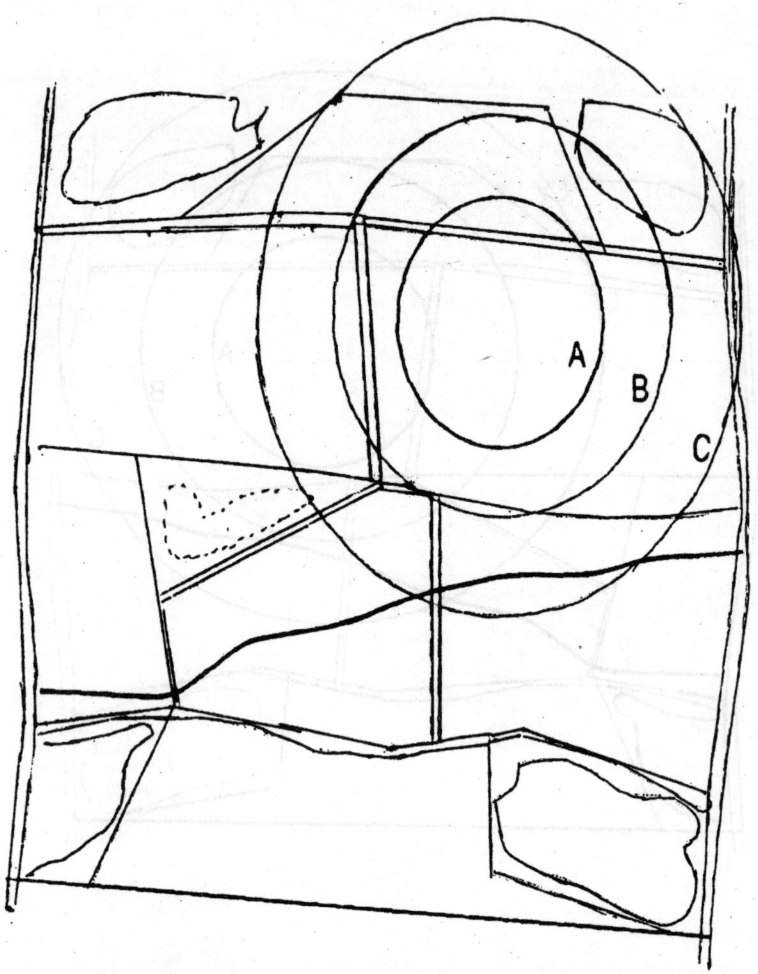

Fig. 9.14: Risk contours indicating the impact area for an accident occurring in kerosene storage vessel; due to severe risk (A), high risk (B), and moderate risk (C).

The applicability of TORAP has been illustrated with a case study of a linear alkyl benzene manufacturing industry situated in Suretgarh, Uttar Pradesh, India. In the first step–hazard identification and ranking - storage and process units involving chlorination, chlorohydration, and quenching were identified as the most vulnerable vis a vis propensity of causing accidents. A detailed study of the credible accidents in storage units and their impacts was then carried out with the help of TORAP. The studies reveal that the storage units of propylene, benzene, chlorine, kerosene, and n-paraffin are the most hazardous, and accidents in these units may cause severe damage to the factory and its surroundings.

To reduce the hazards associated with the storage of chlorine, propylene, benzene and kerosene, proper hazard minimization/mitigation measures should be taken. A few suggestions are made in this context :

- ◊ Instead of one or two large-capacity vessels several vessels of smaller capacity should be used for storage.
- ◊ Adequate space should be kept between the storage vessels and buffers provided between them so that adverse consequences of failure in one of them do not cause second or higher order accidents.
- ◊ Sensitive gas detecting devices for flammable and toxic gases should be installed in the premises of storage area and other units.
- ◊ There should be regular and thorough inspections of electronic control equipment followed by meticulous maintenance.
- ◊ Sufficient quantities of inert gases should be readily available to dilute the concentrations of toxic/flammable gases if they escape to the atmosphere, and to control fire.
- ◊ A thorough emergency preparedness strategy should always be kept in position, fortified by periodic drills or 'dry runs' so that the damage is contained if an accident does occur.

REFERENCES

1. Kletz, T. A., (1986), What went wrong, *(Gulf Publication)*, London.

2. Lees, F. P., Loss prevention in the process industries, *(Butterworths, London)*, 1-3.

3. Abbasi, S. A., Krishnakumari, P., and Khan, F. I. (1999). Hot Topics, (*Oxford University Press, Chennai)* (in press).

4. Khan, F. I., and Abbasi, S.A. (1999). Accident in chemical process industries an analysis of their causes and consequences, J. of Loss Prevention in Process Industires, (in press).

5. The Hindu (1997). Major fire in Vizag refinery, *The Hindu publication*, September 15, 1.

6. Greenberg, H. R. and Crammer, J. J. (1991), Risk assessment and management for chemical process industries, *(Van Nostrand Reinhold*), New York.

7. Khan, F. I. and Abbasi, S. A.,(1985) Risk analysis: a systematic method of hazard assessment and control, *J of Industrial Pollution Control*, 11(2), 89.

8. Van Sciver, G. R., (1990) Quantitative risk analysis in the chemical process industries, *Reliability Engineering and System Safety*, 29, 55.

9. Kayes,P. J., (1986) Manual of industrial hazard assessment technique, (*Technica Ltd.*, London).

10. Green book (1992), Methods for determining of possible damage to people and objects resulting from release of hazardous materials, *Rep CPR 16E*, Voorburg, Warrington.

11. Khan, F. I. and Abbasi, S. A., (1996) Simulation of accidents in a chemical process industry using software TORAP, *Indian J of Chemical Technology*, 3, 338.

12. Khan F. I., and Abbasi, S. A. (1996), Risk analysis of epichlohydrin manufacturing industry using new computer automated tool MAXCRED, *J Loss Prevention In Process Industries*, 10(2), 91.

13. Khan, F. I., and Abbasi, S. A., (1999) MAXCRED: a new software package rapid risk assessment, *Environmental Modeling and Software* (in press)

14. Khan, F. I., and Abbasi, S. A., (1997) , Risk analysis of a cloralkali industry situated in densely populated area, *Process Safety Progress*, 16(3), 172.

15. Khan, F. I. and Abbasi, S. A., (1997), Rapid risk analysis of a typical chemical industry using TORAP, *Indian J of Chemical Technology*, 4,167.

16. Khan, F. I., and Abbasi, S. A., (1997) optHAZOP an optimal and effective procedure to conduct HAZOP study, *J Loss Prevention in Process Industries*, 10(3), 191, 1997.

17. Khan F. I., and Abbasi, S. A.,(1997) Mathematical model for HAZOP study time estimation, *J of Loss Prevention Process Industries*, 10(4), 249.

18. Khan, F. I. and Abbasi, S. A., (1998) TOPHAZOP: a knowledge based software tool for conducting HAZOP in a rapid, efficient yet inexpensive manner, *J Loss Prevention Process Industries*, 11 (2) 321.

19. Khan, F. I., and Abbasi, S. A., (1998) Inherently safer design based on Rapid risk analysis , *J Loss Prevention in Process Industries*, 11(3), 421.

20. Khan, F. I., and Abbasi, S. A., (1998). Models of Domino effect analysis in chemical process industry, *Process Safety Progress*, 17(2), 107.

21. Pasman, H. J., Duxbury, H. A. and Bjordal, T. R., Major hazards in the process industries : achievements and challenges in loss prevention, *Journal of Hazardous Materials*, 30, 1.

22. Pitersen, C. M., (1990). Consequences of accidental release of hazardous material, *Journal of Loss Prevention Process Industries*, 3, 136.

23. AIChE (1994), Dow's Fire and Explosion Index classification guide, AIChE *Technical manual LC 80-29237 :AIChE,* New York, 1994.

24. Khan, F. I. and Abbasi, S. A., (1997), A maximum credible accident analysis based quantitative risk assessment study of a chemical process industry, Indian Chemical Engineer, 39B (2), 90.

25. Contini, S., Amendola, A. and Ziomas, I., (1991) Benchmark exercise on major hazard analysis, *Joint Research Centre-ISPRA*.

26. Reliability Directorate (1992) , Failure frequency of hardware components, Report *No. 87-2981R.27/NVE, Reliability Directorate*.

27. European Community, (1992). Council direction on the major accident hazardous of certain industrial activities, Report No. 82/50/501 EEC, London.

28. Khan, F.I., and Abbasi, S.A. (1999). Accident in chemical process industries an analysis of their causes and consequences, J. of Loss Prevention in Process Industries, (in press).

Chapter 10

MODELS FOR DOMINO EFFECT ANALYSIS IN CHEMICAL PROCESS INDUSTRIES

In the risk assessment parlance, especially with reference to chemical process industries, the term domino effect is used to denote 'chain of accidents', or situations when a fire/explosion/missile/toxic load generated by an accident in one unit in an industry causes secondary and higher order accidents in other units. The multi-accident catastrophe which occurred in a refinery at Vishakhapatnam, India, on 14 September 1997, claiming 60 lives and causing damages to property worth over Rs 600 million, is the most recent example of the damage potential of domino effect. But, even as domino effect has been documented since 1947, very little attention has been paid towards modelling this phenomenon. In this chapter, we have provided a conceptual framework based on sets of appropriate models to forecast domino effects, and assess their likely magnitudes and adverse impacts, while conducting risk assessment in a chemical process industry. The utilizability of the framework has been illustrated with a case study.

INTRODUCTION

The biggest industrial accident of the 1990s - the 'HPCL Disaster' (also known as 'The Vishakhapatnam Disaster') which occurred on 14 September 1997 at the HPCL (Hindustan Petroleum Corporation Limited) refinery near the city of Vishakhapatnam, India, has brought into sharp focus the destructive potential of domino effect *vis a vis* industrial accidents. It was this effect (also variously called 'cascading effect' or 'chain of accidents') which was responsible for a single failure in the HPCL refinery with limited damage potential to escalate into a series of major accidents, eventually claiming over 60 lives, causing damage to property worth over Rs 600 million ($ 20 million), and terrorising a sprawling city of over 2 million inhabitants (Hindu,1997).

On 14 September, 1997, one of the eight Horton spheres, filled with LPG/crude/kerosene and situated near the main gate of the HPCL refinery, caught fire at 6.40 a.m. It, then exploded (Figure 10.1) rocking the entire city and making people think they have experienced an earthquake. The second sphere exploded 15 minutes later and before noon the others also caught fire. The tanks were all full, with crude imports unloaded at the HPCL berth just a few days ago. Fire spread all over the place. People, particularly those living in the vicinity of the refinery, ran for their lives. As the refinery is located in a densely populated industrial belt with several other major industries nearby, a large number of people were directly affected by the happenings at HPCL. Huge tongues of flames and thick black smoke billowed into the sky and joined the hovering monsoon clouds. A sharp shower brought the soot down on the people who saw their white shirts turn black. The rain water flooding the roads was also black and murky.

With both the entrances to the refinery blocked by burning tanks, neither the fire tenders nor the officials could enter the premises for several hours. Only when the contents in the tanks were burnt out, could they venture in.

It rained again later in the evening preceded by strong winds when the flames lingered low and closer to the ground. This again created panic among the public but soon more rain poured down to provide some relief.

The death toll which eventually crossed 60 would have been higher had the fire started half-an-hour later than it did, when the first shift staff would have come in to relieve the night shift. And Sunday being a holiday, the administrative personnel,

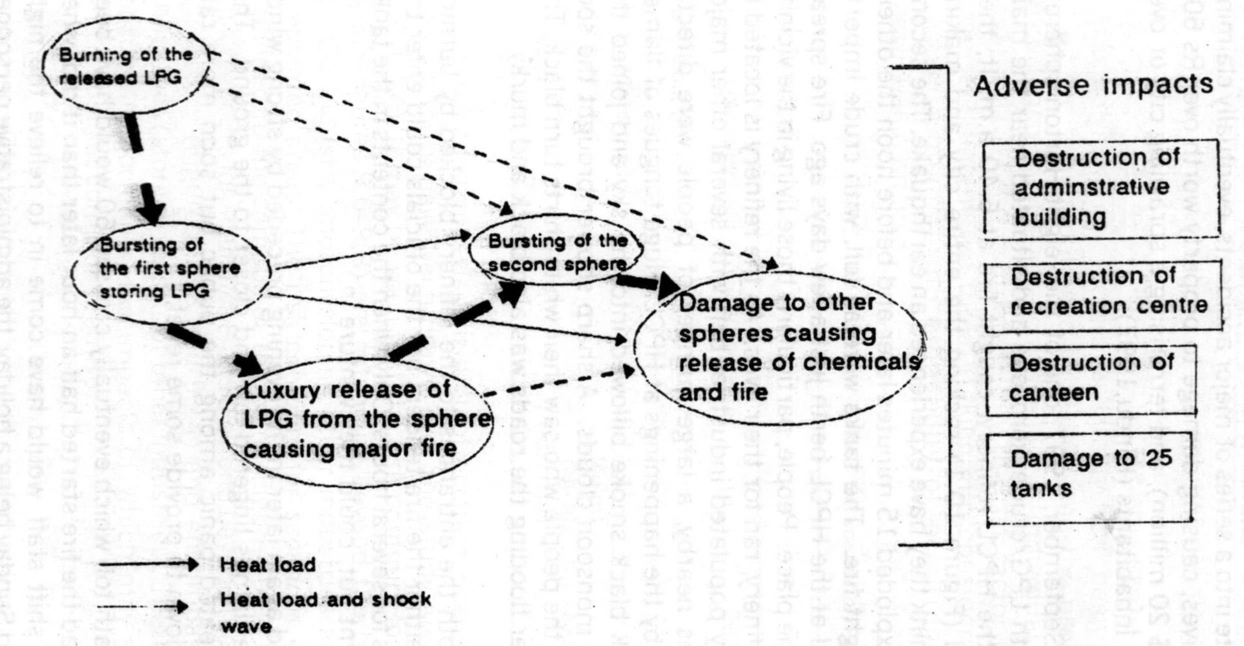

Fig. 10.1: Most probable sequence of events leading to the HPCL's Vishakhapatnam disaster. Thickness of the lines represents the intensity of the heat load impact.

who numbered over 200, were saved as they were not on duty (Hindu,1997).

The deep scars on the psyche of the Vishakhapatnam residents that this series of accidents have left can be gauged from what happened 9 days after the catastrophe. A loud sound created by some innocent activity near a naphtha tank in the HPCL confines,, on 23 September 1997 generated such a panic that workers on duty nearby fled from the scene asking safety personnel of HPCL to raise an alarm which they promptly did. This only helped to spread more panic all over the refinery campus and soon the entire Vishakhapatnam city was gripped with fear. People residing near the refinery complex ran helter-skelter, schools abruptly closed down, shops downed shutters and the entire region was plunged into a war-like situation.

Even as the HPCL's Vishakhapatnam disaster is one of the biggest industrial disaster involving domino effect, it is neither the first instance of its type nor the worst ever. Well-documented instances of chain of accidents in chemical process industries have been occurring since 1940s; the worst such accident–in terms of death toll - had occurred in Mexico city on 19 November, 1984. The accident, which also involved petrochemicals claimed 650 lives (Pietersen, 1986). Table 10.1 presents an illustrative list of some of the major domino accidents that have occurred during the last 30 years.

MODELLING OF DOMINO EFFECTS

In-spite of the destructive potential of domino accidents, and the risk which many industries face from their likelihood all over the world, this phenomenon has received much lesser attention than other aspects of risk assessment.

A thorough scan of primary literature based on CD-ROM, INSPECT, and INDEX searches, as also direct search of books and journals, reveals that the available reports on domino effect are by-and-large confined to qualitative description and interpretations of the events that took place during some of the multiple accidents occurring in the past. For example, Kletz

(1985,1991), Pietersen (1986,1990), Mallikarajunan *et al.* (1988), Pritchard (1989), Bagster and Pitblado (1991), Prugh (1992), and Lees (1996) have described such past multiple accident events and have discussed the various possible ways in which one accident had led to another. The paucity of quantitative studies on this subject can be gauged from the fact that the 3500-page magnum opus by Lees (1996) on risk assessment in process industries contains barely two-page worth of material on domino effect.

A rare quantitative study on domino effect has been reported recently by Latha *et al.* (1992). They have focused on chain of accidents initiated by fire and have conceptualised various ways in which a fire can initiate domino effect. They have also tried to quantify the heat loads generated by various types of fires, the 'strengths' of the primary accidents thus caused, and the mechanisms by which heat load would disperse and cause vessel failures. But, useful as this work is, it is confined to only a few aspects of fire-related domino effects. Our careful study of past accidents has indicated that there can be several other accident events beside fires which can trigger a chain of accidents. Indeed some of the explosions are so powerful that they can simultaneously cause more than one chain of accidents to begin.

In this chapter we have –a) catalogued various types of events which can initiate domino effect, b) presented models for analysing and assessing these events, and c) developed a framework for conducting domino effect analysis in chemical process industries. The applicability of the framework has been demonstrated with an illustrative case study.

Table 10.1. List of the important accidents involving domino effect (Lees, 1996)

Location	*Chemical*	*Date*	*Deaths*	*Injury*
Texas city, USA	Ammonium nitrate	April, 1947	552	3000
Nigata, Japan	Natural gas	June, 1964	3	
Feyzin, France	Propane	January, 1966	18	81
Texas city, USA	Butadiene	October, 1969	13	5
Crescent city, USA	Liquefied petroleum gas (LPG)	June, 1970	0	66
Beek, The Netherlands	Pressurized Naptha	November, 1975	14	104
Westwego, USA	Grain	December, 1977	36	10
Galveston, USA	Grain	December, 1977	15	5
Texas city, Texas	Liquefied petroleum gas (LPG)	November, 1978	7	10
Borger, USA	Light hydrocarbons	January, 1980	0	41
Livingston, USA	Petrochemicals	September, 1982		No details
Mexico city, Mexico	LPG	November, 1984	650	6400
Antwerp, Belgium	Ethylene oxide	July, 1987		No details
Antwerp, Belgium	Ethylene oxide	March, 1989	2	5
Pasadena, USA	Isobutane	October, 1989	23	103
Nagothane, India	Ethane and Propane	November, 1990	31	63
Bradford, UK	Azodiisobutyronitrile (AZDN)	July, 1992	No details	
Vishakhapatnam, India	LPG	September, 1997	60	

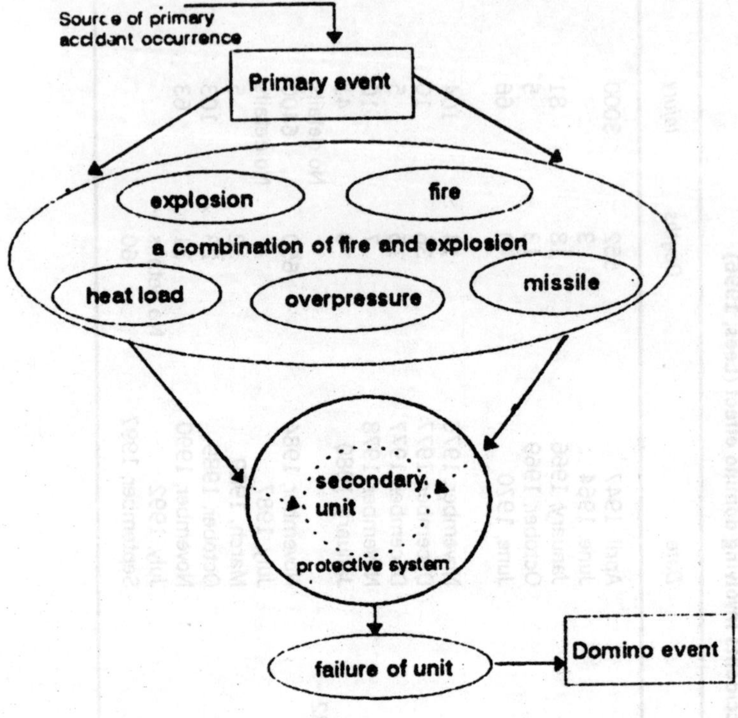

Fig. 10.2: Process sequence in a typical domino event

EVENTS THAT CAN INITIATE DOMINO EFFECT

We have identified the following events which can trigger domino effect:

I. fire,
II. explosions :
   A. blast waves,
   B. missiles,
III. toxic release,
IV. simultaneous and interactive impacts of fire and explosion.

To assess the likelihood of occurrence of these events, and their damage potential, and to forecast whether domino effects would occur and in what sequence, deterministic models have to be used in conjunction with probabilistic analysis. The former type of models are useful in quantifying physical and chemical processes such as size of a leak, chemical release mode and rate, dispersion (including cloud formation), type of explosion and its intensity, heat/missile/shock wave loads generated, etc. The latter type, which are based on careful recapitulation and analysis of past accidents, are needed to work out the probabilities of equipment failure, direction of missiles, direction of fire ball / toxic cloud movement etc.

While choosing the models and conducting analysis based on them one has to be conscious of the following:

a) the computed magnitudes of incident heat and shock-wave effects would have a measure of uncertainty;
b) the stress/strain patterns in vessels effected by fire or fire-explosion combinations would be complex and difficult to quantify with precision;
c) external and transient factors such as wind direction can play a major role;
d) simultaneous occurrence of heat and mass transfer, made complex by transient and rapidly changing nature and magnitudes of the initiating events such as fire due to leakage in a storage vessel, can contribute to the uncertainty.

A typical sequence of a domino accident which indicates the complexities associated with this phenomena, is given in Figure 10.2.

We now present brief reviews of the events which could initiate domino effects vis a vis accidents in chemical process industries.

Fire as initiating event

Fire is a result of uncontrolled oxidation of chemical in the presence of air liberating heat. Fire in a unit generates heat load, which may be large enough to overcome the creep of the material of construction of the nearby units or build up high pressure in the nearby units due to boiling/heating of the chemicals contained in them. Important aspects of modelling domino effects due to fire include prediction of flame heights, flame merging phenomena in fires in close proximity, the effect of wind on flame tilt and length, and heat load.

Fire is among the most common accident that occurs in chemical industries; it is estimated that fire of large or small magnitude occurs at least once a year. There is considerable possibility that one among ten fires can generate heat load large enough to precipitate a significant accident. Study of past accidents indicates that the affects of major fires are significant upto a distance of 200 m from their point of origin (Kletz, 1983; Roberts, 1983; Kayes, 1986; Davies, 1993; Prugh, 1994) while the spacing between hazardous units is generally kept between 50 to 150 m. It, thus, becomes very important to model the phenomena of heat load effecting a nearby unit. Several factors would play important roles including atmospheric conditions (which would determine the direction of maximum flux and rate of heat loss), flame geometry, position and orientation of the receiving unit, the nature and state of chemical contained in the receiving (target) unit etc.

A conceptual model of domino effect triggered by a fire is presented in Figure 10.3. The following combination of events can be initiated by a fire:

fire - explosion - release of chemical ...
fire - release of chemical - fire ...
fire - flammable chemical release - explosion - ...
fire - toxic release ...

Which of these combinations would actually occur, would depend upon the source characteristics (release rate, atmospheric conditions, release conditions, etc.) as well as target characteristics (location, operation, chemical in use, etc.).

The severity of fire in terms of heat load is a function of type of release, flammability and the quantity of the chemical

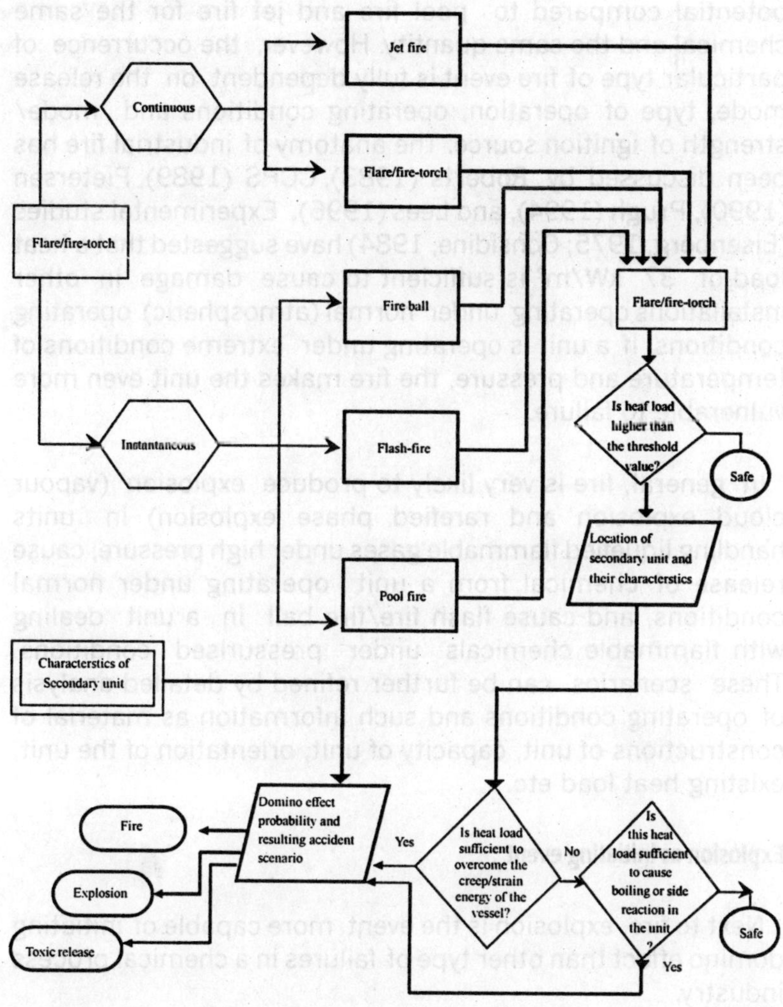

Fig. 10.3: Domino effect model for fire as initiating event

involved, strength of ignition source, and finally the type of fire. For example–flash fire and fire ball have higher damage potential compared to pool fire and jet fire for the same chemical and the same quantity. However, the occurrence of particular type of fire event is fully dependent on the release mode, type of operation, operating conditions and mode/ strength of ignition source. The anatomy of industrial fire has been discussed by Roberts (1983), CCPS (1989), Pietersen (1990), Prugh (1994), and Lees (1996). Experimental studies (Eisenberg, 1975; Considine, 1984) have suggested that a heat load of 37 kW/m^2 is sufficient to cause damage in other installations operating under normal (atmospheric) operating conditions. If a unit is operating under extreme conditions of temperature and pressure, the fire makes the unit even more vulnerable to failure.

In general, fire is very likely to produce explosion (vapour cloud explosion and rarefied phase explosion) in units handling liquefied flammable gases under high pressure, cause release of chemical from a unit operating under normal conditions, and cause flash fire/fire ball in a unit dealing with flammable chemicals under pressurised conditions. These scenarios can be further refined by detailed analysis of operating conditions and such information as material of constructions of unit, capacity of unit, orientation of the unit, existing heat load etc.

Explosion as initiating event

Next to fire, explosion is the event more capable of initiating domino effect than other type of failures in a chemical process industry.

The explosion process

An explosion is a sudden and violent release of energy. In chemical process industries, the energy, thus, released may be chemical or physical; the destructive potential of an explosion depends on the rate at which the energy is released.

Examples of the violent release of physical energy are the explosion of a vessel due to high gas pressure and the sudden rupture of a vessel due to brittle fracture. Thermal energy is also very important in this context - it often creates the conditions for an explosion rather than being a direct source of energy for the explosion itself. In particular, thermal energy may lead to superheating of a liquid under pressure which in turn would cause flashing off of the liquid if it is let to come down to atmospheric pressure.

Certain types of chemical reactions may release energy large enough to cause explosion. For example, explosion may be caused by a violent release of chemical energy in a vessel due to combustion of flammable gas or decomposition of reaction products in a runaway chemical reaction.

An explosion in a vessel due to chemical reaction tends to be 'uniform' while similar reaction in a long pipe gives a 'propagating' explosion.

Deflagration and detonation

Combustion of flammable gas may lead to two kinds of explosions: deflagration or detonation. In a deflagration, the flammable mixture burns relatively slowly. For hydrocarbon-air mixtures the deflagration velocity is typically of the order of 1 m/s.

In a detonation, the flame front travels as a shock wave followed closely by a combustion wave which releases the energy to sustain the shock wave. The detonation front may reach velocities of the order of the velocity of sound and the hot products of combustion may have supersonic velocities. For hydrocarbon-air mixtures the detonation velocity is typically of the order of 2000-3000 m/s.

Detonation generates greater pressures and is more destructive than a deflagration. But, a deflagration may turn into a detonation, particularly when travelling down a long pipe.

Where a transition from deflagration to detonation is occurring, the deflagration velocity naturally exceeds that quoted above.

Table 10.2. Summarized results of consequence analysis of primary event in various units at various locations

Unit/Scenario *Damaging Effect*	**Damaging Potential at Distance(m)** *50*	*75*	*100*	*125*
LDPE plant (ETHYLENE) : CVCE followed by flash fire				
Heat load (kW/m)	50.55	39.35	27.15	18.73
Missile Velocity (m/s)	341.52	250.1	185.3	138.75
Penetration strength (mm) (for 3 kg missile)	32	24	16	10
Peak Over pressure (kpa)	115	95	75	63
Blast wave velocity (m/s)	471	347.3	222.2	162.6
PP Plant (PROPYLENE) : CVCE followed by fire ball				
Heat load (kW/m)	115.32	78.23	51.4	39.71
Missile Velocity (m/s)	421.31	348.45	215.3	158.31
Penetration strength (mm) (for 3 kg missile)	43	35	22	14
Peak Over pressure (kpa)	157	129	98	79
Blast wave velocity (m/s)	757	514.2	368.1	247.3
General storage farm (BUTENE): BLEVE followed by fire ball				
Heat load (kW/m)	128.1	85.42	60.3	42.5
Missile Velocity (m/s)	399.4	287.5	197.1	121.5
Penetration strength (mm) (for 3 kg missile)	37	29	17	11
Peak Over pressure (kpa)	132	98	78	65
Blast wave velocity (m/s)	648.3	368.1	235.2	188.1
Cracker Unit (LPG) : BLEVE followed by fire ball				
Heat load (kW/m)	155.5	133.1	87.55	73.23
Missile Velocity (m/s)	—	—	—	—
Penetration strength (mm) (for 3 kg missile)	—	—	—	—
Peak Over pressure (kpa)	145	110	81	68
Blast wave velocity (m/s)	701	455.6	287.6	204.4
Caustic Plant (CHLORINE) : Instantaneous release followed by *dispersion*				
Heat load (kW/m)	—	—	—	—
Missile Velocity (m/s)	—	—	—	—
Penetration strength (mm) (for 3 kg missile)	—	—	—	—
Peak Over pressure (kpa)	95	65	42	35
Blast wave velocity (m/s)	347.3	188.1	76.6	27.4

Shock / blast waves

Shock wave, or blast wave causes the damaging impact of explosions besides missiles. Blast wave generates overpressure that may injure people and damage equipment and buildings. The anatomy of blast waves and their damaging effects have been discussed by Maurer (1977), Baker *et al.* (1983), Martinsen *et al.* (1986), Medard (1989), Prugh (1991), Greenbook (1992), Cates and Samuels (1993), Venerate *et al.* (1993), Van den Berg and Lennoy (1993), and Davies (1993),

An explosion in air is accompanied by a very rapid rise in pressure leading to the formation of a shock wave. Such wave always travels outwards (from the epicentre of the explosion) with the higher-pressure parts moving at higher velocities. After it has travelled some distance, the shock wave reaches a constant limiting velocity that is greater than the velocity of sound in the air, or in the unburned gas, as in the case of a vapour cloud. The shock wave has a profile in which the pressure rises sharply to a peak value and then gradually tails off. As the shock wave travels outwards the peak pressure at the shock front falls.

At some distance from the explosion centre, a region of negative pressure, or under pressure follows the region of positive pressure, or overpressure, in the shock wave. The underpressure is quite weak and does not exceed above 4 psi.

The peak overpressure p^o, given by the Friedlander equation (Davies, 1993), is related to overpressure at different instants after explosion:

$$p = p^o(1 - t/td) \exp(-\alpha t/td) \quad (1)$$

Where p is overpressure, p^o is the peak overpressure, t is the time, td is the duration time and a is the decay parameter.

A study of past accidents (Clancey, 1972; Eisenberg *et al.* 1975; Pietersen, 1986; Prugh, 1991; Davies; 1993; Khan and Abbasi, 1996b) suggests that an over pressure of 0.7 atm is necessary to cause damage and fatality. In industrial explosions, this order of overpressure is observed over a distance of several hundred meters.

Based on careful studies of past industrial accidents and considering such factors as properties of materials used in the construction of the equipment and the mechanisms of material failure these authors believe that the probabilities of an explosion causing a secondary accident is as much a function of the characteristics of target units, such as material of construction, chemical involved, degree of conjunction, operating conditions, etc., as it is of the overpressure/blast wave (Figure 10.4). Once we take this into consideration, it would appear that domino effect may occur even when overpressure is less than 0.7 atm if the secondary unit has been weakened in any manner due to corrosion, creep, or fatigue. Recent reports of the continuing post-mortom of the HPCL's Vishakhapatnam disaster (Shankar, 1997), have implicated corrosion as being the prime cause behind the primary accident (vessel failure) leading to domino effect. However, the value of 0.7 atm as the lower limit of damaging overpressure is a useful index because this level of overpressure if reached, would damage an otherwise 'fit' unit operating under normal conditions of temperature and pressure.

Missiles

When a vessel or a pipe explodes, its walls are shattered into fragments, which are hurled outwards as missiles by the blast.

Missiles are generally classified as primary and secondary. Primary missiles are those resulting from the bursting of a containment so that energy is imparted to the fragments which become missiles. Secondary missiles occur due to the passage of a blast wave which imparts energy to objects in its path, turning them into missiles.

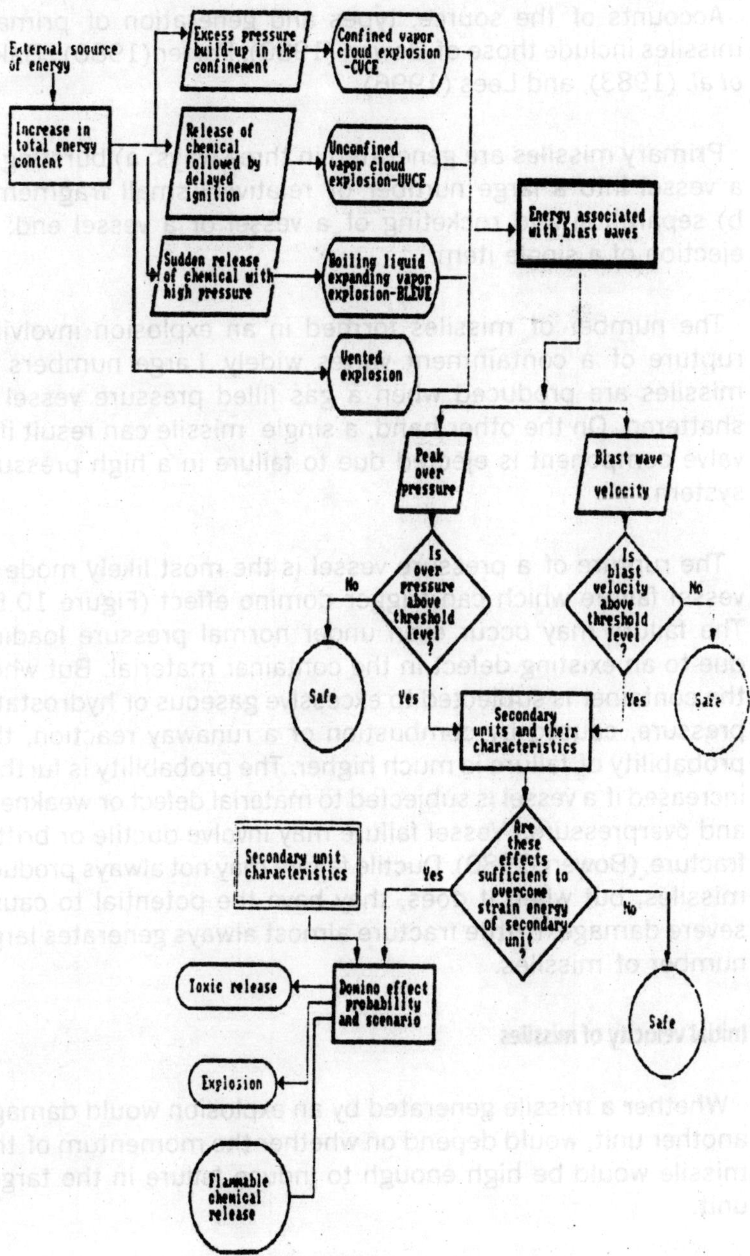

Fig. 10.4: Domino effect model for last wave as initiating event

Accounts of the source, types and generation of primary missiles include those of Bowen (1980), Porter (1980), Baker *et al.* (1983), and Lees (1996).

Primary missiles are generated in three ways: a) bursting of a vessel into a large number of relatively small fragments; b) separation and rocketing of a vessel or a vessel end; c) ejection of a single item.

The number of missiles formed in an explosion involving rupture of a containment varies widely. Large numbers of missiles are produced when a gas filled pressure vessel is shattered. On the other hand, a single missile can result if a valve component is ejected due to failure in a high pressure system.

The rupture of a pressure vessel is the most likely mode of vessel failure which can trigger domino effect (Figure 10.5). The failure may occur even under normal pressure loading due to an existing defect in the container material. But when the container is subjected to excessive gaseous or hydrostatic pressure, caused by combustion or a runaway reaction, the probability of failure is much higher. The probability is further increased if a vessel is subjected to material defect or weakness and overpressure. Vessel failure may involve ductile or brittle fracture. (Bowen, 1980). Ductile failure may not always produce missiles, but when it does, they have the potential to cause severe damage. Brittle fracture almost always generates large number of missiles.

Initial velocity of missiles

Whether a missile generated by an explosion would damage another unit, would depend on whether the momentum of the missile would be high enough to induce failure in the target unit.

Table 10.3. Results of domino effect analysis due to an accident in General storage farm

Unit (Chemicals)	Damaging effects observed	Chances of occurrence of domino effect (%)	Conclusion (secondary accident scenario)	Net probability of occurrence (/yr)
Primary Event : BLEVE followed by fire ball in			4.5E-05	
(Butene)	General storage farm			
Domino Effect : Caustic plant (Chlorine)	shock wave heat load	7.0 13.0	A domino scenario is instantaneous release followed by dispersion.	8.5E-06
PP Plant (Propylene) by fire ball.	shock wave heat load	12.0 11.0	A domino scenario is BLEVE followed	9.8E-06
LDPE plant (Ethylene) by fire ball/flash fire.	shock wave heat load	12.0 15.0	A domino scenario is BLEVE followed	1.1E-05
Cracker plant (LPG)	shock wave heat load	15.0 20.0	A domino scenario is an instantaneous release followed by fire ball/flash fire.	1.4E-05
2nd Butene tank (Butene)	shock wave heat load	17.0 23.0	A domino scenario is CVCE followed by fire ball.	1.6E-05

The initial velocity of a missile is a function of: a) the force or pressure on the missile, b) the transfer of momentum to the missile, and c) the transfer of energy to the missile.

The force acting on a missile depends on the manner of its generation. In case of the rupture of a pressure vessel containing gas, the overpressure may occur due to a slow and gradual rise in pressure or due to a sudden rise as in an explosion. Two types of forces act on the fragments to eject them from the vessel. One is the differential between the gas pressure and the ambient pressure. The other is the dynamic pressure, or wind. In practice, the pressure differential acts only for a very short time, and the acceleration of the fragments is essentially due to the dynamic pressure (Lees, 1996; Khan and Abbasi, 1996b).

In case of rupture of a partition wall by an explosion, too, the forces acting on the fragments are the pressure differential (between the face of the wall and the ambient pressure) and the dynamic pressure. If the pressure incident on the wall is that of a rapidly rising blast wave and the wall configuration is such that the pressure differential across it persists until the wall ruptures, the relevant pressure is the reflected pressure. If, on the other hand, the pressure on the wall rises relatively slowly, the pressure differential is simply that between that pressure and the ambient pressure. In either case, the pressure differential usually acts only for very short time and the acceleration of the fragments is mainly due to the dynamic pressure.

The momentum of the missiles generated from the bursting of a gas filled vessel can also be estimated as the fraction of the available energy which is converted into the kinetic energy of the fragments.

To estimate the force of ejection of an item, such as valve spindle, by a high pressure jet of fluid, an approach based on conservation of momentum may be used.

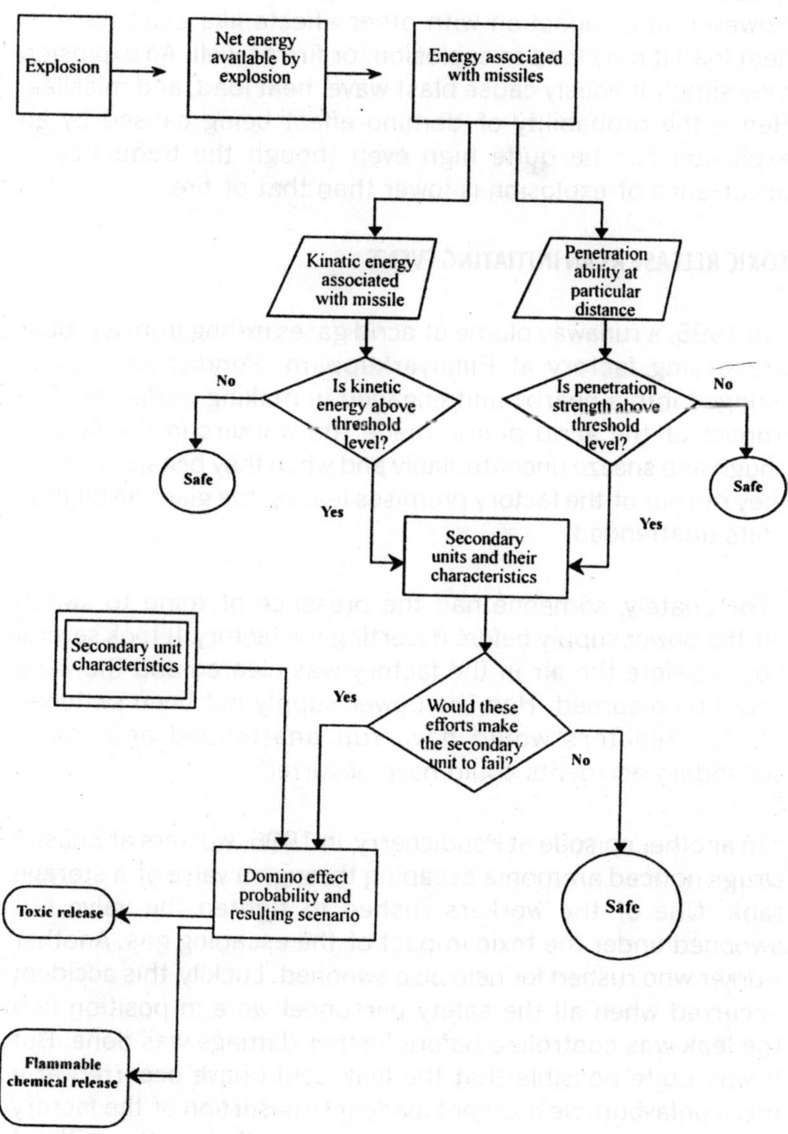

Fig. 10.5: Domino effect model for missile effect as initiating event

The impact of a missile in causing domino effect is primarily related to release of chemical (Pasman *et al.* 1992; Khan and Abbasi, 1996b; 1997) from the target vessel (Figure 10.5). However, in conjunction with other effects like blast wave or heat load it may load to explosion, or fire as well. An explosion may simultaneously cause blast wave, heat load, and missiles. Hence the probability of domino effect being caused by an explosion can be quite high even though the frequency of occurrence of explosion is lower than that of fire.

TOXIC RELEASE AS AN INITIATING EVENT

In 1995, a runaway plume of acrid gases exiting from a rubber processing factory at Pillaiyarkuppam, Pondicherry, India, plunged into a nearby unit engaged in making perfumes. The impact of the acrid plume made the workers in the factory cough and sneeze uncontrollably and when they began to choke, they ran out of the factory premises leaving the giant distillation units unattended.

Fortunately, someone had the presence of mind to switch off the power supply before deserting the factory. It took several hours before the air in the factory was cleared and the work could be resumed. Had the power supply not been switched off the reactors would have run unattended and major secondary accidents could have occurred.

In another episode at Pondicherry, in 1996, workers at Shasun Drugs noticed ammonia escaping through a valve of a storage tank. One of the workers rushed to tighten the valve but swooned under the toxic impact of the escaping gas. Another worker who rushed for help also swooned. Luckily, this accident occurred when all the safety personnel were in position and the leak was controlled before further damage was done. But it was quite possible that the leak could have occurred at a more unfavourable moment leading to desertion of the factory premises and the consequent operation of the reactors without human checks.

Secondary accidents due to the harm caused by toxic release to supervisory staff have not been a common phenomena in the past. But increasing automation, as a consequence of which lesser number of persons are being deployed for supervising process controllers in an industry, toxic release can harm the supervisory personnel to the extent of making them ineffective thereby forcing a situation in which the processes run without human controls. Major chain of accidents can result in such situations. In developing countries like India, the extent of automation has increased but has not reached the stage common in developed countries where process control rooms are carefully constructed and located so as to minimise danger to them from likely accidents.

To quantify the probability of occurrence of domino effect duc to the release of toxic chemicals, the following parameters need to be considered in addition to process details, PIDs, etc. required for all other type of initiating events:

i) control systems employed in the various units in the plant;
ii) detailed description of the human duties and their relative positions in the plant;

The probability of occurrence of domino effect due to this event is likely to be far lower than the other initiating events.

COMBINATION OF INITIATING EVENTS

It has been revealed by various case studies (Kletz, 1985; Marshall, 1987; Raghavan and Swaminathan, 1996; Lees, 1996; Khan and Abbasi, 1995;1997) that a real-life industrial accident is more often than not a combination of different accidental events (explosion, fire and toxic release) rather than a single event. Therefore, in order to study domino effect, probabilities of occurrence of each type of event should be estimated and the final probability of occurrence of domino effect computed accordingly.

Table 10.4. Results of domino effect analysis due to an accident in PP plant.

Unit (Chemicals)	Damaging effects observed	Chances of occurrence of domino *effect* (%)	Conclusion (secondary accident scenario)	Net probability of occurrence (/yr)
Primary Event : CVCE followed by flash fire in (Ethylene) PP plant				5.0E-05
Domino Effect :				
Caustic plant (Chlorine)	shock wave heat load missile	11.0 13.0 5.0	A domino scenario is instantaneous release followed by dispersion	1.3E-05
General storage (Butene)	shock wave heat load missile	15.0 14.0 9.0	A domino scenario is CVCE followed by fire ball.	1.7E-05
LDPE plant (Ethylene)	shock wave heat load missile	11.0 12.0 7.0	A domino scenario is BLEVE followed by fire ball/flash fire.	1.4E-05
Cracker plant (LPG)	shock wave heat load missile	8.0 11.0 5.0	A domino scenario is an instantaneous release followed by fire ball/flash fire.	1.1E-05

PROCEDURE FOR DOMINO EFFECT ANALYSIS (DEA)

As a first step towards DEA, a detailed consequence assessment should be conducted for the primary event (missiles, heat load, overpressure) at the location of secondary unit using latest models for studying the impact generated at the source of the primary event, directional probabilities (in case of missiles and fire jets), damage radii, scenarios of vessel failure etc. The sequence of steps to conduct such a domino effect analysis is presented in Figure 10.6. This procedure recommends use of computer software to conduct detailed consequence analysis. For this purpose, authors (Khan and Abbasi, 1996a) have developed a software package MOSEC (MOdelling and Simulation of Explosion in Chemical process industries). This software enables the detailed study of damaging characteristics of industrial fires and explosions. The software MOSEC incorporate four different models of explosion namely confined vapour cloud explosion (CVCE), unconfined vapour cloud explosion (UVCE), boiling liquid expanding vapour cloud explosion (BLEVE), and vented explosion. It also incorporates three models for fires, namely flash fire, pool fire, and fire ball. To estimate the damage potential most of these models use multi-energy method (Van den Berg and Lennoy, 1993).

In subsequent steps of DEA, the probability of occurrence of domino effect is estimated on the basis of the location, contents, orientation, and material characteristics of the likely target units. The analysis employs, *inter alia*, vessel failure and probit models (Clancey, 1972; Eisenberg *et al.* 1975; Pietersen, 1986; Greenbook, 1992; Davies, 1993; Lees, 1996; Khan and Abbasi, 1997). If the probability of occurrence of a 'hit' is sufficiently high then an accident scenario for the secondary accident is developed followed by an estimation of its damaging potential. This procedure is repeated till the full chain of likely accidents is covered.

To facilitate DEA we propose a two-tier sequence of studies.

Level I study

In this section, a screening is done in order to identify the units which may come under the spell of domino effect. For this purpose threshold values of different damaging effects reported in references Eisenberg *et al.* (1975), Clancey (1972),Pietersen (1986), Pasman *et al.* (1992), Lees (1996), and Khan and Abbasi (1997) are used. For example, it is reported that an overpressure of 0.7 atm can destroy a unit by blast wave impact, a heat load of 37 kW/m^2 is sufficient to induce vessel failure, and a missile (sharp edged) having a velocity higher than 75 m/s has sufficient potential to penetrate the target unit provided that it collides with the unit. If the estimated values of these parameters at the location of the target unit are higher than the threshold values, a detailed analysis or level II study is then performed.

LEVEL II STUDY

At level II, more detailed analysis has to be conducted to verify the existence of domino effect, using the damage potential of the primary event and the characteristics of the secondary unit. The characteristics to be considered are:

- material of construction of the unit,
- process chemicals with operating conditions,
- quantities of chemicals in use,
- properties of the chemicals,
- location of the unit in terms of distance and degree of conjunction, and
- prevailing wind direction.

Using these characteristics and damage potential of primary event, the verification and probability estimation for a secondary event (domino effect) can be worked out as follows:

Table 10.5. Results of domino effect analysis due to an accident in LDPE plant

Unit (Chemicals)	Damaging effects observed	Chances of occurrence of domino effect (%)	Conclusion (secondary accident scenario)	Net probability of occurrence (/yr)
Primary Event : (Propylene)	CVCE followed by fire ball in LDPE plant			3.0E-06
Domino Effect :				
Caustic plant (Chlorine)	shock wave heat load	9.0 10.0	A domino scenario could be toxic release through relief valve or leakage.	5.4E-07
General storage (Butene)	shock wave heat load missile	9.0 13.0 10.0	A domino scenario could be BLEVE followed by fire ball/flash fire.	8.6E-07
PF Plant (Propylene)	shock wave heat load missile	7.0 11.0 9.0	A domino scenario could be CVCE followed by fire ball/flash fire.	7.4E-07
Cracker plant (LPG)	shock wave heat load missile	6.0 9.0 5.0	A domino scenario could be a instantaneous release followed either by UVCE or fire ball.	5.6E-07

Heat load

Some 60% of the total heat radiation incident over a vessel would be absorbed by the vessel shell and the remaining 40% would be absorbed by the chemicals in the vessel to raise the temperature/pressure of the contents (Latha *et al.* 1992)

i) Vessel failure due to high pressure build-up

The probability of secondary accident due to heat absorbed by the chemical:

$$0.40*q*Ar = M* Cp *DT \qquad (2)$$

where, Cp represents specific heat of chemical (kJ/m²/°C), DT is temperature difference, M is mass of chemical (kg), and Ar is area of vessel (m²).

The rise in temperature can be translated in terms of pressure rise by using ideal gas law; assuming the content (in case of pressurised gas) as near to ideal gas. For liquefied gases the energy absorbed by the chemical would lead to the formation of vapour.

At temperature T_2, the fraction of vapour generated is governed by (Kayes, 1986; Greenbook, 1992).

$$f = 1\text{-}\exp(\text{-}Cp*(T_2 \text{-} T_1)/Hv) \qquad (3)$$

and

$$Vvap = f * V \qquad (4)$$

where, Hv is heat of evaporation (kJ/kg), V is total volume of chemical (m³), f is fraction of vapour and Vvap is volume of vapour generated (m³).

The pressure developed in the vessel:

$$P_2 = n*R*T_2/Vvap + P_1 \qquad (5)$$

If P_2 is greater than the relief pressure, it may lead to release of the chemical contained in the vessel and if it is greater than the design pressure then the vessel may burst. The degree

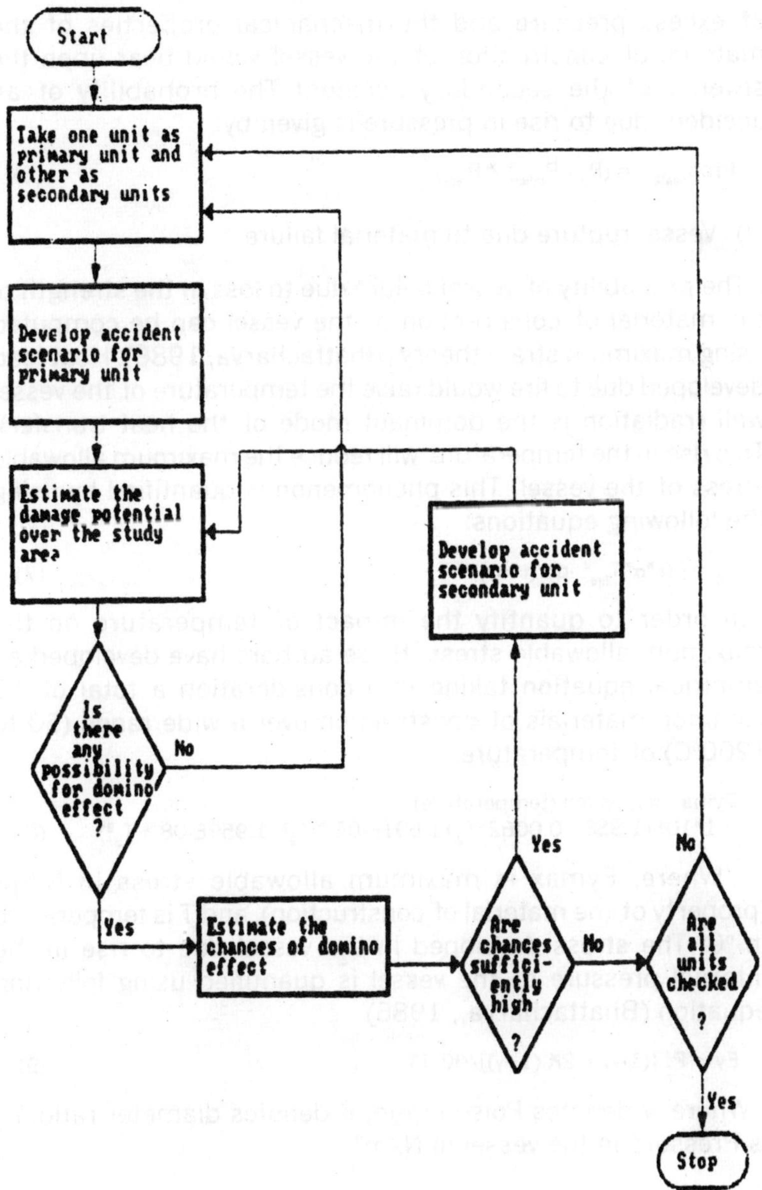

Fig. 10.6: Procedure for domino effect analysis

of excess pressure and the mechanical properties of the material of construction of the vessel would bear upon the severity of the secondary accident The probability of an accident due to rise in pressure is given by:

$$Prob_{heat2} = (P_2 - P_{relief}) / P_{relief} \quad (6)$$

ii) Vessel rupture due to material failure

The probability of vessel failure due to loss in the strength of the material of construction of the vessel can be computed using maximum strain theory (Bhattacharya, 1986). Heat load developed due to fire would raise the temperature of the vessel wall (radiation is the dominant mode of the heat transfer). This rise in the temperature will reduce the maximum allowable stress of the vessel. This phenomenon is quantified by using the following equations:

$$T_2 = [(\varepsilon^*\sigma^*T_{fire}^{\ 4} - q)/(\varepsilon^*\sigma)]^{1/4} \quad (7)$$

In order to quantify the impact of temperature on the maximum allowable stress, these authors have developed an empirical equation taking into consideration a total of 15 common materials of construction over a wide range (50 to 1200°C) of temperature.

Eymax = *function* (temperature)

$$= 1^*10^8 (1.858 - 0.0062^*T_2 + 1.691E\text{-}05 * T_2^{\ 2} - 1.959E\text{-}08 * T_2^{\ 3}) \quad (8)$$

Where, Eymax is maximum allowable stress in N/m^2 (property of the material of construction), and T is temperature in °C. The stress developed in the vessel due to rise in the internal pressure of the vessel is quantified using following equation (Bhattacharya,, 1986)

$$Ey = P_2 [(1-\gamma) + 2K (1+\gamma)]/(K^2-1) \quad (9)$$

Where, v denotes Poison ratio, K denotes diameter ratio, P_2 is Pressure in the vessel in N/m^2.

Subsequently the probability of vessel failure due to failure of the material of construction is estimated as:

$$Prob_{heat2} = Ey_{max}/Ey \quad (10)$$

Table 10.6. Results of Domino effect analysis due to an accident in cracker unit

Unit (Chemicals)	Damaging effects observed	Chances of occurrence of domino *effect (%)*	Conclusion (secondary accident scenario)	Net probability of occurrence (/yr)
Primary Event : (LPG)	BLEVE followed by fire ball in Cracker unit			1.0E-04
Domino Effect :				
Caustic plant (Chlorine)	shock wave heat load	17.0 21.0	A domino scenario is BLEVE followed by dispersion.	3.4E-05
General storage (Butene)	shock wave heat load	19.0 25.0	A domino scenario is BLEVE followed by flash fire.	3.9E-05
PP Plant (Propylene)	shock wave heat load	13.0 17.0	A domino scenario is CVCE followed by fire ball/flash fire	2.8E-05
LDPE plant (Ethylene)	shock wave heat load	10.0 15.0	A domino scenario is CVCE followed by fire ball.	2.4E-05

Finally the probability of secondary accident due to heat load is computed using the function;

$$Prob_{heat} = Prob_{heat1} \cup Prob_{heat\,2} \quad (11)$$

$$Prob_{heat} = minimum[1, \{1-(1-Prob_{heat1})*(1-Prob_{heat2})\}] \quad (12)$$

Blast load (overpressure)

When a blast wave developed due to an explosion hits an object, it causes a diffraction/drag type of impact. The likely damage due to such impact is measured by effective overpressure at the wave front.

$$Pd = Cd * p^o \quad (13)$$
$$Pe = p^o + Pd \quad (14)$$

Where,

Pd is dynamic pressure, p^o is peak overpressure, and Cd is drag coefficient (= 0.1 for sphere, 1.2 for cylinder, 2.0 for square; Lees, 1996).

$$Pr_{blast} = -23.8 + 2.92 * \ln Pe \quad (15)$$

Using equations 15 and 16 (Pasman *et al.* 1992) the probability of damage due to explosion can be computed:

$$Prob_{blast} = fr\,(Pr_{blast}) \quad (16)$$

Where *fr* represents probit function (function that transforms probit values to percent chance), and Pe is explosion pressure in Pa.

Missile load

The impact of missiles on the secondary units is assessed using three parameters:

i) penetration strength, A:

$$Prob_A = (\text{penetration strength} - \text{thickness of vessel})/\text{penetration strength} \quad (17)$$

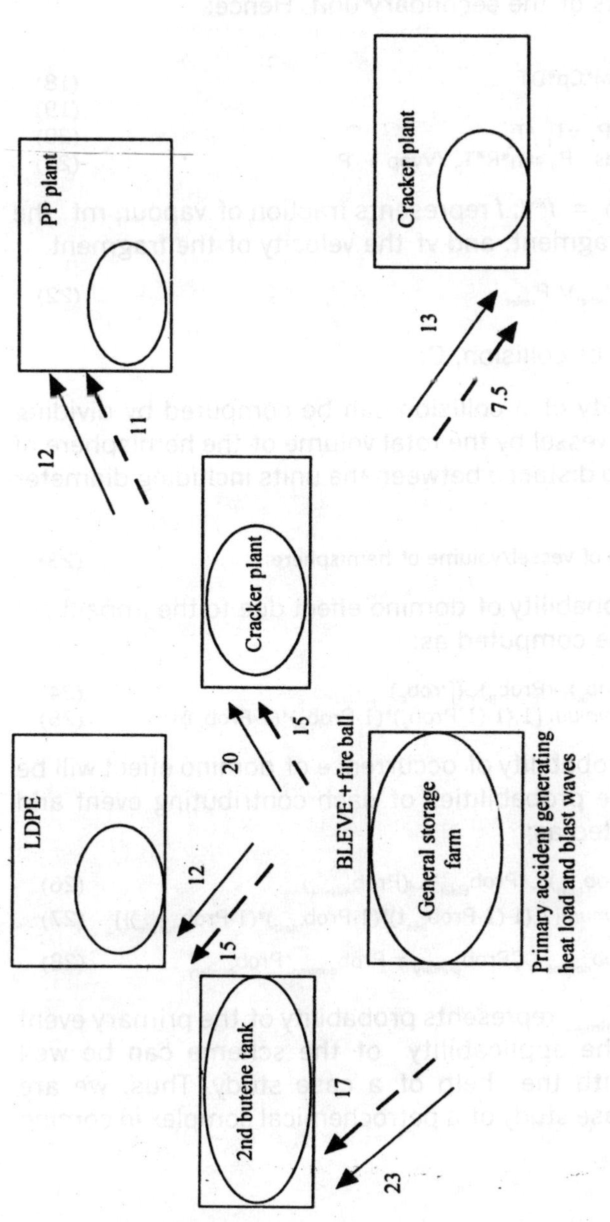

Fig. 10.7: Simplified diagram showing locations of units and impact of primary accident (accident in butene tank) over other units.

ii) impact energy, B:

As reported by Davies (1993) and Lees (1996) about 40-50% of total impact energy is utilized in raising the temperature of the contents of the secondary unit. Hence:

$$1/4 * mf*vf^2 = M*Cp*DT \quad (18)$$

$$T_2 = T_1 + DT \quad (19)$$

for gases $P_2 / P_1 = T_1 / T_2$ (20)

for liquefied gas $P_2 = n*R*T_2 / Vvap + P_1$, (21)

Where, Vvap = f*V; f represents fraction of vapour, mf the mass of the fragment, and vf the velocity of the fragment.

$$Prob_B = (P_2 - P_{relief}) / P_{relief} \quad (22)$$

iii) Probability of collision, C:

The probability of a collision can be computed by dividing the volume of vessel by the total volume of the hemisphere of radius equal to distance between the units including diameter of the vessel.

$Prob_C$= volume of vessel/volume of hemisphere (23)

Thus, the probability of domino effect due to the impacts of missiles can be computed as:

$$Prob_{missile} = (Prob_A) \cup (Prob_B) \cup (Prob_C) \quad (24)$$

$$Prob_{missile} = minimum\ [1, \{1-(1-Prob_A)*(1-Prob_B)*(1-Prob_C)\}] \quad (25)$$

Finally the probability of occurrence of domino effect will be the sum of the probabilities of each contributing event and can be estimated as;

$$Prob_{domino} = (Prob_{heat}) \cup (Prob_{blast}) \cup (Prob_{missile}) \quad (26)$$

$$Prob_{domino} = minimum[1, \{1-(1-Prob_{heat})*(1-Prob_{blast})*(1-Prob_{missile})\}] \quad (27)$$

$$Prob_{domino} = Prob_{domino}\ ÇÇ Prob_{primary} = Prob_{domino} * Prob_{primary} \quad (28)$$

Where $Prob_{primary}$ represents probability of the primary event occurrence. The applicability of the scheme can be well understood with the help of a case study. Thus, we are presenting a case study of a petrochemical complex in coming section.

APPLICATION OF DEA - AN EXAMPLE

We now illustrate the use of the two-tier DEA procedure with reference to its application to a petrochemical complex situated at Ramgarh, India. It is a multi-product petrochemical industry engaged in the manufacture of low density polymer (LDP), high density polymer (HDP), caustic soda and other related products.

As a prelude to DEA, a screening was carried out of all the units for their hazard potential. This led to the identification of these five units as highly vulnerable units: general storage farm, cracker unit, poly propylene (PP) plant, low density poly ethylene (LDPE) plant, and caustic plant. The screening also helped to identify the five most hazardous chemicals: butene, liquefied petroleum gas (LPG), propylene, ethylene and chlorine. The most credible accident scenarios of the primary event were then developed for each unit and a detailed consequence analysis was carried out. The consequences of the primary event were then visualised at the secondary unit. The summary of the study is presented below.

Primary event : general storage farm (butene)

Butene is highly flammable and as it is stored in a liquid form, the most credible accident scenario visualised for this chemical is BLEVE followed by fire ball. The consequence analysis of this scenario, conducted using the software MOSEC, is presented in Table 10.2. Similar analysis for other units is also presented in Table 10.2. The potentiality of damage, if an accident occurs in this unit, to other units situated 50 to 100 meters away has been assessed. The resulting domino effects are described below:

Domino effect triggered by the primary event

Chlorine tank (caustic soda unit situated 70 meters away)

Level I study suggests moderate probability of domino effect due to shock wave and thermal effect.

Level II study verifies the same. The probability of domino effect occurrence due to blast waves is 7.5 % and due to heat load is ~13% (Table 10.3).

The scenario for the secondary accident has been visualised as a sudden release of chlorine. It is because large quantity of chlorine is stored in liquid form under high pressure, hence a leak or build-up of high pressure would cause the chemical to be released instantly. The resulting damage would be caused by toxic load with slight overpressure.

General storage farm (second butene tank situated 60 meters away)

Level I study suggests high probability of domino effect.

Level II study suggests 17% probability due to blast wave and 23% chances due to thermal effect (Table 10.3).

The scenario for the secondary accident in this case has been visualised as: confined vapour cloud explosion (CVCE) followed by fire ball. It is because, heat load would create high internal pressure (about 30% higher than the design value) adding to the already high pressure at which the chemical was stored. As the chemical is highly flammable, it would catch fire soon after release, turning into a fire ball.

LDPE Plant (ethylene tank situated 65 meters away)

Level I study suggests moderate probability of domino effect.

Level II analysis suggests that domino effect probabilities due to the blast and the heat load are 12% and 15% respectively (Table 10.3). The secondary accident would be of a nature similar to the one that may occur in the butene tank: confined vapour cloud explosion followed by fire ball.

PP Plant (propylene tank situated 90 meters away)

Level I study suggests moderate probability of domino effect.

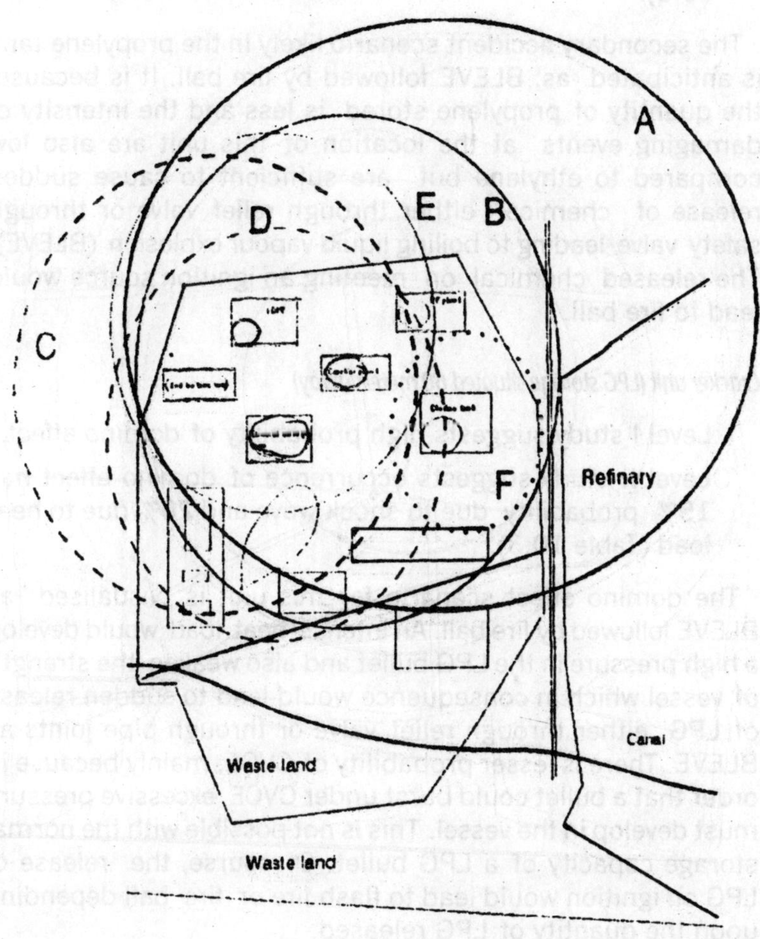

Fig. 10.8: Plot showing severe risk contour due to PP plant (A), Cracker plant (B), General storage farm (C), 2nd butene tank (D), LDPE plant (E), and Chlorine storage vessel (F).

Level II study suggests 12% probability of domino effect due to shock wave and 11% due to heat radiation (Table 10.3).

The secondary accident scenario likely in the propylene tank is anticipated as; BLEVE followed by fire ball. It is because, the quantity of propylene stored is less and the intensity of damaging events at the location of this unit are also low compared to ethylene but are sufficient to cause sudden release of chemical either through relief valve or through safety valve leading to boiling liquid vapour explosion (BLEVE). The released chemical on meeting an ignition source would lead to fire ball.

Cracker unit (LPG storage situated 60 meters away)

Level I study suggests high probability of domino effect.

Level II study suggests occurrence of domino effect has 15% probability due to shock wave and 20% due to heat load (Table 10.3).

The domino effect scenario for this unit is visualised as BLEVE followed by fire ball. An intense heat load would develop a high pressure in the LPG bullet and also weaken the strength of vessel which in consequence would lead to sudden release of LPG either through relief valve or through pipe joints as BLEVE. There is lesser probability of CVCE, mainly because in order that a bullet could burst under CVCE excessive pressure must develop in the vessel. This is not possible with the normal storage capacity of a LPG bullet. Of course, the release of LPG on ignition would lead to flash fire or fire ball depending upon the quantity of LPG released.

A simplified block diagram showing the occurrence of domino effect in various units triggered by a primary accident in the butene tank is presented in Figure 10.7. It is evident that the second butene tank situated near the primary accident site is highly vulnerable. The chlorine tank is comparatively safe.

Table 10.7. Results of domino effect analysis due to an accident in caustic plant.

Unit (Chemicals)	Damaging effects observed	Chances of occurrence of domino effect (%)	Conclusion (secondary accident scenario)	Net probability of occurrence (/yr)
Primary Event (Chlorine)	Instantaneous release followed by dispersion in Caustic soda plant			1.0E-06
Domino Effect : Cracker plant (LPG)	shock wave	3.0	Very low probability of LPG release followed by flash fire.	3.0E-08
General storage	shock wave	1.0	Very low probability of (Butene) butene release followed by flash fire.	1.0E-08
PP Plant (Propylene)	shock wave	7.0	A domino scenario may be propylene release followed by flash fire.	7.0E-08
LDPE plant (Ethylene)	shock wave	0.00	A domino scenario may be ethylene release followed by flash fire.	—

Similar studies were conducted to analyse the domino effect due to accidents in either one or other of these units PP plant, LDPE plant, cracker unit, and caustic plant. The results are summarised in Tables 10.4-10.7.

Consequence analysis covering all the hazardous units reveals that, as far as the primary event is concerned, PP plant has the maximum damage potential as it causes lethal overpressure over the largest area. Risk contours, signifying an individual risk factor greater than $1*10^{-4}$ /yr due to the cumulative effect of the primary and secondary accidents, are plotted over the industry's map in Figure 10.8. It is evident that an accident in PP plant would endanger an area of more than 300 m radius, while the risk contour for cracker plant would encompass an area of about 200 m radius. An accident in the chlorine storage vessel would bring an area of about 70 m radius under severe risk, the lowest area-of-lethal-impact of the scenarios considered.

In summary, the cracker plant has the potential of causing the most damaging primary accident and *also* has the maximum probability to lead to domino effect. The intensities of blast wave, missiles, and heat load would be lethal over an area of 200 meters radius if an accident occurs in this plant. The plant has several vulnerable units close to it which would facilitate a long chain of accidents. On the other hand an accident in caustic plant has the least damage potential and also the least probability of causing secondary accidents.

SUMMARY AND CONCLUSIONS

1. Study of the probability of occurrence of domino effect and forecasting the impacts if such chain of accident occurs, should be an integral part of any risk assessment study. This is borne out by the history of several past accidents including the most recent one—and one of the biggest ever—which occurred at the HPCL's refinery at Vishakhapatnam, India, on 14 September, 1997.

2. Domino effect can be initiated by five types of primary events - fire, overpressure/shock waves, missiles, toxic load, and a combination of one or more of these.
3. In this chapter we have presented details of the mechanisms by which domino effect is initiated by the five types of primary accidents mentioned above. We have then presented models for determining the probabilities of occurrence of these events, developing secondary (and higher order) accident scenarios, forecasting their consequences, and developing disaster prevention/ management strategies based on these studies.
4. A two-tier procedure for conducting a typical domino effect study has been presented. The applicability of the procedure has been demonstrated with the help of a case study.
5. One of the interesting and revealing observations that emerges from the case study is that it is not necessary that the unit which has the maximum damage potential (if it meets with an accident) would also have the maximum probability of causing domino effect. It is because the probability of occurrence of domino effect depends not only upon the damage potential of the primary accident but also on a number of others factors such as characteristics of the secondary unit, location of the unit, and site characteristics.

NOMENCLATURE

Ar = cross sectional area of the unit, m^2
Cd = drag coefficient
Cp = specific heat of the chemical, ($kJ/m^2/\ ^\circ C$)
DT = temperature difference, °C
Eymax = maximum allowable stress of the material of construction of the unit, N/m^2
Ey = stress developed in unit due to the rise in temperature, N/m^2
f = fraction of vapours
Hv = heat of evaporation (kJ/m^2)
K = diameter ratio
M = mass of the chemical, kg
mf = mass of fragment generated due to explosion, kg

n = number of moles of chemical
P = pressure, N/m^2
p° = peak over pressure, N/m^2
p = over pressure, N/m^2
Pr = Probit value
Prob = probability (/yr)
q = heat flux, kJ/m^2
R = gas constant (N m/°C)
T = temperature, °C
t = time , s
td = decay time, s
V = volume of the unit, m^3
vf = velocity of fragment generated due to explosion, m/s
Vvap = volume of chemical transformed to vapour, m^3

Symbols

ε = emissivity of the atmosphere
σ = Stefen-Boltzman's constant
γ = Poisson ratio
α = decay parameter

Subscript

1 = initial condition
2 = final condition
blast = shock wave load
d = dynamic
e = explosion
fire = average fire temperature, K
heat = heat load
missile = missile load
relief = pressure relief condition

REFERENCES

1. Bagster, D. F. and Pitblado, R. M., (1991). The estimation of domino incident frequencies - an approach, *Process Safety Environ*, 69B, pp 196-210.

2. Baker, W. E., Cox, P. A., Westin, P. S., Kulesz, J. J., Strehlow, R. A., (1983). Explosion hazards and evaluations , *Elesevier Science Ltd.*, Amsterdam.

3. Bhattacharya, B. C., (1986). Chemical equipment design , *CBC publication*, Delhi, pp 214–16.

4. Bowen, J. H., (1980). Missile problems in the chemical and nuclear power industries , *Nuclear Engineering*, 19, p 149.

5. Cates, A. T. and Samuels, B., (1993). A simple assessment methodology for vented explosions , *J Loss Prevention in the Process Industries*, 4, pp. 287–96.

6. Centre for Chemical Process Safety, (CCPS), (1989). Guidelines for chemical process quantitative risk analysis , *CCPS Publication G-8*, pp.8, 87, 108, 114, 116, and 123–33.

7. Clancey, V. J., (1972). Diagnostic features of explosion damage , *Proc. Sixth International Meeting of Forensic Science*, Edinburgh.

8. Considine, M., (1984). Thermal radiation hazards ranges from large hydrocarbon pool fires , *SRD report number R297*, U. K. atomic energy authority.

9. Davies, P. A., (1993). A guide to the evaluation of condensed phase explosions , *J of Hazardous Materials*, 33, pp 1-18.

10. Eisenberg, N. A., Lynch, C. J., and Breeding, R. J., (1975). Vulnerability model–A simulation system for assessing damage resulting from marine spills, *Report G-D-136-75*, Washington DC.

11. Greenbook , (1992). Methods for determining the possible damage to people and objects resulting from release of hazardous materials , *Rep CPR 16E, Voorburg*, Warrington.

12. Hindu, (1997). Major fire in Vizag refinery, *The Hindu publication*, September 15, pp 1.

13. Kayes, P. J., (1986). Manual of industrial hazard assessment technique, *Technica Ltd.*, London.

14. Khan, F. I., and Abbasi, S. A., (1995). An anatomy of industrial accidents, *Report CPCE/RA 011/95*, Pondicherry University.

15. Khan, F. I., and Abbasi, S. A., (1996a). MOSEC -Modelling and Simulation of Explosion in Chemical industries, *Report CPCE/RA 013/96*, Pondicherry University.

16. Khan, F. I., and Abbasi, S. A., (1996b). Accident simulation in chemical process industries using software MAXCRED , *Indian J of Chemical Technology*, 3, pp. 338–44.

17. Khan, F. I., and Abbasi, S A, (1997). Risk analysis of a cloralkali industry situated in densely populated area , *Process Safety Progress*, 16(3), pp. 172–85.

18. Kletz, T. A., (1983). After the investigation of fire. *Fire Prevention*, 162, pp 16-22.

19. Kletz, T. A., (1985). What Went Wrong, *Gulf Publication Company*, England.

20. Kletz, T. A., (1991). Process Safety: An Achievement , *Jr. of Institution of Mechanical Engineers*, 11, 34.

21. Latha, P., Gautam, G., and Raghavan, K. V., (1992). Strategies for quantification of thermally initiated cascade effects , *J of Loss Prevention Process Industries*, 5(1),.

22. Lees, F. P., (1996). Loss prevention in process industries , (*2nd edition*), *Butterworths*, 1-3, London.

23. Mallikarajunan, M. M., Raghavan, K. V., and Pietersen, C. M., (1988). An approach to maximum Credible Accident Analysis of a cluster of Chemical industries , *EnviroTec International Conference*, September 21–24, Bombay.

24. Martinsen, W. E., Johnson, D. W., and Terrell, W. F., (1986). BLEVEs : Their causes, effects and prevention, *Hydrocarbon Processing*, 65, p 11.

25. Marshall, V. C., (1987). Major chemical hazards , *John Wiley and Sons*, New York.

26. Maurer, B., (1977). Modelling of vapour cloud dispersion and deflagration after bursting of tanks filled with liquefied gas, *2nd International Symposium on Loss Prevention in Process Industries*, Heidelberg, pp.VI–305.

27. Medard, L. A., (1989). Accidental explosions, *Ellis Horwood*, England.

28. Pasman, H. J., Duxbury, H. A., Bjordal, P, (1992). Major hazards in the process industries, : achievements and challenges in loss prevention, *J of Hazardous Materials*, 33, pp. 1-33.

29. Pietersen, C., M., (1986). Analysis of the LPG disaster in Mexico City, *Loss Prevention and Safety Promotion*, 5, pp. 21.

30. Pietersen, C. M., (1990). Consequence of accidental release of hazardous materials, *J of Loss Prevention Process Industries*, 3, pp. 136–41.

31. Porter, W. H. L., (1980). Generation of missiles and destructive shock fronts and their consequences , *Nuclear Engineering*, 19, pp. 171–78.

32. Pritchard, D. K., (1989). A review of methods for predicting blast damage from vapour cloud explosions, *J of Loss Prevention Process Industries*, 2(4), pp.193–99.

33. Prugh, R. W., (1991). Quantitative evaluation of BLEVE Hazards, *Chemical Engineering Progress*, 87(2), pp. 60-71.

34. Prugh, R. W., (1992). Modelling and mitigation of the consequence of accidental release of hazardous materials, *Plant Operation Progress*, 11 (1), pp.19-25.

35. Prugh, R. W., (1994). Quantitative evaluation of fire ball hazards , *Process Safety Progress*, 13(2), pp. 83-89.

36. Raghavan, K. V., and Swaminathan, G., (1996). Hazard assessment and disaster mitigation in petrochemical industries, *Oxford Publishing Company*, Chennai.

37. Roberts, A. F., (1983). Thermal radiation hazards from pressurised storage, *Fire Safety*, 4, pp.197-212.

38. Scielly, N. F. and High, W. G., (1986). The blast effects of explosion, *Proceeding of International Symposium on Loss Prevention and Safety Promotion*, Cannes.

39. Shankar R., (1997). Corrosion, *Science Express : Indian Express Publication*, pp.1-3, (30 September), Chennai.

40. Van den Berg, A. C., and Lennoy, A., (1993). Methods for vapour cloud explosion blast modelling , *J of Hazardous Materials*, 34, pp 171–79.

41. Venerate, J E S, Rutledge, G A, Sumathipala, and Sollows, K, (1993). To BLEVE or not to BLEVE: Anatomy of Boiling Liquid Expanding Vapour Explosion, *Process Safety Progress*, 12(2), pp 67–70.

Chapter 11

STUDIES ON THE PROBABILITIES AND LIKELY IMPACTS OF CHAINS OF ACCIDENT (DOMINO EFFECT) IN A FERTILISER INDUSTRY

Fertilizer industries are among the chemical process industries where hazardous chemicals are handled and accidents relating to heat/mass/momentum transfer involving such chemicals can occur, leading to explosions, fires, and toxic releases.

Based on the experience gained from past accidents in fertilizer industries, techniques and tools of risk assessment are often employed to identify hazards and forecast likely disasters but such efforts have almost always been focussed on single events or 'stand alone' accidents. The interaction of an accident with other hazardous units which can cause secondary, tertiary, or higher order accidents is seldom studied. But in real-life situations, chain of accidents have a high probability to occur. This has happened in the past and the likelihood of it occurring in future is forever increasing as more and more industries are located close to each other due to increasing population pressure in most countries. In view of this we have developed models and computer-automatic tools to study chain of accidents or 'domino effects'. In this Chapter, which is a sequel to Chapter 10, we have presented an illustrative case study of chain of accidents in a major fertiliser industry situated near the city of Chennai (Madras).

INTRODUCTION

Most techniques of risk assessment in chemical process industries handle stand-alone accidents and their likely consequences. However, in reality, it often happens that accident in one unit causes a secondary accident in a nearby unit which in turn sometimes triggers a tertiary accident, and so on. The probability of occurrence and adverse impacts of such *domino* or *cascading* effects are increasing due to increasing congestion in industrial complexes and increasing density of human population around such complexes.

Well-documented instances of chain of accidents in chemical process industries have been occurring since 1940s; the worst such accident–in terms of death toll - had occurred in Mexico city on 19 November, 1984. The accident, which also involved petrochemicals claimed 650 lives (Petersen, 1986,1990). Other well-known instances of chain of accidents are: Texas, 1947; Westwego, 1977; Mexico city, 1984; Antwerp, 1987, 1989; Pasadena, 1989; Nagothane, 1990; Bradford, 1992 and the Vishakhapatnam Disaster (Khan and Abbasi 1999b).

As a part of an ongoing programme of developing techniques and tools for conducting risk assessment in chemical process industries, we have introduced a computer automated tool MAXCRED (Khan and Abbasi, 1997a 1999a;), and its advanced version MAXCRED-II (Khan and Abbasi, 1997b; 97c 99a) for conducting a rapid risk assessment systematic, optimising HAZOP (hazard and operability study) procedure (Khan and Abbasi, 1997c), and the methods of safety-based design (Khan and Abbasi, 1997c). All these procedures essentially deal with accidents in a single industry, more so in one of the units of an industry. Recently, we have developed a set of models to study domino effort (Khan and Abbasi, 1998a), and have developed a computer automated tool, DOMIFFECT based on the models, (Khan and Abbasi, 1998b). In this paper we present a brief description of the occurrence of domino effect and the application of the tools and techniques developed by us to a typical fertiliser industry.

Table 11.1. List of different toxic release and dispersion models used in the analysis module of DOMIFFECT

Event	Model incorporated	Important references
Toxic release		
Light gas	Gaussian dispersion model with Erbink modification	Greenberg and Crammer,1992 Erbink,1993,1995
Heavy gas	Box model, modified plume path theory for heavy gas	Deaves,1992; Khan and Abbasi,1997f,1997g
Explosion		
BLEVE	thermodynamic and heat transfer models	Pietersen,1990, Pasman et al., 1992; Prugh,1991;Venerate et al., 1993
VCE	multi energy model	Ven den Berg and Lennoy,1993; Prugh,1987; Prugh,1992; Lees, 1996
CVCE	vapour cloud explosion in confinement (vessel or building)	Singh,1988; Pietersen,1990; Lees,1996; Pritchard,1989; CCPS,1994
Fire		
Flash fire	flare model, fire torch and spontaneous combustion	Kayes,1986, Greenberg and Crammer,1992; CCPS,1994

Pool fire	combustion of liquid pool	Roberts,1983; Mourehouse,1982, Davies,1993; Greenbook,1992
Fire ball	spontaneous combustion of vapour cloud	Roberts, 1983; Davies,1993 Prugh, 1994
Domino effect		
Heat load	thermodynamic models, fluid expansion models, probit function, heat transfer models.	Pietersen,1990; Bagster and Pitblado,1990; Prugh,1994 Latha et al. 1992; Khan and Abbasi, 1997h
Blast wave load	shock wave model, pressure-impulse model pressure loading, pressure-temperature interac-·tion, empirical probit models.	Pritchard,1989; Pietersen,1990, Davies,1993; Khan and Abbasi,1997h,Lees,1996
Missile load	missile impact model, thermo-dynamic models, heat transfer models, and empirical models for probability estimation.	Bowen,1980; Davies,1993, Khan and Abbasi, 1997h

OCCURANCE OF DOMINO/CASCADDING EFFECTS

For a chain of accidents to occur in a chemical process industry, it is necessary that–a) there are sensitive units within 'striking distance' of the missiles or fires generated by an accident, b) the missiles or fires are powerful enough to cause failure in the units they strike, and c) the probability of both the 'hits' occurring, and with sufficient destructive potential, is high.

In order to forecast domino effects it is essential to understand factors such as mechanism of accidents, site characteristics, and such interactions which may initiate a chain of accidents. A typical pattern of domino effect in a chemical process industry is presented in Figure 11.1,a-c. Figure 11.1a shows building-up of high pressure in the vessel, which causes explosion and fire, generating heat load, shock waves, and missiles. The nearby unit (sphere, Figure 11.1b) is damaged, in turn generating fresh heat load, shock waves, and missiles. The chain continues through to the third unit (Figure 11.1c) In a real-life situation such a cascading effect would propagate until no more units are present within the danger zone created by the previous accidents.

Four main types of primary accidents can trigger domino effect:

I. fire,
II. explosions :
   A. blast waves,
   B. missiles,
III. toxic release,
IV. simultaneous and interactive impacts of fire and explosion.

One has to employ deterministic models in conjunction with probabilistic analysis to forecast the occurrence and assess the consequences of domino effect. (Latha *et al* .1992; Khan and Abbasi, 1998c,98d,98e,99a). Whereas probabilistic models based on previous histories are useful in the assessment of 'event frequencies', equipment reliability,

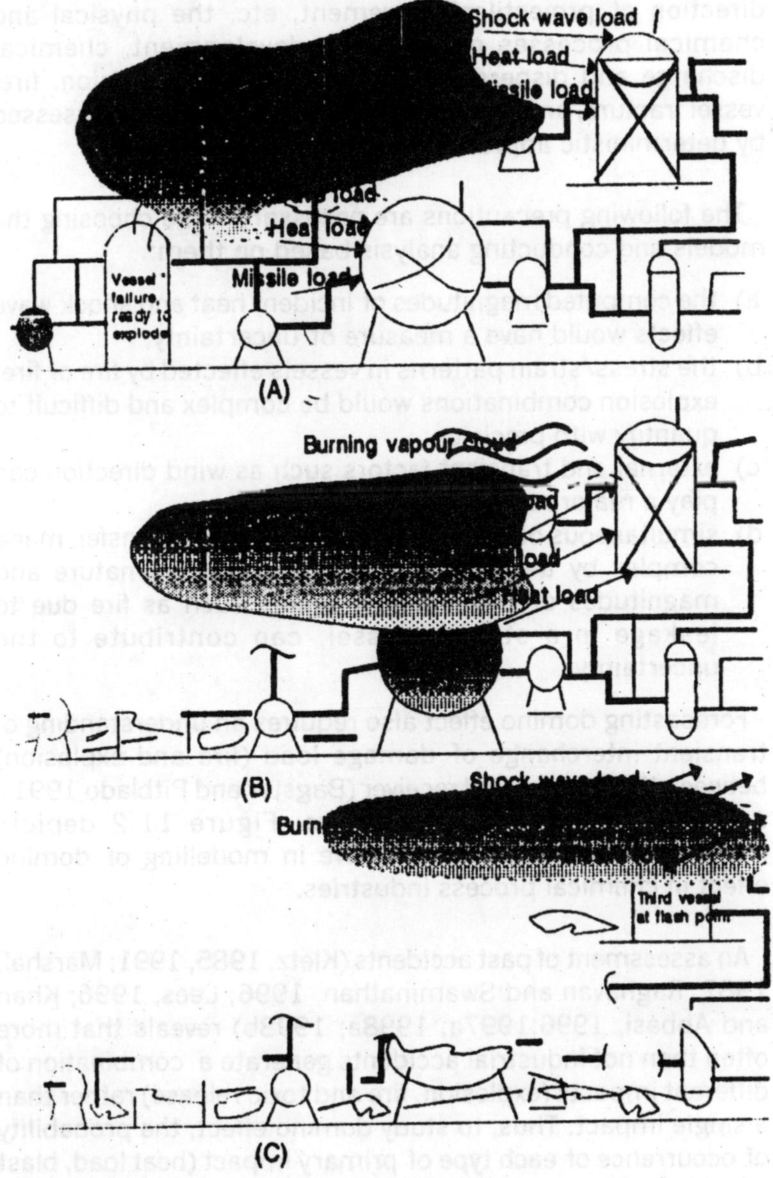

Fig. 11.1: Typical chain of accidents in a chemical process industry

direction of projectiles' movement, etc, the physical and chemical processes namely leak development, chemical discharge and dispersion, cloud formation, explosion, fire, vessel fracture, and related phenomena, can best be assessed by deterministic approaches.

The following precautions are necessary while choosing the models and conducting analysis based on them:

a) the computed magnitudes of incident heat and shock-wave effects would have a measure of uncertainty;
b) the stress/strain patterns in vessels effected by fire or fire-explosion combinations would be complex and difficult to quantify with precision;
c) external and transient factors such as wind direction can play a major role;
d) simultaneous occurrence of heat and mass transfer, made complex by transient and rapidly changing nature and magnitudes of the initiating events such as fire due to leakage in a storage vessel, can contribute to the uncertainty.

Forecasting domino effect also requires an understanding of transient interchange of damage load (fire and explosion) between the initiator and receiver (Bagster and Pitblado,1991), and mechanics of vessel failure. Figure 11.2 depicts mechanism and difficulties involve in modelling of domino effect in chemical process industries.

An assessment of past accidents (Kletz, 1985, 1991; Marshal, 1987; Raghavan and Swaminathan, 1996; Lees, 1996; Khan and Abbasi, 1996;1997a; 1998a; 1998b) reveals that more often than not industrial accidents generate a combination of different impacts (explosion, fire and toxic release) rather than a single impact. Thus, to study domino effect, the probability of occurrence of each type of primary impact (heat load, blast wave, and missiles) should be estimated and combined according to the anticipated sequence of events in order to forecast the nature and intensity of the secondary accident (Khan and Abbasi, 1997a, 1997b; 1998a, 98b, 98c,99b).

Table 11.2. Main units of urea plant

Units	*Reference used in Figure 9*	*operation/purpose*
Hydro-de-sulphuriser – I (HDS-I)	A	Hydro- de-sulphurisation
Hydro-de-sulphuriser – II (HDS-II)	B	Hydro- de-sulphurisation
Hydro-de-sulphuriser stripper	C	Hydrogen sulphide stripping
Primary reformer	D	Reforming
Secondary reformer	E	Reforming
Primary CO converter	F	CO to CO_2 conversion
Secondary CO converter	G	CO to CO_2 conversion
Methanator	H	Methanation
Catacab reboiler	I	Reboiler
CO_2 absorber	J	CO_2 absorption
Ammonia reactor	K	Ammonia formation
Ammonia absorber	L	Ammonia absorption

A COMPUTER-AUTOMATED TOOL *THE DOMIFFECT*

We have developed DOMIFFECT (DOMIno eFFECT) as a computer-automated tool for conducting risk assessment in chemical process industries (Khan and Abbasi 1998b). The sequence of steps involved in DOMIFFECT is shown in Figure 11.3. DOMIFFECT is menu-driven and interactive, capable of the following:

- estimation of all possible hazards from toxic release to explosion;
- handling of interaction among different accidental events (generation of domino or cascading accident scenarios);
- estimation of domino effect probability;
- estimation of domino effect consequences.

The software incorporates state-of-the-art models for studying–a) the hazards of fire, explosion, toxic release, or combination of these present in a chemical process industry; b) the damage potential of likely accidents, assessed on the basis of credible scenarios the tool develops; c) the likelihood of second accident being triggered by the first, d) the scenarios of the second accident, their damage potential, and the probability of their causing a third accident (steps similar to a–c above), and so on. A major feature of DOMIFFECT is the

Table 11.3. The result of DOMIFFECT for accident in HDS-I unit and subsequent domino effect in other units

Parameters			Values
Primary Event			
Fire : Fire ball			
Radius of the fire ball	(m)	:	87.4
Duration of the fire ball	(s)	:	54
Energy released by fire ball	(kJ)	:	5.4E+06
Radiation heat flux	(kJ/sq.m)	:	735.6
Secondary Effect : HDS –II unit			
Domino Checking			
Distance of study for secondary unit	(m)	:	45.0
Domino effect due to heat load :			
Probability of domino effect due to heat load	(%)	:	7.5
Domino effect due to over pressure			
Probability of domino effect due to explosion is	(%)	:	13
Domino effect due to missile :			
Probability of domino effect due to missile	(%)	:	2

{Cont.}.....

Domino analysis : *result for scenario instantaneous release followed by fire ball*

Fire : Fire ball

Radius of the fire ball	(m)	:	95.5
Duration of the fire ball	(s)	:	61
Energy released by fire ball	(kJ)	:	7.2E+06
Radiation heat flux	(kJ/sq.m)	:	825.2

Secondary Effect : Primary reformer unit

Domino Checking			
Distance of study for secondary unit	(m)	:	60.0
Domino effect due to heat load :			
Probability of domino effect due to heat load	(%)	:	23
Domino effect due to over pressure			
Probability of domino effect due to explosion is	(%)	:	17
Domino effect due to missile :			
Probability of domino effect due to missile	(%)	:	9

Domino analysis : *result for scenario BLEVE followed by fire ball*

Explosion : BLEVE			
Total energy released	(kJ)	:	12890.0
Peak over pressure	(kPa)	:	65.8
Variation of over pressure in air	(kPa/s)	:	0.67
Shock velocity of air	(m/s)	:	389.0
Duration of shock wave	(ms)	:	300

{Cont.}...

No missile effect

Fire : Fire ball

Radius of the fire ball	(m)	:	88.1
Duration of the fire ball	(s)	:	36
Energy released by fire ball	(kJ)	:	1.0E+06
Radiation heat flux	(kJ/sq.m)	:	745.6

Secondary Effect : Secondary reformer unit

Domino Checking			
Distance of study for secondary unit	(m)	:	75.0
Domino effect due to heat load :			
Probability of domino effect due to heat load	(%)	:	10
Domino effect due to over pressure			
Probability of domino effect due to explosion is	(%)	:	5
Domino effect due to missile :			
Probability of domino effect due to missile	(%)	:	9

Domino analysis : *result for scenario BLEVE followed by fire ball*

Explosion : BLEVE			
Total energy released	(kJ)	:	11540.0
Peak over pressure	(kPa)	:	87
Variation of over pressure in air	(kPa/s)	:	0.57

{Cont.}.....

Shock velocity of air	(m/s)	:	341.0
Duration of shock wave	(ms)	:	256
No missile effect			
Fire : Fire ball			
Radius of the fire ball	(m)	:	72
Duration of the fire ball	(s)	:	33
Energy released by fire ball	(kJ)	:	7.7E+05
Radiation heat flux	(kJ/sq.m)	:	634.5
Secondary Effect : Catacarb reboiler unit			
Domino Checking			
Distance of study for secondary unit	(m)	:	90.0
Domino effect due to heat load :			
Probability of domino effect due to heat load	(%)	:	10
Domino effect due to over pressure			
Probability of domino effect due to explosion is	(%)	:	7
Domino effect due to missile :			
Probability of domino effect due to missile	(%)	:	0
Domino analysis : *result for scenario continuous release followed by pool fire*			
Fire : Pool fire			
Radius of the pool fire	(m)	:	47.0
Burning area	(sq.m)	:	91781.8

{Cont.}.....

Burning rate	(kg/min)	:	574.8
Heat flux	(kJ/sq.m)	:	332

Secondary Effect : Ammonia converter unit

Domino Checking			
Distance of study for secondary unit	(m)	:	70.0
Domino effect due to heat load :			
Probability of domino effect due to heat load	(%)	:	25.0
Domino effect due to over pressure			
Probability of domino effect due to explosion is	(%)	:	30
Domino effect due to missile :			
Probability of domino effect due to missile	(%)	:	11

Domino analysis : *result for scenario BLEVE followed by toxic load build-up*

Explosion : BLEVE			
Energy released during explosion	(kJ)	:	2.4E+08
Peak over pressure	(kPa)	:	626.4
Variation of over pressure in air	(kPa/s)	:	420.4
Shock velocity of air	(m/s)	:	763.6
Duration of shock wave	(ms)	:	41
Missile characteristics:			
Kinetic energy associated with fragment	(kJ)	:	3.9E+06
Initial velocity of fragment	(m/s)	:	252.9

{Cont.}.....

Penetration at distance of 250 meter :			
Concrete structure	(m)	:	0.127
Brick structure	(m)	:	0.162
Steel structure	(m)	:	0.024
Toxic release and dispersion			
Box instantaneous model : Elevated source			
Concentration at distance of 200 (m),	(kg/cu.m)	:	3.452E-04
Heavy gas puff Characteristics			
Ground level concentration of puff	(kg/cu.m)	:	3.452E-04
Ground level concentration on puff axis	(kg/cu.m)	:	3.452E-03
Cloud radius	(m)	:	1.167E+02
Maximum distance travelled by the cloud	(m)	:	7.874E+02
Maximum ground level concentration	(kg/cu.m)	:	4.3

module automatic analysis. This module uses a knowledge-base built into DOMIFFECT, estimates the detailed consequences, and checks the probability of occurrence of secondary or higher order accidents at different locations.

DOMIFFECT is a PC-based software with object oriented architecture coded in C++. It consists of six main modules DATA, ACCIDENT SCENARIO, ANALYSIS, DOMINO, GRAPHICS and USER INTERFACE.

The data and accident scenario modules of DOMIEFFECT have several features similar to the ones included in MAXCRED II, with the following additions:

i) automatic scenario generation sub-module of DOMIFFECT consists of more elaborate knowledge-base than the corresponding MAXCRED-II sub-module;
ii) a different and faster inference engine is used for retrieving the knowledge from knowledge-base;
iii) user-defined scenario generation sub-module consists of more advanced models for explosion and fire . It also treats gaseous dispersion in much more elaborate fashion incorporating density effects (heavier than air), atmospheric conditions, and terrain characteristics.

The analysis module has the features contained by the corresponding module of MAXCRED-II. In addition it contains newer models (see Table 11.1).

The Domino module assess the damage potential of the primary event at the point of location of the secondary unit and checks for the likelihood of the occurrence of the secondary accident. If the probability of the secondary accident is sufficiently high then the appropriate accident scenarios are developed and analysed for consequences.

This module has an independent set of information, which should be provided by the user. The information pertains to the operational details of the secondary unit, chemicals used, meteorological conditions, and topological characteristics. This

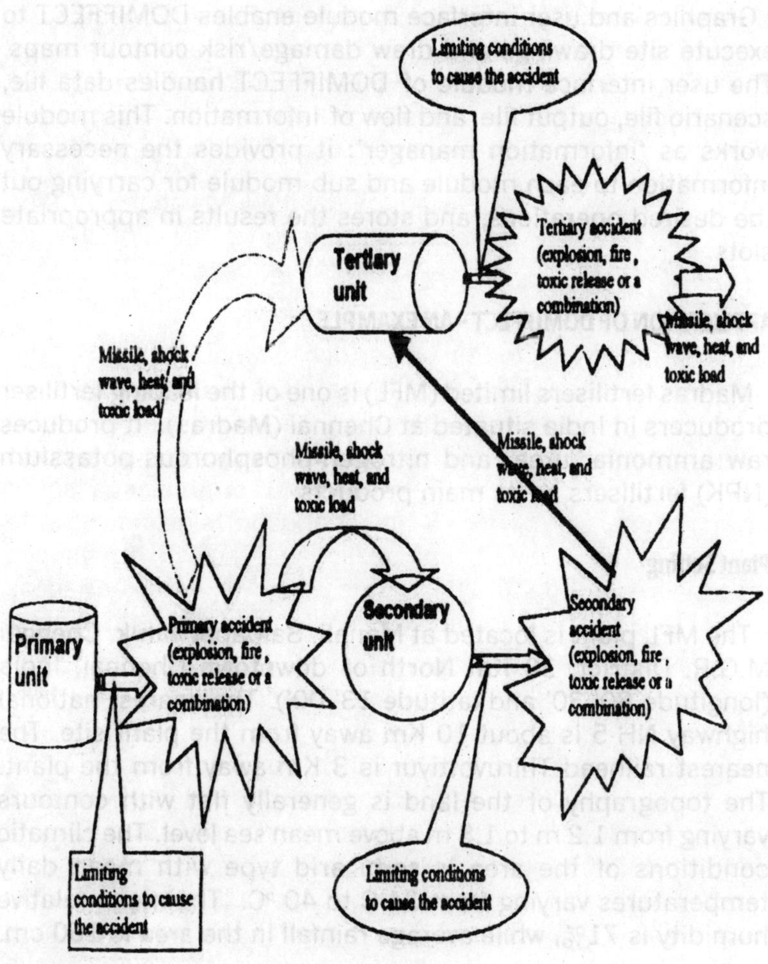

Fig. 11.2: The mechanism of domino/cascading effect pertaining to modelling of chain of accidents in chemical process industries.

module consists of various sets of mathematical models for estimating the damage consequences, and probability of occurrence of secondary accidents due to the effect of the primary ones (Table 11.1).

Graphics and user interface module enables DOMIFFECT to execute site drawings and draw damage/risk contour maps. The user interface module of DOMIFFECT handles data file, scenario file, output file, and flow of information. This module works as 'information manager': it provides the necessary information to each module and sub-module for carrying out the desired operations, and stores the results in appropriate slots.

APPLICATION OF DOMIFFECT - AN EXAMPLE

Madras fertilisers limited (MFL) is one of the leading fertiliser producers in India situated at Chennai (Madras). It produces raw ammonia, urea, and nitrogen-phosphorous-potassium (NPK) fertilisers as its main products.

Plant Setting

The MFL plant is located at Manali, Saidapet Taluk, Chengai M.G.R. District, 30 Km North of downtown Chennai, India (longitude 80' 30' and latitude 13' 00'). The nearest national highway NH-5 is about 10 Km away from the plant site. The nearest railhead Thiruvottiyur is 3 Km away from the plant. The topography of the land is generally flat with contours varying from 1.2 m to 1.8 m above mean sea level. The climatic conditions of the area is semi arid type with mean daily temperatures varying from 24°C to 40 °C. The mean relative humidity is 71%, while average rainfall in the area is 160 cm.

As illustrated in Figure 11.4, the plant is surrounded by a number of major industries, several of which process highly hazardous chemicals. Several populous villages are also located close to the industry.

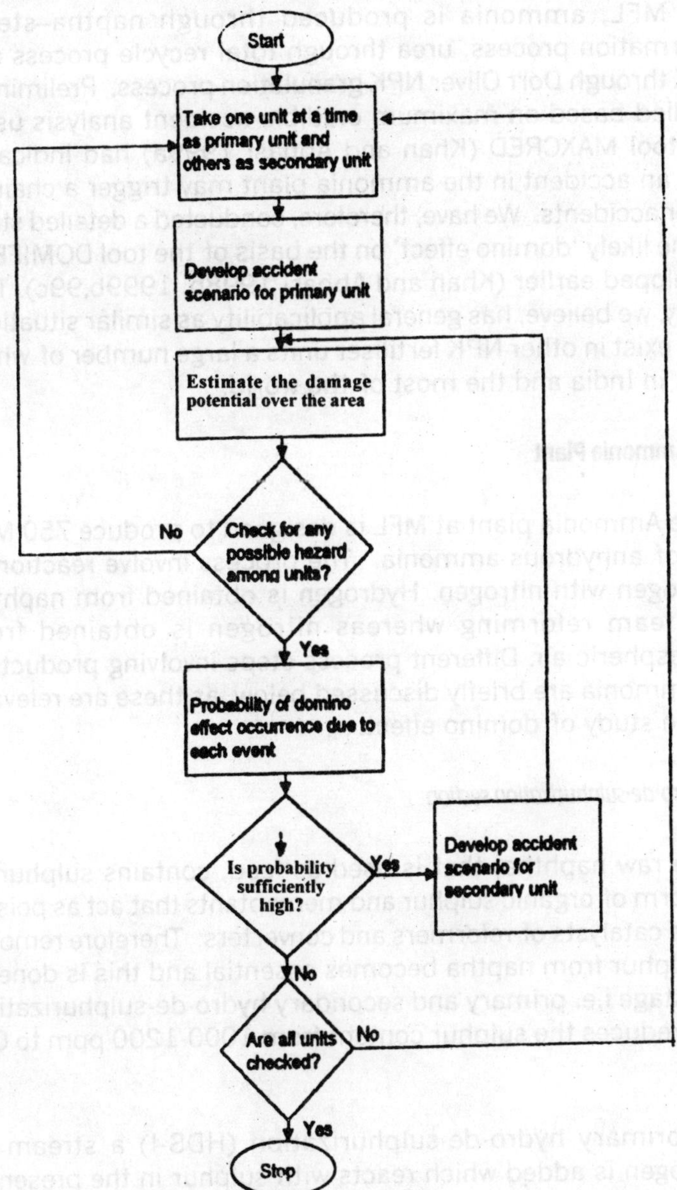

Fig. 11.3: Steps involve in DOMIFFECT

At MFL, ammonia is produced through naptha–steam reformation process, urea through total recycle process and NPK through Dorr Oliver NPK granulation process. Preliminary studied based on maximum credible accident analysis using the tool MAXCRED (Khan and Abbasi 1999a) had indicated that an accident in the ammonia plant may trigger a chain of other accidents. We have, therefore, conducted a detailed study of the likely 'domino effect' on the basis of the tool DOMIFECT developed earlier (Khan and Abbasi 1998b, 1999b,99c). This study, we believe, has general applicability as similar situations may exist in other NPK fertiliser units a large number of which exist in India and the most of the world.

The Ammonia Plant

The Ammonia plant at MFL is designed to produce 750 MT/day of anhydrous ammonia. The process involve reaction of hydrogen with nitrogen. Hydrogen is obtained from naphtha by steam reforming whereas nitrogen is obtained from atmospheric air. Different process steps involving production of ammonia are briefly discussed below, as these are relevant to the study of domino effect.

i. Hydro-de-sulphurization section

The raw naphtha, that is used as feed, contains sulphur in the form of organic sulphur and mercaptants that act as poison to the catalysts of reformers and converters. Therefore removal of sulphur from naptha becomes essential and this is done in two stage i.e. primary and secondary hydro-de-sulphurization that reduces the sulphur content from 1000-1200 ppm to 0.2 ppm.

In primary hydro-de-sulphurization (HDS-I) a stream of hydrogen is added which reacts with sulphur in the presence of Co-Mo catalyst to form H_2S which is stripped off from naphtha. In secondary hydro-de-sulphurization (HDS-II), again in the presence of Co-Mo catalyst, H_2S is formed which is absorbed as ZnS in a sulphur absorbent vessel containing ZnO.

ii. Gas reforming section

The de-sulphurised naphtha so formed is reformed in primary and secondary reformers. In primary reformers steam is added which reacts with naphtha at 797°C in the presence of nickel catalyst to form H_2, CO and CO_2:

$$C_xH_y + H_2O \Leftrightarrow H_2 + CO + CO_2 - \text{Heat}$$

The gases formed in primary reformer now enters the secondary reformer where preheated air is introduced. The H_2 and other combustible gases burn with oxygen leaving nitrogen required later to make ammonia:

$$2H_2 + O_2 + N_2 \rightarrow 2H_2O + N_2 + \text{Heat}$$

Due to the above reaction, the gas temperature reaches upto 1315°C. Methane is formed and the temperature drops down. After cooling the reformed gas enters high temperature shift converter which operates at 368°C in the presence of iron as catalyst. Here CO reacts with H_2O to form CO_2 and hydrogen:

$$CO + H_2O \Leftrightarrow CO_2 + H_2 + \text{Heat}$$

Heat of reaction again increases the gas temperature to 424°C so it is again cooled in a heat exchanger and the exit gases after cooling are sent to a low temperature shift converter which operates at 216°C in the presence of copper as catalyst. CO_2 and H_2 are produced.

iii. CO_2 removal section

Gases coming out from gas reforming section contain not only H_2 and N_2 but also steam, CO_2, CO, CH_4 and Argon (Ar). All oxygen-containing compounds are poisonous to ammonia synthesis catalyst (iron): it becomes essential to remove or convert them into other harmless compounds. Steam is removed by condensing it at low temperature while CO_2 is removed by absorbing in catacarb solution (K_2CO_3) at high pressure and low temperature; the solution is regenerated by boiling it into another vessel, stripping it of CO_2.

Table 11.4. Summarized results of consequence analysis of primary event in various units at various locations

Unit/Scenario *Damaging Effect*	**Damage Potential at Distance(m)** *50*	*75*	*100*	*125*
HDS-I unit : fire ball				
Heat load (kW/m)	150	92	67	7
Missile Velocity (m/s)	341	250	185	138
Penetration strength (mm) (for 3 kg missile)	32	24	16	10
Peak Over pressure (kpa)	115	95	75	63
Blast wave velocity (m/s)	471	347.3	222.2	157.6
Ammonia converter unit : CVCE followed by dispersion				
Heat load (kW/m)	115	78	51	39
Missile Velocity (m/s)	421	348	215	158
Penetration strength (mm) (for 3 kg missile)	43	35	22	14
Peak Over pressure (kpa)	257	171	110	83
Blast wave velocity (m/s)	757	514.2	368.1	247.3
Primary reformer : BLEVE followed by fire ball				
Heat load (kW/m)	128	85	60	42
Missile Velocity (m/s)	399	287	197	121
Penetration strength (mm) (for 3 kg missile)	37	29	17	11
Peak Over pressure (kpa)	132	98	78	49
Blast wave velocity (m/s)	648	368	235	188

{Cont.}.....

Secondary reformer : BLEVE followed by fire ball				
Heat load (kW/m)	167	101	75	54
Missile Velocity (m/s)	415	302	214	156
Penetration strength (mm) (for 3 kg missile)	41	33	21	16
Peak Over pressure (kpa)	163	103	79	57
Blast wave velocity (m/s)	689	371	227	179
HCS-II : BLEVE followed by fire ball				
Heat load (kW/m)	174	141	91	79
Missile Velocity (m/s)	—	—	—	—
Penetration strength (mm) (for 3 kg missile)	—	—	—	—
Peak Over pressure (kpa)	175	110	88	62
Blast wave velocity (m/s)	572	375	256	161
Catacarb reboiler : fire ball				
Heat load (kW/m)	155	133	87	73
Missile Velocity (m/s)	—	—	—	
Penetration strength (mm) (for 3 kg missile)	—	—	—	
Peak Over pressure (kpa)	—	—	—	
Blast wave velocity (m/s)	—	—	—	

Often it is cheaper to convert the residual gases (CO and CO_2) rather than to remove them. The conversion process occurs in a methanator:

$$CO + CO_2 + 7H_2 \text{ ÕÕ } CH_4 + 3H_2O + \text{Heat}$$

iv. Compression, synthesis and refrigeration section

In compression section purified synthesis gas coming out from gas reform section is compressed from 24.6 kg/cm^2 absolute pressure to 240 kg/cm^2 absolute pressure in three stages.

In synthesis section compressed nitrogen reacts with compressed hydrogen at high pressure to give ammonia:

$$N_2 + 3H_2 \Leftrightarrow 2NH_3 + \text{Heat}$$

Subsequently in refrigeration section the ammonia is liquefied and separated.

HAZARD IDENTIFICATION

As a prelude to domino effect analysis (DEA), of all the units of the ammonia plant mentioned above were screened for their hazard potential. A number of these units involve hazardous chemicals and extreme operating conditions. Towards this we have used hazard identification and screening analysis (HIRA) technique (Khan and Abbasi, 1997c,1998f). The results are presented in Table 11.2 and Figure 11.5. It evident from the figure that units such as: HDS-I, II, primary and secondary reformer, reboiler, and ammonia converter, are most hazardous, while the units: primary and secondary CO converter, CO_2 absorber, etc. are comparatively less hazardous. Therefore, for the saps of brevity, we have limited the subsequent discussions to the more hazardous units.

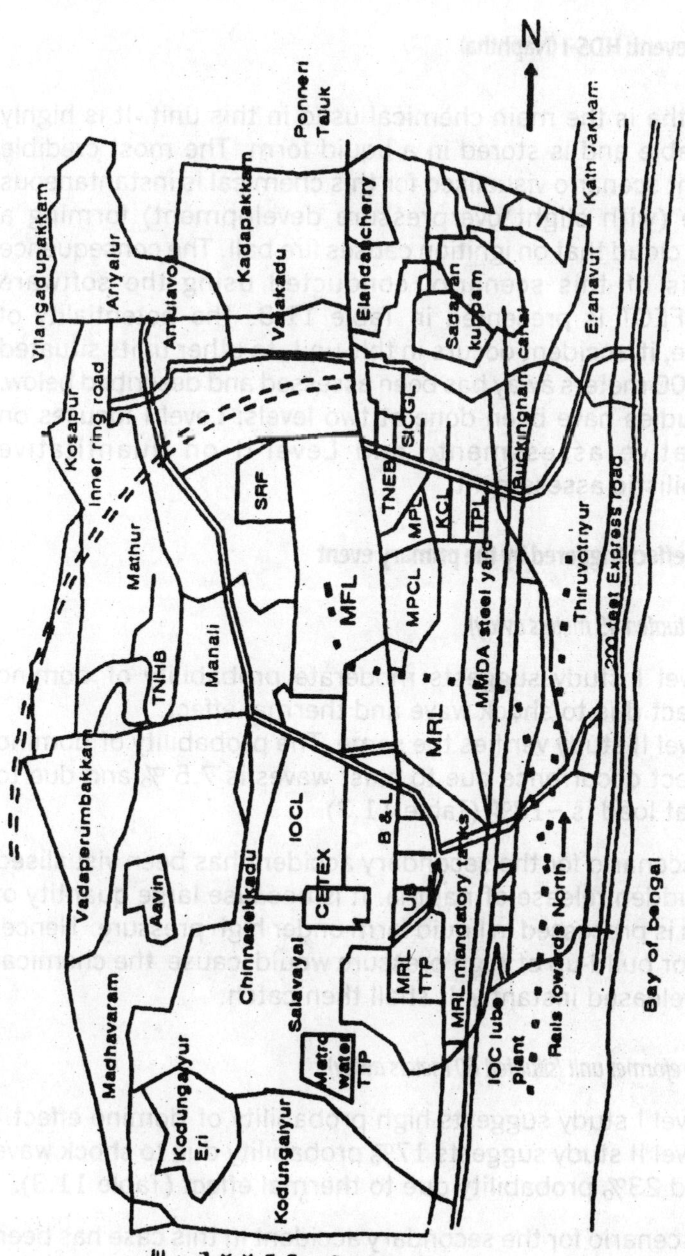

Fig. 11.4: Map showing location of MFL and its surroundings

HAZARD ASSESSMENT

Primary event: HDS-I (Naphtha)

Naphtha is the main chemical used in this unit. It is highly flammable and is stored in a liquid form. The most credible accident scenario visualised for this chemical is instantaneous release (with slight overpressure development) forming a vapour cloud that on ignition causes fire ball. The consequence analysis of this scenario, conducted using the software DOMIFFECT is presented in Table 11.3. The potentiality of damage, if accident occurs in this unit, to other units situated 45 to 100 meters away has been assessed and described below. The studies have been done at two levels; Level I focuses on qualitative assessment and Level-II on quantitative probabilistic assessment.

Domino effect triggered by the primary event

HDS-II (situated 45 meters away)

Level I study suggests moderate probability of domino effect due to shock wave and thermal effect.
Level II study verifies the same. The probability of domino effect occurrence due to blast waves is 7.5 % and due to heat load is ~13% (Table 11.3).

The scenario for the secondary accident has been visualised as a sudden release of naptha. It is because large quantity of naptha is processed in liquid form under high pressure. Hence, a leak or build-up of high pressure would cause the chemical to be released instantly. It shall then catch.

Primary reformer unit (situated 60 meters away)

Level I study suggests high probability of domino effect.
Level II study suggests 17% probability due to shock wave and 23% probability due to thermal effect (Table 11.3).

The scenario for the secondary accident in this case has been visualised as: boiling liquid expanding vapour explosion

Table 11.5. Results of domino effect analysis due to an accident in primary reformer unit

Unit (Chemicals)	Damage effects observed	Probability of occurrence of domino effect (%)	Conclusion (secondary accident scenario)	Net probability of occurrence (/yr)
Primary Event :	BLEVE followed by fire ball in Primary reformer			4.5E-05
Domino Effect :				
HDS-I	shock wave heat load missile load	7.0 13.0 2.0	A domino scenario is instantaneous release as BLEVE followed by flash/fire ball.	9.3E-06
HDS-II	shock wave heat load missile load	12.0 11.0 5.0	A domino scenario is BLEVE followed by fire ball.	1.1E-05
Ammonia converter	shock wave heat load missile load	12.0 15.0 5.0	A domino scenario is CVCE followed by formation of toxic load.	1.3E-05
Secondary reformer	shock wave heat load missile load	15.0 20.0 10.0	A domino scenario is an instantaneous releaseas BLEVE followed by fire ball/flash fire.	1.7E-05
Catacarb reboiler	shock wave heat load missile load	7.0 3.0 1.0	A domino scenario is pool fire	4.8E-06

(BLEVE) followed by fireball. It is because heat load from the primary accident would create high internal pressure (about 30% higher than the design value) adding to the already high pressure at which the chemical was stored. As the chemical is highly flammable, it would catch fire soon after release, turning into a fire ball.

Secondary reformer unit (situated 75 meters away)

Level I study suggests moderate probability of domino effect.
Level II analysis suggests that domino effect probabilities due to the shock and heat load are 5% and 10% respectively (Table 11.3). The secondary accident would be similar to the one that may occur in the previous unit (instantaneous release –BLEVE- followed by fire ball).

Catacarb reboiler (situated 90 meters away)

Level I study suggests moderate probability of domino effect.
Level II study suggests 7% probability of domino effect due to shock wave and 10% due to heat radiation (Table 11.3).

The secondary accident scenario likely in the catacarb reboiler is visualised as: continuous release that on ignition would give pool fire. This shall happen because the quantities of hydrogen and naptha processed are lesser and the intensity of damaging events at the location of this unit are also likely to be lower. However, they may still be strong enough to cause release of chemical either through relief valve or through safety valve. The exited chemical on meeting ignition source would lead to pool fire.

Ammonia converter (situated 70 meters away)

Level I study suggests high probability of domino effect.
Level II study suggests probability of occurrence of domino effect 30% due to shock wave and 25% due to heat load (Table 11.3).

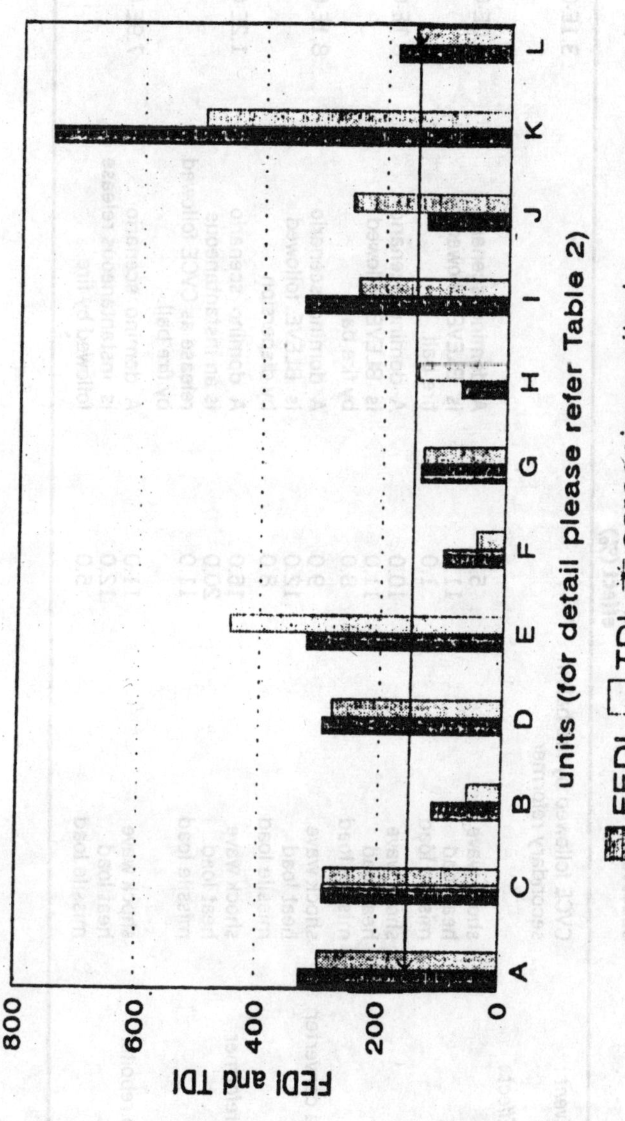

Fig. 11.5: The results of hazard identification steps: FEDI stands for fire and explosion damage index, while TDI stands for toxic damage index.

Table 11.6. Results of domino effect analysis due to an accident in secondary reformer unit

Unit (Chemicals)	Damage effects observed	Probability of occurrence of domino effect (%)	Conclusion (secondary accident scenario)	Net probability of occurrence (/yr)
Primary Event :	CVCE followed by fire ball in secondary reformer			3.1E-05
Domino Effect :				
HDS-I	shock wave heat load missile load	5.0 11.0 1.0	A domino scenario is BLEVE followed by fire ball.	5.0E-06
HDS-II	shock wave heat load missile load	10.0 11.0 8.0	A domino scenario is BLEVE followed by fire ball.	8.1E-06
Ammonia converter	shock wave heat load missile load	9.0 12.0 8.0	A domino scenario is BLEVE followed by dispersion	8.1E-06
Primary reformer	shock wave heat load missile load	15.0 20.0 11.0	A domino scenario is an instantaneous release as CVCE followed by fire ball.	1.2E-05
Catacarb reboiler	shock wave heat load missile load	11.0 12.0 5.0	A domino scenario is instantaneous release followed by fire.	7.9E-06

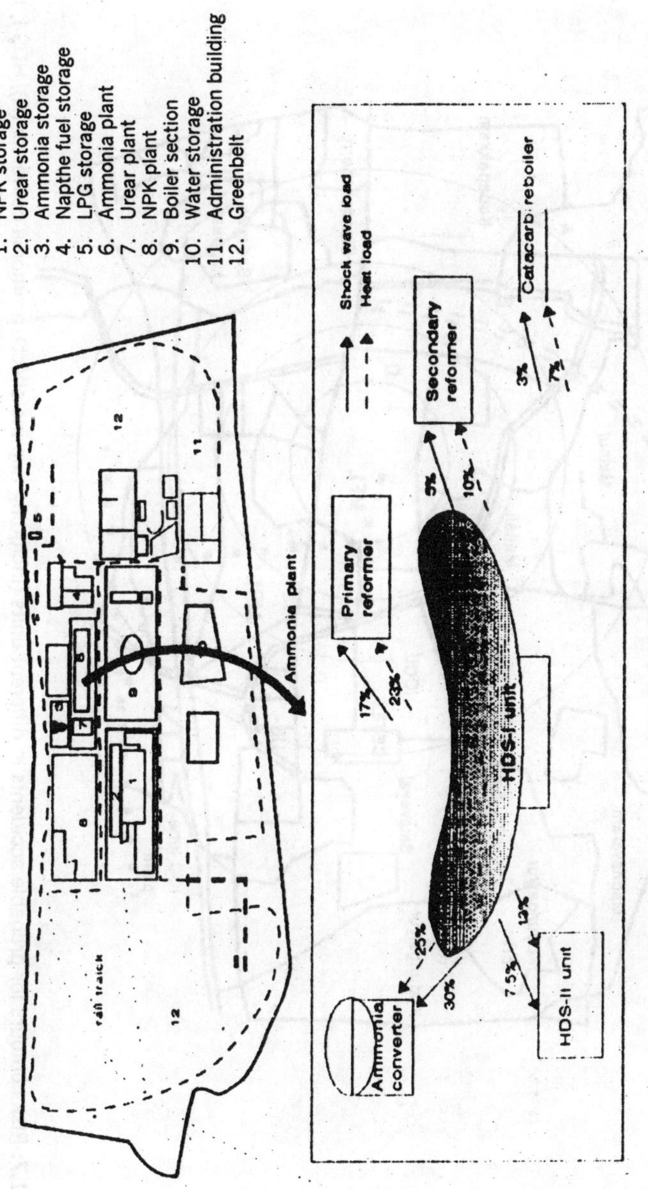

Fig. 11.6: Layout of ABCFL the plant and probability of domino effect due to HDS-I unit.

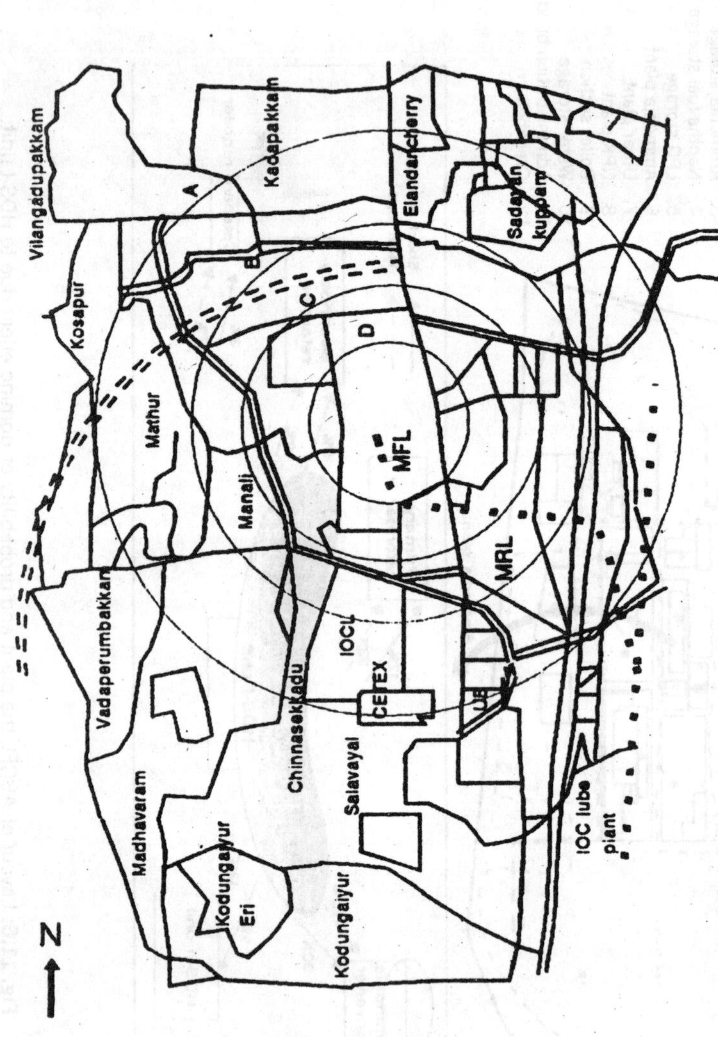

Fig. 11.7: Risk contours for probable accidents in different units (including domino effect), ammonia converter (A), HDS-I, II (B), primary and secondary reformer (C), catacarb reboiler (D).

Table 11.7. Results of domino effect analysis due to an accident in HDS-I unit

Unit **(Chemicals)**	***Damage*** **effects observed**	***Probability*** **of occurrence of domino effect (%)**	***Conclusion*** **(secondary accident scenario)**	***Net probability*** **of occurrence (/yr)**
Primary Event : Fire ball in HDS-I unit				7.5E-05
Domino Effect :				
HDS-II	shock wave heat load missile load	7.0 13.0 2.0	A domino scenario is instantaneous release followed by fire ball.	1.5E-05
Primary reformer	shock wave heat load missile load	17.0 23.0 9.0	A domino scenario is BLEVE followed by fire ball.	3.1E-05
Ammonia converter	shock wave heat load missile load	30.0 25.0 11.0	A domino scenario is BLEVE followed by dispersion of toxic gas	3.9E-05
Secondary reformer	shock wave heat load missile load	5.0 10.0 9.0	A domino scenario is an instantaneous release as BLEVE followed by fire ball.	1.6E-05
Catacarb reboiler	shock wave heat load missile load	3.0 10.0 0.0	A domino scenario is continuous release leading to pool fire	9.5E-06

Table 11.8. Results of domino effect analysis due to an accident in HDS-II unit

Unit (Chemicals)	Damage effects observed	Probability of occurrence of domino effect (%)	Conclusion (secondary accident scenario)	Net probability of occurrence (/yr)
Primary Event :	BLEVE followed by fire ball in HDS-II unit			7.5E-05
Domino Effect :				
Primary reformer	shock wave heat load missile load	9.0 15.0 3.0	A domino scenario is instantaneous release as BLEVE followed by fire ball.	1.8E-05
HDS-I	shock wave heat load missile load	15.0 18.0 3.0	A domino scenario is BLEVE followed by fire ball.	2.4E-05
Ammonia converter	shock wave heat load missile load	20.0 15.0 3.0	A domino scenario is CVCE followed by dispersion.	2.5E-05
Secondary reformer	shock wave heat load missile load	11.0 15.0 3.0	A domino scenario is an instantaneous release as BLEVE followed by flash fire.	1.9E-05
Catacarb reboiler	shock wave heat load missile load	2.0 5.0 3.0	A domino scenario is pool fire	7.2E-06

Table 11.9. Results of domino effect analysis due to an accident in Ammonia converter

Unit (Chemicals)	Damage effects observed	Probability of occurrence of domino effect (%)	Conclusion (secondary accident scenario)	Net probability of occurrence (/yr)
Primary Event :	BLEVE followed by dispersion in Ammonia converter			1.5E-05
Domino Effect :				
HDS-I	shock wave heat load missile load	12.0 21.0 3.0	A domino scenario is instantaneous release as BLEVE followed by fire ball	4.8E-06
HDS-II	shock wave heat load missile load	6.0 17.0 3.0	A domino scenario is BLEVE followed by fire ball.	3.6E-06
Primary reformer	shock wave heat load missile load	24.0 35.0 3.0	A domino scenario is BLEVE followed by fire ball/flash fire.	7.8E-06
Secondary reformer	shock wave heat load missile load	35.0 46.0 3.0	A domino scenario is an instantaneous release followed by fire ball/flash fire.	9.8E-06
Catacarb reboiler	shock wave heat load missile load	11.0 16.0 3.0	A domino scenario is instantaneous release followed by fire ball.	4.1E-06

Table 11.10. Results of domino effect analysis due to an accident in Catacarb reboiler

Unit (Chemicals)	Damage effects observed	Probability of occurrence of domino effect (%)	Conclusion (secondary accident scenario)	Net probability of occurrence (/yr)
Primary Event :	fire ball in Catacarb reboiler unit			2.2E-05
Domino Effect :				
HDS-I	shock wave	—	A domino scenario is instantaneous release followed by fire.	2.4E-06
	heat load	11.0		
	missile load	—		
HDS-II	shock wave	—	A domino scenario is release followed by fire ball.	1.9E-06
	heat load	9.0		
	missile load	—		
Ammonia converter	shock wave	—	A domino scenario is release followed by dispersion.	2.4E-06
	heat load	11.0		
	missile load	—		
Secondary reformer	shock wave	—	A domino scenario is an instantaneous/ continuous release followed by flash fire.	2.2E-06
	heat load	10.0		
	missile load	—		
Primary reformer	shock wave	—	A domino scenario is fire ball.	2.8E-06
	heat load	13.0		
	missile load	—		

The domino effect scenario for this unit is visualised as BLEVE followed by creation of toxic zone. An intense heat load would develop a high pressure in the reactor and also weaken the strength of vessel, which in consequence would lead to sudden release, either through relief valve or through pipe joints as BLEVE. CVCE may also occur, but there is lesser probability of CVCE, mainly because to cause a CVCE excessive pressure must develop in the vessel, which is not possible with the impact of observed damaging effect (heat and shock wave). Of course, the release of ammonia on ignition would lead to flash fire depending upon the quantity released. Subsequent dispersion of unburned ammonia in the atmosphere would cause toxic load. The combustion of ammonia is difficult, whereas its toxicity impact is highly likely. Therefore we have given more emphasis on toxic effect and ignored the effect of fire.

A simplified block diagram showing the occurrence of domino effect in various units triggered by a primary accident in the HDS-I is presented in Figure 11.6. It is evident that the ammonia reactor and primary reformer are highly vulnerable to accidents which may trigger domino effect. It is interesting to note that eventhough ammonia converter is placed at larger distance than primary/secondary reformer, it still has high probability of causing domino effect. It is because most (more than 50%) of the time, the atmospheric conditions favour conveying higher heat and shock wave load to ammonia reactor. Further, ammonia converter involves larger quantity of chemical operating at extreme conditions of temperature and pressure than other units, making it more vulnerable.

Similar studies were also conducted to analyse the domino effect due to accidents in either one or other of these units: HDS-II, primary and secondary reformer, ammonia reactor, Catacarb reboiler, and CO_2 absorber. The results are summarised in Tables 11.4 –11.10.

RISK ASSESSMENT AND CONCLUSION

A detailed consequence analysis covering all the hazardous units reveals that, as far as the primary event is concerned, ammonia converter has the maximum damage potential as it causes lethal overpressure, and toxic load over the largest area. Risk contours, signifying an individual risk factor greater than $1*10^{-4}$ /yr due to the cumulative effect of the primary and secondary accidents, are plotted over the industry's map in Figure 11.7. It is evident that an accident in ammonia converter would endanger an area of more than ~400 m radius, while the risk contour for primary/secondary reformer units would encompass an area of about ~250 m radius. An accident in the HDS- I and II would bring an area of about ~150 m radius under severe risk.

In summary, the ammonia converter has the potential of causing the most damaging primary accident and *also* has the maximum probability to cause domino effect. The intensities of blast wave, missiles, heat and toxic load would be lethal over an area of ~400 meters radius if an accident occurs in this plant. The plant has several vulnerable units close to it, which would facilitate a long chain of accidents. On the other hand an accident in Catacarb reboiler has the least damage potential and also the least probability of causing secondary accidents.

REFERENCES

1. Bagster, D. F., and Pitblado, R. M., (1991). The Estimation of Domino Incident Frequencies–an approach, *Process Safety Environ*, 69B, 196.

2. Hindu, (1997). Major fire in Vizag refinery, *The Hindu publication*, September 15, pp 1.

3. Khan, F. I., and Abbasi, S. A., (1996). Accidents in a chemical industry usin¸ the software package MAXCRED, *Indian journal of chemical technology*, 3, 338-44.

4. Khan, F. I., and Abbasi, S. A., (1997a). Risk analysis of epichlohydrin manufacturing industry using new computer automated tool MAXCRED, *J Loss Prevention In Process Industries*, 10(2), 91–100.

5. Khan, F. I., and Abbasi, S. A., (1997b). Risk analysis of a cloralkali industry situated in densely populated area, *Process Safety Progress*, 16(3), 172-185.

6. Khan, F. I., and Abbasi, S. A. (1997c), Rapid quantitative risk analysis of using a software package MAXCRED-II,*Indian journal of chemical technology*, 4, 196-210.

7. Khan, F. I., and Abbasi, S. A., (1997d) optHAZOP : an effective and optimal HAZOP study, *Journal of loss prevention in process industries*, 10(3), 191–204.

8. Khan, F. I., and Abbasi, S. A., (1997e). Accident hazard index : a multi-attribute scheme for process industry hazard rating, *Institution of Chemical Engineers (IChemE) of UK (Environmental protection and safety)*, IChemE, 75B, 217–24.

9. Khan, F. I., and Abbasi, S. A., (1998a). Models for domino effect analysis in chemical process industries, *Process Safety Progress*, 17(2), 107–23.

10. Khan, F. I., and Abbasi, S. A., (1998b). DOMIFFECT (DOMino eFFECT) : a new software for domino effect analysis in chemical process industries, *Environment modelling and software*, 13, 163–77.

11. Khan, F. I., and Abbasi, S. A. (1998c), Inherently safer design based on rapid risk analysis. *Journal of loss prevention in process industries*, 11, 361–72.

12. Khan, F. I., and Abbasi, S. A., (1998d) Rapid quantitative risk analysis petrochemical industry using new software MAXCRED, *Journal of cleaner production*, 6, 9-22.

13. Khan, F. I., and Abbasi, S. A. (1998e). Accident simulation as a tool for assessing and calculation environmental risk in CPI : a case study, *Korean journal of chemical engineering*, 11(2), 124-35.

14. Khan, F. I., and Abbasi, S. A., (1998f) Multivariate hazard identification and ranking system, *Process safety Progress*,–AIChE, USA , 17 (3), 157–71.

15. Khan, F. I., and Abbasi, S. A., (1999a). MAXCRED–A new software package for rapid risk assessment in chemical process industries, Environment modelling and software, 14, 11-25.

16. Khan, F. I. and Abbasi, S. A. (1999b). The World's Worst Energy Industry Accident of 1990s–what happened and what might have been: a quantitative study, *process safety progress* 17 (4), 311–25.

17. Khan, F. I. and Abbasi, S. A. (1999c). *Risk analysis of chemical process industries: advanced techniques*, Discovery publishing House, New Delhi 435 pages.

18. Kletz, T. A., (1985). What went wrong , *Gulf Publication Company*, England.

19. Kletz, T. A. (1991). The Philips explosion - this is where I came in, *Chem Engr*, , 493, 47.

20. Latha, P., Gautam, G., and Raghavan, K. V., (1996). Strategies for quantification of thermally initiated cascade effects, (In Disaster management in process industries; *Ed:* Raghvan, and Sawminathan), Oxford press, Chennai.

21. Lees, F. P., (1996) Loss Prevention in chemical process industries, Butterworts, London.

22. Marshall. V. C. (1987). *Major Chemical Hazards*, Ellis Horwood, Chichester.

23. MFL (1999). Risk analysis of Madras Fertiliser Limited, Centre for Pollution Control and Energy Technology, *report No: RA/01/99*, 167 pages.

24. Petersen, C. M., (1986). Analysis of the LPG disaster in Mexico City, *Loss Prevention and Safety Promotion*, 5, 21.

25. Petersen, C. M., (1990). Consequence of accidental release of hazardous materials, *J of Loss Prevention Process Industries*, 3, 136.

26. Raghavan, K. V., and Swaminathan, G., (1996). Hazard assessment and disaster mitigation in petrochemical industries, *Oxford Publishing Company*, Chennai.

27. Shankar, R., (1997). Corrosion, *Science Express* : *Indian Express Publication*, pp.1-3, (30 September), Chennaip.